Anonymous

Narrative of Messrs. Moody and Sankey's Labors in Great Britain and Ireland

With eleven addresses and lectures in full

Anonymous

Narrative of Messrs. Moody and Sankey's Labors in Great Britain and Ireland
With eleven addresses and lectures in full

ISBN/EAN: 9783337325084

Printed in Europe, USA, Canada, Australia, Japan

Cover: Foto ©berggeist007 / pixelio.de

More available books at **www.hansebooks.com**

NARRATIVE OF

Messrs. Moody and Sankey's

LABORS IN

GREAT BRITAIN AND IRELAND,

WITH

ELEVEN ADDRESSES AND LECTURES IN FULL.

NEW AND COMPLETE EDITION.

No other published account of the GREAT AWAKENING *is more complete, while no other has Verbatim Reports of* ADDRESSES.

ANSON D. F. RANDOLPH & COMPANY,
770 BROADWAY, NEW YORK.
1875.

COPYRIGHT, 1875, BY
ANSON D. F. RANDOLPH & COMPANY

ROBERT RUTTER,
BINDER,
34 BEEKMAN STREET, N. Y.

NOTE.

I.

The following Narrative of the Religious Awakening in Great Britain and Ireland has been compiled from the voluminous correspondence of the BRITISH EVANGELIST, and THE CHRISTIAN, two weekly journals published in London.

In the preparation of this Narrative it was found necessary to condense the original matter, in order to bring the volume within reasonable limits. The aim of the editor, however, has been to present a consecutive account of the development and progress of the work in each of the places where it was prosecuted, and it is believed that the American reader will, in these pages, obtain something like a comprehensive view of this wonderful movement, and its immediate and marvelous results.

It is the design of the present publishers to issue in April or May, a supplementary pamphlet, which shall embrace an account of the subsequent labors of Messrs. Moody and Sankey in other parts of England.

FEBRUARY, 1875.

II.

The Supplementary Issue, which now follows page 122 of this volume, has been enlarged, and brings down the Narrative to the close of the work in England.

OCTOBER, 1875.

INTRODUCTION.

MESSRS. MOODY AND SANKEY.

THE HISTORY OF THEIR WORK.

IT is for obvious reasons desirable, when strangers like Messrs. Moody and Sankey are exciting so much interest, and producing so great an impression, that the public should be made acquainted with the circumstances that have brought them into the position which they now occupy. It ought to be known that they have not run unsent, and have not taken upon themselves, without due cause, the responsibility of the work which they are now carrying on. A brief sketch of the chief facts in Mr. Moody's life will show clearly how this matter stands.

Mr. Moody was born in the year 1837, in the State of Massachusetts, in the district which was the scene of the great awakening, under Jonathan Edwards, about a hundred years before. But so far from his inheriting anything from that remarkable movement, he had not even heard the Gospel of the grace of God till he was about seventeen years of age. Going about that time to Boston, to be trained for business in the establishment of an uncle, he one day went into the church of Dr. Kirk, a Congregational minister in that city. There, for the first time, he listened to an evangelical sermon. It had the effect of making him uncomfortable, and he resolved not to go back. He felt that his heart had been laid bare, and he wondered who had told the preacher about him. Something, however, induced him to go back next Sunday, and the impression was renewed. A Sunday-school teacher in whose class he had been, having come to see him and ask for him at his place of business, he opened up his mind to him, and he was enabled to enter into that peace and joy in believing to which he has been the instrument of introducing so many.

Not very long after this, Mr. Moody left Boston and proceeded to Chicago, where he entered into business for himself. Being full of the desire to be useful, he went into a Sunday-school, and asked the superintendent if he would give him a class. In this school there were twelve teachers and sixteen pupils; and the answer to his application was that if he could gather a class for himself he would be allowed to teach them. Mr. Moody went out into the streets, and by personal application, succeeded in bringing in a score of boys. He enjoyed so much the work of bringing in recruits, that instead of teaching the class himself, he handed it over to another teacher, and so on, until he had filled the school. Then he began to entertain the notion of having a school of his own. He went to work in a neglected part of the city. Sunday is the day devoted by many to concerts, balls, and pleasure generally. Mr. Moody saw that to succeed in such a population, a school must be exceedingly lively and attractive, and as he observed that the Germans made constant use of music in their meetings, he was led to consider whether music might not be employed somewhat prominently in the service of Christ. Not being himself a singer, he got a friend who could sing to help him, and for the first few evenings the time was spent between singing hymns and telling stories to the children, so as to awaken their interest and induce them to return. A hold having in this way been established, the school was divided into classes, and conducted more in the usual way.

This school became the basis of wider operations. After a time a lively interest in divine things began to appear among the children. This led to the holding of meetings every night, and to the offering of prayers and delivery of addresses suitable to the circumstances of the children. These meetings began to be attended also by the parents, some of whom shared the blessing. It may be stated here that some of those young persons who were converted at this time, remain to the present day the most valuable and active coadjutors in the work with which Mr. Moody is associated in Chicago.

In most cases neither the children nor their parents had hitherto been connected with any Christian Church. Mr. Moody began to find himself constrained to supply them with spiritual food. At first he encouraged them to connect themselves with other congregations. But it was found that in these they were next to lost or swallowed up: they felt themselves strangers, sometimes unwelcome strangers, while they lost all the benefit of neighborhood, mutual interest, and combination in the worship of God. Gradually, therefore, Mr. Moody felt shut up to taking charge of them, and supplying them with Christian instruction. Both school and church continued to increase, the school amounting to about a thousand, and suitable buildings were erected through the liberality of friends. Mr. Moody had by this time given up business, so that he might be free to give his whole time and attention to the work. As he felt himself called by the Lord to this step, he resolved to decline all salary or allowance from any quarter, and trust for his maintenance solely to what it might be put into the hearts of God's people to contribute. Being quite destitute of private means, this resolution showed that his faith in a divine call to give himself to Christian work was capable of bearing a great strain. At the same time, while adopting this course for himself, he has never pressed it upon others, unless they should clearly see it to be their duty. And while believing himself called to a kind of supplementary work in the ministry, he is very far from prescribing the same *rôle* to others. On the contrary, he is the steady friend of a regular ministry, being fully persuaded that in "ordaining elders in every city," the apostles meant to set up the permanent platform of the Christian Church. Mr. Moody is delighted to obtain the co-operation of the clergy; and among the many subjects of congratulation and thankfulness which his visit to Edinburgh has supplied, one of the most important has been, that from the very first he has been received most readily by the ministers, and has obtained from all denominations very cordial support. The clergy have shown by this that they take the same view substantially as Mr. Moody himself of the relation of the regular to the irregular ministry. While believing the ministry to be a divine institution, they do not believe that it monopolizes the grace of God for the conversion of men and the upbuilding of the Church. God may move out of the regular course, and often does to show His sovereignty and to impress the truth. "Not by might, nor by power, but by my Spirit, saith the Lord."

Mr. Moody had acquired a position of much influence in the United States in connection with Sunday-school and mission work, when the war broke out between North and South. This led to a new turn being given to his labors. There was a large camp in the neighborhood of Chicago, to which he gave much attention, going there night after night and striving to bring the soldiers under the influence of divine grace. When the Christian Commission was organized, under the presidency of his friend, George H. Stuart, of Philadelphia, Mr. Moody became one of his most energetic coadjutors. He did not go into the army as an agent of the Commission, but he was President of the Executive branch for Chicago, and nine different times he went to one or other of the scenes of warfare, remaining some weeks and working with all his might. These services with the army were of no little use, not only in producing direct fruit, but also in developing that prompt and urgent method of dealing with men, that strenuous endeavor to get them to accept immediate salvation, which is still so conspicuous a feature of his mode of address. With wounded men hovering between life and death, or with men in march, resting for an evening in some place which they were to leave to-morrow, it was plainly, so far at least as he was concerned, the alternative of "now or never;" and as he could not allow himself or allow them to be satisfied with the "never," he bent his whole energies to the "now."

Mr. Moody's labors in the army were often much blessed. Of all his campaigns of this kind there was none on which he looked back with more pleasure than one in the State of Tennessee, in connection with troops under the command of General Howard. That General being in the fullest sympathy with Mr. Moody, their work together was very earnest and much blessed.

The war being ended, Mr. Moody had more time to develop his work in Chicago.

To set others to work in the vineyard had long been one of his chief aims, and by means of the Young Men's Christian Association, in which he took a great interest, he was highly successful. It is to be observed, that in America these associations are much larger and more influential than they usually are in this country, and that their operations ramify over a much more extensive sphere. Mr. Moody strove to inspire the Chicago Association with his own spirit, and to send them to work in the vineyard. The hall of the association became one of the stated scenes of his own labors. The association was very unfortunate in the matter of fires—its first building having been burnt down in 1867, and its second in the great fire of Chicago in 1871. According to Mr. David Macrae, "the lightning city" showed such activity of movement that the money for the second building was all subscribed before the fire had completed the destruction of the first. This, we believe, is somewhat hyperbolical; but in sober truth, the arrangements for the restoration of the building after the first fire were made with wonderful rapidity. The new building contained a hall of enormous size. Mr. Moody was accustomed to preach to his own people in the morning, to superintend a Sunday-school of about a thousand in the afternoon, and to preach again in the evening in the hall of the Young Men's Association.

In October, 1871, occurred the terrible fire which destroyed a great part of Chicago. Mr. Moody, with his wife and two children, was roused in the middle of the night to find the fierce fire approaching their dwelling, and, leaving his house and household gear to their fate (all the property he possessed), had to hurry along to seek shelter in the houses of friends. Mr. Moody's school and church, as well as the buildings of the Young Men's Christian Association, perished likewise in the conflagration. The feelings of himself and his fellow-citizens, on going to see the ruins, can hardly be conceived. But after the first stunning sensation was got over, faith and hope revived. In one month after the fire, a temporary erection was completed! No small energy must have been required to accomplish this, amid the confusion, the bustle, and the infinity of things that had to be attended to. But reared the wooden building was, and it has served the purpose of church and school till now, when a new and substantial building is sufficiently advanced to allow the basement story to be used for public services.

Besides what he did for his own place of meeting, Mr. Moody took an active part in putting things in order after the fire. As President of the Young Men's Association, and having under him a staff of active workers, he received the contributions of many friends. Among the rest, the sum contributed by Edinburgh, approaching, if we remember rightly, to £2,000, was consigned to the care of the Mayor and himself. He fully shared in the feeling of affectionate gratitude which the ready help of this country on that occasion inspired. Many were moved to tears by that token of good-will and sympathy; it was hardly credible to them that Great Britain should be hastening to their help. There is less danger now of such men misunderstanding the real feeling of England towards the United States.

It was shortly before the fire that Mr. Moody and Mr. Sankey began to work together. Mr. Sankey was in business somewhere in Pennsylvania, and Mr. Moody, happening on some public occasion to sit near him, was attracted by his beautiful voice. The thought struck him that Mr. Sankey would be a valuable assistant to him in many ways, in the Sunday-school, in the church, and in the training of the Young Men's Christian Association. He accordingly entered into an engagement with him, and he has come with Mr. Moody to this country to help him in his work by conducting "The Service of Song." Mr. Moody has always been eager to secure music—and especially good music—as an aid in preaching the gospel. It is his belief that the gospel may be presented in song as well as in speech, and that while the song has a marvelously attractive power, it is also fitted to express better than plain speech the emotion suitable to the truths of the gospel. Abhorring the notion of providing a musical entertainment merely to please those who are not in the kingdom of God, he seeks to move their hearts and win them to Christ by truth expressed in the most winning tones. The idea of profaning the worship of God by uttering sacred words not felt by the singer, would be revolting; but it must occur to every one who has heard Mr. Sankey that the charm of his service is in the blending of his heart with his song. There is not only no similarity, but an absolute contrast between Mr. Sankey's hymns and the performances of paid professional singers, who will often sing the most sacred words, not because they are feeling them, but to let the audience hear how well they can sing. It is also in subserviency to spiritual ends that Mr. Sankey uses the harmonium. It has been found quite compatible with spiritual and hearty worship; and the probability is, that at extra meetings, for Sabbath-schools, congregational soirees, children's churches, and the like, its use will now become more common among us.

When things had settled down after the Chicago fire, Mr. Moody began to think of permanent premises for his school and church. A suitable site was secured, and it was resolved to proceed with the erection of a large and commodious building, which, besides accommodation for the schools, will have a hall or church, containing sittings for 2,500. The cost of the whole will be about £20,000. Mr. Moody, by his disinterested labors, has made so many friends all over his country that the contributions have flowed freely from all parts. Among the most interesting was a colossal subscription from 500,000 Sabbath-school children, of five cents each, all anxious to have a brick in Mr. Moody's tabernacle. From Pekin he received a contribution of $300 from an unknown friend. A few converted Chinamen collected a few dollars even from their Pagan countrymen. A little while ago it seemed likely that the whole sum necessary would be provided, but the collapse in business which has occurred may deprive the enterprise of some of the expected contributions.

We are not aware what were the deeper reasons that induced Mr. Moody to devote the time which he is now giving to evangelistic work in this country. We should suppose that he was influenced by the feeling that the churches here stand specially in need of the application of those brisker, livelier, more direct modes of appeal which are more characteristic of America. He may have thought that there was a great amount of solid knowledge and doctrinal orthodoxy here, and that if it could only be kindled up by a spark from heaven, the results would be very remarkable. The immediate cause of his coming over to spend a year was that he was invited by two gentlemen— Mr. Pennefather, of London, and Mr. Bainbridge, of Newcastle. It was a singular circumstance that both these gentlemen died before or about the time of his arrival. The time selected for his visit to this country was very characteristic of the man. His new church had begun to be built, and his schools and congregation were soon to be transferred from the temporary building to the basement story (all that is yet ready) of the new one. Most pastors would have thought that at such a time there was a special reason for their staying at home. Mr. Moody, however, felt that were he to stay, the burden of a thousand little things would be thrown on him, which others could arrange as well as he could, and which in his absence they would have to arrange. Not having had a house of his own since the fire, he was less tied to home, and his family being, like all Americans, fond of traveling, his wife and children have come with him. In regard to the spiritual superintendence of the congregation, it is supplied in a large measure by members of the flock, with occasional help from other pastors. Mr. Moody trains his people to be independent in fact, as they are Independent in name. It may be stated, however, that in one respect the congregation is Presbyterian; it is governed by a session, not by the whole membership. We understand that Mr. Moody has found no reason to regret that his congregation and schools have been left under this kind of arrangement, the reports of their welfare and progress being very satisfactory.

On arriving in this country in midsummer of last year, Messrs. Moody and Sankey's first field was York. Their progress there was slow. They had to win their way to the confidence of the people, and that by slow degrees, as at first they had none of the clergy to back them, and there was a general suspicion or uncertainty with regard to them. The other towns visited in the north of England were Sunderland, Newcastle, and Carlisle. In some of these the impression produced was very great. Newcastle especially responded in a wonderful way. The work of grace seemed to advance there wonderfully, and the power of Heaven fell on the hearts of the people. Some friends in Edinburgh, hearing of what was doing in Newcastle, invited Messrs. Moody and Sankey to pay a visit to Edinburgh.

Mr. Kelman, of Leith, went twice to Newcastle to judge of the work for himself, and returned full of joy and expectation. Accompanied by Mr. Sankey, Mr. Moody came to Edinburgh about the middle or towards the end of November. Here they were received with much cordiality by influential members both of the clergy and the laity. Our readers are familiar with the progress of the work in this city and in Leith. Mr. Moody has taken a remarkable hold of the people of Edinburgh; and of Mr. Sankey's influence, if there were no other evidence of it, it would be enough that his hymns have become popular melodies, and that they are being sung or hummed everywhere by old and young.

What are the elements of Mr. Moody's power? He is not a man of much education or culture; his manner is abrupt and blunt; his speech bristles with Americanisms; his voice is sharp, rapid, and colloquial; and he never attempts anything like finished or elaborate composition. But he is in downright earnest. He believes what he says; he says it as if he believed it, and he expects his audience to believe it. He gets wonderfully near to his hearers, without any apparent effort. Whatever size the audience may be, he is at home with them at once, and he makes them feel that they are at home with him. He is gifted with a rare sagacity, an insight into the human heart, a knowledge of what is stirring in it, and of what is fitted to impress it. He has in his possession a large number of incidents and experiences well-fitted to throw light on the points he employs them to elucidate, and to clench the appeals which he uses them to enforce. In addition to all this, he has a deeply pathetic vein, which enables him to plead very earnestly at the very citadel of the heart. At first his tone may seem to be hard. He will take for his text, "There is no difference," and press the doctrine of universal condemnation as if the worst and the best were precisely alike. Possibly the antagonism of his audience is somewhat roused. But by and by he will take them with him to some affecting deathbed, and his tone will show how profoundly his own heart is stirred by what is happening there. The vein of pathos comes out tenderly and beautifully. He seems as if he were lying on the ground pleading in tears with his hearers to come to Christ. But, most important of all, he seems to rely for effect absolutely on divine power. Of course, every true preacher does, but in very different degrees of conscious trust and expectation. Mr. Moody goes to his meetings, fully expecting the divine presence, because he has asked it. He speaks with the fearlessness, the boldness, and the directness of one delivering a message from the King of kings and Lord of lords. And he takes pains to have his own heart in the spirit of the message. He tries to go to his audience loving them, and actively and fervently longing for their salvation. He says that if he does not try to stir up this spirit of love beforehand, he cannot get hold of an audience; if he does, he never fails. He endeavors to address them with a soul steeped in the corresponding emotion. He seems to try, like Baxter, never to speak of weighty soul concerns without his whole soul being drenched therein.

With all this, there is in Mr. Moody a remarkable naturalness, a want of all approach to affectation or sanctimoniousness, and even a play of humor which spurts out sometimes in his most serious addresses. Doubtless he gets the tone of his system restored by letting out the humor of him after a long day's hard and earnest work. For children he has obviously a great affection, and they draw to him freely and pleasantly. We should fancy him a famous man to lead a Sunday-school excursion party to the country, and set them agoing with all manner of joyous and laughing games. We are sure he himself would be the happiest of the party, enjoying the fun himself as well as pleased at their enjoyment of it. The repression of human nature, or the running of it into artificial moulds, is no part of his policy. We are sure he must agree with the late Dr. Guthrie, that there is nothing bad in human nature except its corruptions, and that our aim should be not to destroy it or any part of it, but to get it restored, as God at first made it. His instincts of sagacity make him recoil from all onesidedness, and desire that men and women, under God's grace, should hide no true accomplishment, and lose no real charm.

At public meetings, Mr. Sankey seldom goes beyond the singing, except to say a few words connected with his hymns, or to give some little incident fitted to encourage and stimulate. The feeling thrown into his singing and the beauty of the singing itself are his great charms. Mr. Sankey is very particular about distinctness of articulation, and in his solos every word and syllable may be heard as distinctly by his audience as if he were speaking. In the after meetings, Mr. Sankey takes a more prominent part. He converses with the anxious, and gives them suitable instruction and counsel. Mr. Moody's mode of dealing with the anxious is marked by great urgency. He shuts them up to a decision, and will hardly let them out of his hands till they have announced their purpose to give themselves to Christ.

It would be difficult to say if ever Edinburgh has been moved to a similar extent in connection with spiritual things. The Reformation, no doubt, was a great spiritual movement, and there were other great movements in the seventeenth century. In July, 1741, George Whitefield came to Edinburgh, after having been in many parts of Scotland, and his movement was wonderful. A week had not elapsed before concern about salvation began to break out. The broken in heart would come to him about the dawn of day, and at seven o'clock every evening he had a service in the open air in the park of the Orphans' Hospital, where the North British Railway Station now is. Later in the day he had another service, and in the evening he was again visited by the anxious. Three weeks after he came he writes like one amazed, and says he verily believes there are three hundred in the city anxious about their souls. He does not know how he is to tear himself away from Scotland. But Edinburgh then had probably not more than one-sixth the population it has now, and as they were all living closely and compactly in the Old Town, a spiritual movement had fewer physical obstacles and difficulties. The movement now going on is probably of much larger dimensions, even proportionally, than that under Whitefield. Thirty years ago William Burns held large meetings in Edinburgh, but the community was not roused in anything like the degree in which it is roused now. One great effect of the movement now going on in Edinburgh is, that other parts of the country are getting into the spirit of hope and expectation. Whenever this spirit is awakened, spiritual results are likely to follow. Whitefield ended his Scotch campaign where Moody and Sankey have begun theirs. We trust that no less a blessing may rest on their labors than on his, and that they too may find, when the term for their return draws nigh, that they cannot yet tear themselves from Scotland.—*Ab. from The Daily Review, Jan. 6th.*

MR. SANKEY'S MUSIC AND SINGING.

No stranger who has ever visited Glasgow has been privileged to sing to such numerous, crowded, and attentive audiences, as Mr. Sankey. He has introduced amongst us a style of music which to a great extent is new in public worship. In Scotland, our service of praise has been hitherto chiefly confined to the use of *psalms*. In many of our churches, *hymns* have been used to a considerable extent, and gradually this style of music is finding its way among all denominations. Mr. Sankey has given us a clearer understanding of what is meant under the third division of the apostle's classifications, viz., *spiritual songs*. He literally "sings the gospel," just as truly and not less powerfully than his friend Mr. Moody preaches it. This element of solo singing in public worship is quite new to us in Scotland, and has proved to be so effective, so attractive, and has been so much blessed amongst us, that it is to be hoped that many who are gifted with the power of song, may take courage, and be induced to follow Mr. Sankey's example, and use this power of song as a new means of bringing the truths of the gospel before the masses of our people. The charm and power of Mr. Sankey's singing are its intense earnestness, and the clear, plain, simple enunciation of every word sung. It is manifest to every one that he feels intensely the truth he is singing, and that he is determined that every one shall hear it and feel it also. He comes with a divine message to his audience, and sets himself to make it known to every one and to be realized by all; hence the deep impression produced by his singing. What he sings is nothing new to us, but how he sings is very new indeed. His songs are simple. The subject is the old, old story. The words are plain and pleasant, but nothing extraordinary; often not to be compared to those of our well-known psalms and hymns. The music is generally pretty and pleasant, but little more; a small portion of it has any claim to originality. Much of it is so Scottish and Irish in its construction that to our people familiar with such music, it is sometimes difficult to realize that what we hear is sacred song. Usually short turns and strains remind us irresistibly of something we know, but cannot recall. In some of the melodies the effect is more marked. Who does not feel the sweetness of familiar Irish melody in "Sweet by-and-by," and the "Valley of Blessing," and the thorough Scottish ring in such songs as "Hold the Fort," "Sweet Hour of Prayer," "The Gate Ajar," "Here am I, send me," and many others. It takes us by surprise to hear gospel truth wafted in the strains of our national music; but is it not possible that this may be the true though unexpected reason why these simple songs have found such a direct and wonderful entrance to the Scottish heart? Mr. Sankey has a fine, full, soft, baritone voice, well trained, and over which he has complete mastery—the organ he uses as a mere accessory, though sometimes its help is not beneficial—for instance, in the song "I am so glad," the effect of the compound triple time is very striking, being quite unknown to us in church music, but the organ having no accent cannot mark this effect, and the first line is always heard in a monotonous staccato style, which unfortunately our audience are too ready to follow. When Mr. Sankey sings clearly out, so as to drown the organ, it is all right, but when our choirs and congregations are left alone to sing this song, they do so in an undecided common time greatly marring its beauty. In the chorus the effect is generally better, for the marked accent of the words helps to keep them right. Mr. Sankey's singing has not the least pretension to be artistic: nothing can be more plain and natural. The music with him is a secondary matter: the words are of the first importance. He sings the words and brings out their full meaning and expression. The music is made subservient, and in time and accent is constantly varied, so as *to fit the words.*

THE WORK IN SCOTLAND.

EDINBURGH.

I.

PRAYER AND FASTING.

Nov. 28, 1873.—We are having a very good time here just now, under the preaching of Mr. Moody and the singing of Mr. Sankey.

We are all delighted with them; ministers of all denominations are joining cordially in the work, and God is indeed working graciously. About 2,000 are out every night hearing; many more come and cannot get into the church. Two churches are to be opened simultaneously each night next week.

The singing of Mr. Sankey lays the gospel message and invitation very distinctly and powerfully on the consciences of the people; and Mr. Moody's gospel is clear, earnest, distinct, and well illustrated—telling of death and resurrection—the "gospel of God." He is a first-rate workman, and very practical, and God has been blessing his preaching.

Every evening there has been a number of souls coming into the inquiry rooms; but last night, when preaching on "the Son of man came to seek and to save that which was lost," the *Spirit* seemed to be working in special power, and old *Formality* got his neck broken, and the wounded and weeping souls came into the inquiry meeting in droves. I had to speak at one time to seven all at once, because there was more corn than reapers; and others were similarly circumstanced. I saw Mr. Moody all the evening with generally more than one. Three rooms were open for inquirers, and I don't know what they had in the others, but we had about forty names on the paper at the close, of those we conversed with in our room. Mr. Moody keeps with us in the elders' vestry. Others, who are less susceptible and can stand at doors, do so, and lay hold of the people as they retire. About one hundred, I should think, were spoken with privately last night, and numbers of them decided for Christ. About ten did so (or professed to do so), in conversation with myself. May the Divine Spirit make it a grand reality to their souls that Christ is theirs! On Tuesday night I had seven who professed conversion.

On Wednesday I fought away with two only, both chronic cases, deep in the mire of their own thoughts, and feelings, and reasonings, and I left them very much the same as I found them. (One of them has been saved.) This was, I suppose, to teach me this lesson, that it is altogether God's work to save, and man is powerless.

This experience made me go out next night with Jesus' word on my lips, "This kind goeth not out but by *prayer* and *fasting;*" *prayer* is the symbol of our dependence upon God, and fasting is the symbol of "no confidence in the flesh"—or self-renunciation. No devil has so powerful a hold of an anxious soul but that *prayer* and *fasting* will cast him out in the name of Jesus.

Our noon prayer-meeting is well attended; about 700 are out daily, and there is a remarkable quickening and earnestness among ministers and Christians generally. I know Edinburgh well, and I am safe to say that I never knew a time when there was a greater appearance of harmony among Christians; unity among the Lord's workers; and humble, prayerful waiting upon God for blessing.

II.

"I HAVE FOUND JESUS."

Dec. 2, 1873.—The work here still makes good progress. On Friday there was much blessing to Christians, and numbers of souls were also brought in. On that evening we had delightful work in the inquiry meeting, and, I think, I had about half-a-dozen I had good hope of. One was specially interesting, a stranger from beyond Stirling. She was passing through, came to the meeting, heard, was awakened, came into the inquiry meeting, and into my hands, along with a girl of twelve, and both professed to see the way of salvation in Acts xiii. 38, 39. This woman was astonished to hear that she had just to believe what she read there to be saved. She said, "Is that all? have I only to believe it?" "Just to believe that forgiveness is yours as a gift from God." "Then I do believe." "Then God says you are justified from all things."

On the Lord's-day morning Mr. Moody preached very effectively to Christians in Dr. Andrew Thompson's Church, and in the evening there were three meetings, one in the Barclay Church, another in Viewforth Free Church, and a third in Fountainbridge Church. About 3,000 heard the Word from Mr. Moody that day.

Mr. Moody preached first in Barclay Church at six o'clock, then Viewforth at 7:45. Then he went down to Fountainbridge Church, where Dr. Bonar had been preaching till he came, but they had despaired of Mr. Moody coming; and just as we were at the church door we met the people coming out, which seemed rather provoking, but Mr. Moody said, "We'll go back and get the inquiry meeting at Viewforth Church," and just as we got back the inquiry rooms were filled—there looked like 200—but there were many workers among them—about one-half were Christians.

One soul, we saw, took Christ there; but I had quite a number at the Barclay Church, where I had conversed with the anxious for one hour previously. It was now ten o'clock, and we left off and came home filled with joy in the Lord over new-born souls. We are thankful for the interest manifested, and the first drops of the plentiful rain.

Large numbers were out again last night, and we had a meeting for inquirers at the Free Assembly Hall. About forty confessed that they were new converts, and about forty stood up as anxious to be saved, and were asked to go to the other side of the hall, where they were conversed with.

I got down beside a young lady whom I saw anxious last night in the inquiry meeting, but did not have the opportunity of speaking to her, and kept at work for an hour with her over the Word of God. I could not tell you at length the deep interest of this case; but at the close I had some hope that she has divine life and will yet get liberty.

A beautiful incident happened as I was speaking to her. A young girl bounded up to us and said, with an overflowing joy, "I am the girl you spoke to at the Barclay Church and gave the book to; now I am just going, but could not leave without coming to tell you that *I have found Jesus.*"

We had a very sweet meeting at noon to-day. Mr. Moody gave us the prayers that God does not answer—Moses, Elijah, Paul. I pointed out to him afterwards, to his great delight, that Moses' prayer was answered, to see the land 1483 years afterwards, but not as in the midst of Israel, but in better company, with Jesus in the midst, on the mount of transfiguration; and he saw the land in the light of the glory of Christ. And when he returned he did not care a bit for the land. He was all taken up with Christ, and instead of speaking of it or the goodly mountain and Lebanon, he and Elias spake to Him of His decease that He should accomplish at Jerusalem, the thing nearest his heart. That is the sight we, too, shall get of it (if we do not see it now) when He comes in His glory, and all His saints with Him.

III.

THE FIRST NIGHT IN BROUGHTON PLACE CHURCH.

Tuesday Night, Dec. 2d.—We have had a most impressive address from Mr. Moody this evening in Broughton Place Church, on the text, "*Where art thou?*" He spoke very solemnly to Christians, and said if they were to wake up, Edinburgh would be filled with awakening from one end to the other, inside of forty-eight hours. Then he spoke to sinners, and it was most alarming. The three steps to hell, he said, were—

1. *Neglect;* 2. *Refuse;* 3. *Despise.*

He told them, even weeping, of their danger, and besought them to get the question settled now. Ah, it is that tender, weeping power in dear Mr. Moody, that is so overwhelming to sinners. He is now preaching in one of the best and largest churches of the New Town, and yet he has been quite as faithful as when among the poor last week in the Old Town; and there have been some marked cases of awakening. Mr. Sankey's singing of "Jesus of Nazareth" had a fine effect upon them. I saw it striking in upon the hearts of many; and many weeping eyes told of its power. A widow in front of me, with her little boy by her side, was moved deeply, and publicly addressed by

Mr. Moody, listened with very wistful eyes; and both of them came to the second meeting. I was anxious about the result of the inquiry meeting in that church, and they were rather long in coming in, but it turned out nearly as good as before. About fifty were conversed with this first night, and there seemed to be quite a number that believed.

The first I got hold of was a working man; and after showing him in the Word the way of life and peace, and getting him to decide, he said:

"My wife's here."

"Where?"

"Sitting there by herself."

"Please bring her here."

She, too, professed faith in the Lord Jesus Christ, and they went home together believing.

Then I got a youth about eighteen in a terrible state of anxiety, and wrought with him a long time, and though hopeful, I do not know that he sees clearly; but he lives near me, and I hope to see him again to-morrow. The life is in, I believe, but he wants liberty.

Then I got a word with about a dozen besides, and gave them books.

I saw three all at once profess Christ in Mr. Moody's hands.

But there were chronic cases that baffled the whole of us, and after ten o'clock there was a man in a corner to whom Mr. Gall had spoken all night, who was all but desperate with conviction. Mr. Moody prayed with him, and he was bowed down and weeping, but we had to leave him still in bondage, showing how entirely it is God's work to set a soul free.

A thing of some special interest to me happened as I was giving in the names. The recorder said to me, "Mr. Reid, you are my wife's spiritual father."

I said I was not aware of it. "When was it?"

"It was in the last revival about twelve years ago, and she used to attend all your meetings when you lived in Edinburgh."

I said I was glad to hear it, and, most likely, I would know her if I saw her.

It is delightful to find fruit "after many days;" it gives one more confidence and hopefulness that what we are now seeing in blossom will ripen into precious fruit."

IV.

FAITH.

TO-NIGHT (*Thursday, Dec.* 4,) we have had a good time. Mr. Moody preached on faith. But you will know what it was when I tell you that it was scriptural, and it was very convincing. Mr. Sankey sang "*Jesus of Nazareth passeth by.*" There was a power in it; many wept. At the close I had three or four anxious sinners, and about as many anxious saints. Mr. Moody had a goodly number professing faith in his hands. Others also were busy. I had some interesting cases of saints in darkness who again got light. Just as I was leaving, Mr. Moody put into my hands a young lady who had been conversed with by one and another all the evening; and just as I spoke the *very last word I intended to speak to her*, her face was lighted up with joy, and she said, "I now trust in the Lord Jesus." Dr. Thomson remarked as we were coming out, that he thought it had been a night of more solid work than any we have yet had. One good thing in being in one of the New Town churches is, that "the poor rich," as a noble worker calls them, have got a chance for their souls. The most respectable men and women have been plentiful in the meeting, and not absent from the inquiry rooms. The poor have far more privileges and opportunities of being saved than the better classes. But they, too, are getting a chance now; and we have seen some marked instances of salvation among them. We returned home, praising God for His grace and blessing.

V.

THE INQUIRERS' MEETING.

Dec. 5, 1873.—I desire to give you an inkling of the work in the inquiry meeting as I have seen it for the past fortnight. I have observed that Mr. Moody speaks to inquirers with an open Bible in his hands, fixing them down to the Word of God, and anchoring their souls on the living rock of the Holy Scriptures. He also gets them to their knees in prayer; and I have seen them rising from his side by twos and threes, wiping their weeping eyes, and smiling through their tears, confessing Christ.

My conviction is, that in the inquiry room in the Barclay Church, where we were, not fewer than fifty souls appeared in a hopeful way to continue their journey heavenward, during the first week.

This week, and especially on Friday night, there have been, we believe, as many in the rooms below the Broughton Place Church; and Dr. Thomson said on Friday night before we left, "I think there could not have been fewer than one hundred inquirers here to-night, and I think more have professed faith in Christ than any night." It was very cheering to see the great heartiness with which Dr. Thomson entered into the work of the inquirers' meeting; and also to see other ministers there, in considerable force, from his own church and other churches engaged in pointing sinners to Christ. Having been every night at work for an hour and a half in the inquiry meeting, and judging of the work from seeing about forty come to Christ in my own hands, I judge that the Lord is doing marvelous things among us, whereof we are glad.

Seven professed faith in Christ all at one time in one company, and we had a conviction that it was reality in at least four of them. On Friday night, after Mr. Moody's solemn word, there seemed to be a great smashing up of souls (as Mr. Radcliffe used to call it), and among others, a lady came into my hands from San Francisco, California, here for the healing of her body; and her trouble was that the Spirit, she thought, had left her. We showed that her anxiety to be saved and her clinging to Christ were evidences to the contrary; and she left after ten minutes' conversation in a state of blessed emancipation and comfort. She was brought to me by one who got out of bondage the night before; and I said, "Perhaps you will be bringing two each on Sunday night." Thus the work of the inquirers' meeting is self-propagating.

VI.

WEEPING FOR A NIGHT, JOY IN THE MORNING.

THE last case we dealt with on Friday night was the most solemn we have seen, except that man who was specially prayed for in the noon-day meeting the other day, and saved that night. This was a young woman weeping floods of tears. She complained of a hard heart, and feared the scorn of the ungodly when she went home; she faintly professed faith in Christ.

I felt such an interest in this girl that I could not sleep without sending her a line by post, inviting her to come next day that my wife might read the Scriptures with her, and tell her more about the Lord Jesus. She came: I was at a meeting I have on Saturday evening. We made special prayer for her, and the person who led us seemed to get near to God, and we had a conviction that we were heard. It was so; for on my return home, I was met with the cheering intelligence, "The girl has been here: I have read with her for nearly two hours; and she has just left, saved and happy. She said she faintly believed last night, as you said, but she is now at liberty, and says she never saw the fullness and freeness of salvation as she sees it now. Her eyes were red and swollen with weeping last night; but she was looking bright and smiling; and the only tears she wept were tears of expressed gratitude that Jesus had received her, and that we had been so interested in her as to care for her for Jesus' sake as we had done." We have seen her since, and she is looking unto Jesus; but her demeanor is quiet and subdued, and she looks as one would do who had just escaped from drowning, or from a terrible railway collision.

VII.

THE SATURDAY MEETING.

Dec. 6.—We have had a meeting to-day for parents and children. It assembled—about 2,000 were present; the parents got a good word. Our dear brother Sankey's singing happily gave the gospel to the children in a number of gospel hymns.

Mr. Moody addressed parents from Deut. iv. 5–11; v. 29; vi. 7. Some young people think they hear too much about Christ and salvation from their parents, but here they have authority from God to speak of them, morning, noon, and night; when lying down and rising up; when sitting in the house and walking by the way. There should be the most diligent instruction of the young by parents, storing their minds with the Word of God.

Then from Mark x. 13–17 he addressed the children, and said that this is the only time when Christ was said to be much displeased. He told of the daughter of an infidel dying in peace, after being only five weeks at the Sunday school. Also, of a boy of twelve, who heard Dr. Chalmers preach, and came, at the close of the service, and said he had nothing to give, but he would give himself to Christ. He did so, and has been the means in our country of establishing many Sabbath-schools, with tens of thousands of scholars, and out of them have grown as many as thirty-eight churches, in which are many precious souls saved and happy, all through this boy coming to Christ and giving himself to Him.

VIII.

MESSRS. MOODY AND SANKEY IN BROUGHTON PLACE CHURCH.

AFTER the labors of Messrs. Moody and Sankey in Broughton Place Church for a week, I am sure you will give me some space in your paper for the statement of a few facts and impressions.

I should consider it a very superfluous work to say anything of the trustworthiness of these excellent men. They have come amongst us not as unknown adventurers without "letters of commendation," but as long-tried and honored laborers in the fields of evangelism in their own country, and more recently in Newcastle and other towns in the north of England, where there appears to have been a pentecostal blessing in which every denomination of Christians has shared. And the ministers, and elders, and deacons of our different churches that have gathered around them every evening, and shared with them in their blessed work, prove the confidence in which they are held by those in whom the Christian people of Edinburgh are accustomed to place confidence.

The service of song conducted by Mr. Sankey, in which music is used as the handmaid of a gospel ministry, has already been described in your columns. I have never found it objected to except by those who have not witnessed it. Those who have come and heard, have departed with their prejudices vanquished and their hearts impressed. We might quote, in commendation of this somewhat novel manner of preaching the gospel, the words of good George Herbert:

"A verse may win him who the gospel flies,
And turn delight into a sacrifice."

There is nothing of novelty in the doctrine which Mr. Moody proclaims. It is the old gospel—old, yet always fresh and young, too, as the living fountain or the morning sun—in which the substitution of Christ is placed in the centre and presented with admirable distinctness and decision. It is spoken with impressive directness, not as by a man half convinced and who seems always to feel that a skeptic is looking over his shoulder, but with a deep conviction of the truth of what he says, as if, like our own Andrew Fuller, he could "venture his eternity on it," and with a tremendous earnestness, as if he felt that "if he did not speak the very stones would cry out." The illustrations and anecdotes, drawn principally from his strangely-varied life, are so wisely chosen, so graphically told, and so well applied as never to fail in hitting the mark.

I wish once more to call attention to one essential feature in the action of these good men—the daily noon-day meeting for prayer. It began some weeks ago in an upper room in Queen Street Hall. That was filled after a few days. Next it was transferred to Queen Street Hall, which is capable of holding 1,200 persons. It was not long ere this became overcrowded, and now there are full meetings every day in the Free Assembly Hall, which is capable of holding some hundreds more. It is a fact with a meaning in it, that simultaneously with the increase in the noon-day meeting for prayer has been the increase in attendance in Broughton Place Church at the evening addresses, and also in the number of inquirers afterwards. Before the end of last week every inch of standing-ground in our large place of worship was occupied with eager listeners, and hundreds were obliged to depart without being able to obtain so much as a sight of the speaker. The number of inquirers gradually rose from fifty to a hundred per night, and on Monday evening this week, when the awakened and those who professed to have undergone the "great change," were gathered together in our church hall, to be addressed by Mr. Moody, no other persons being admitted, there were nearly three hundred present, and even these were only a part of the fruits of one week. I wish to give prominence to the statement that the persons who conversed with the perplexed and inquiring were ministers, elders, and deacons, and qualified private members of our various churches; and also Christian matrons and Bible-women, as far as their valuable services could be secured.

And now, at the close of the week of special services in Broughton Place Church, I wish to repeat the statement in your paper which I made on Monday in the Assembly Hall, that there is no week in my lengthened ministry upon which I look back with such grateful joy. I would not for the wealth of a world have the recollection of what I have seen and heard during the past week blotted out from my memory. When Howe was Chaplain to Cromwell at Whitehall, he became weary of the turmoil and pomp of the palace, and wrote to his "dear and honored brother," Richard Baxter, telling him how much he longed to be back again to his beloved work at Torrington. "I have devoted myself," he said, "to serve God in the work of the ministry, and how can I want the pleasure of hearing their cryings and complaints who have come to me under convictions." I have shared with many beloved brethren during the past week in this sacred pleasure, and it is like eating of angels' bread, first to hear the cry of conviction, and yet more to hear at length the utterance of the joy of reconciliation and peace!

I was much struck by the variety among the inquirers. There were present from the old man of seventy-five to the youth of eleven, soldiers from the Castle, students from the University, the backsliding, the intemperate, the skeptical, the rich and the poor, the educated and the uneducated; and in how many instances were the wounded healed and the burdened released!

It may be encouraging to Christian parents and teachers to be told that very much of this marvelous blessing, when once begun in a house, has spread through the whole family, and those who already had the knowledge of divine truth in their minds by early Christian education, formed by far the largest proportion of the converts. The seed was there sleeping in the soil, which the influence from above quickened into life.

There was a considerable number of skeptics among the inquirers, but their speculative doubts and difficulties very soon became of no account when they came to have a proper view of their sins. Some have already come to tell me of their renunciation of unbelief, and their discipleship to Christ. One has publicly announced that he can no longer live in the ice-house of cold negations, and has asked Mr. Moody to publish the address which brought light to his heart, and to circulate it far and wide over the land.

I witnessed no excesses in the inquiry rooms, but there was often deep and melting solemnity, sometimes the sob of sorrow, and the whispered prayer of contrition or gratitude. There must, however, occur at times imprudent things and excesses in connection with even the best works that have imperfect though good men employed about them. But cold criticism that is in search of faults, or ultra-prudence that attempts nothing from fear of making mistakes, is not the temper in which to regard such events. I would not dare to take either of these positions, "lest haply I should be found to be fighting against God."

I have already expressed my high appreciation of Mr. Moody's manner of addressing. If some think that it wants the polished elegance of certain of our home orators, it has qualities that are far more valuable; and even were it otherwise, the great thing is to have the gospel of the grace of God clearly and earnestly preached to the multitudes who are crowding every night to listen to him. When the year of jubilee came in ancient times among the Jews, I suspect the weary bond-slave or the poor debtor cared little whether it was proclaimed to him with silver trumpets or with rams' horns, if he could only be assured that he was free.—I am, etc.,

ANDREW THOMSON.

63 Northumberland Street,
    Edin., *Dec.* 9, 1873.

IX.

December 7th.—There was a meeting at Broughton Place Church—two thousand present. Another at Free Lady Glenorchy's Church—one thousand present. Mr. Reginald Radcliffe was in the Free Assembly Hall—fully one thousand present; so that upwards of four thousand heard evangelistic preaching. The meetings have been full of interest—hundreds cut up by the truth, and the gospel preached in the power of the Holy Ghost. Christians seem filled with joy, and ready to take each other to their bosoms; there is such a fine spirit of love and unity pervading their minds.

Mr. Moody was speaking of the sinners' excuses in Luke xiv. this evening at both the meetings.

A number of blind children were awakened. There were also some good cases here too among men and women, as well as at the other church. I have been preaching in a neighboring town, but I am told that Mr. Moody has preached like a giant to-night, and there was great power. Mr. Sankey had also much power in singing, "Prodigal Child" and "Free from the Law."

SECOND MEETING WITH YOUNG CONVERTS.

Monday, Dec. 8th.—This evening there was a prayer-meeting in Dr. Thomson's church, and the inquirers met Mr. Moody in the room below along with those who had been recently converted. About seventy stood up and told of the blessing they had received through Mr. Moody's preaching and the Word of God as it had been brought before their minds. This indicates a considerable awakening; for I judge that for every one who comes to the inquiry meeting to be conversed with, there must be nine who go home with the arrow of conviction in their souls. Few could summon up courage to go there and face strangers. The most part go home to weep, and pray, and read, and ponder alone.

There has been some very blessed work this evening. There seemed to be a goodly number of inquirers, and men and women well qualified to speak with them. We were occupied with four young men for the greater part of the evening, and they all professed faith in Christ, but we fear they only saw men as trees walking; but if there is life the liberty will come by-and-by. On going to ask Mr. Moody to come to speak with them, I found him at the door trying to find out the condition of all that went out. Just as I went up to him, he was saying to three ladies, "Oh, surely you will not think of leaving without Christ. He will converse with you."

And so saying, he got them down on a seat, and me beside them, and left. I could judge from their Bibles, that were well marked, that they were not careless persons, but probably Christians who would not like to commit themselves by saying they were "saved," but who had a secret trust in Christ; and I think I was right, for no sooner did I bring before them in an earnest and personal way one or two texts, than they seemed deeply interested; and as they were troubled that they had not sufficient conviction of sin, they appeared to be greatly helped by being told that I had no deep conviction of sin—in fact no appreciable conviction of sin at all; but that I felt a want, and was drawn to Christ by His personal loveliness, and that the sin-crisis came a year afterwards. After this I took them to the precious Word, in Romans iii. 24–26, and they all professed faith in Jesus; and I gave them back into Mr. Moody's hands, and they all left confessing Christ. At this very moment a lady came to me and said, "Dear sir, will you come and see a girl over in yon corner that nobody can make anything of? She says she came to hear Mr. Moody preach; she has never been able to get here before, and he has not preached, and she is disappointed and angry, and says she did not come here to be spoken to." I went at once, asked her to come to a quiet place where I could see her alone; but she sat like a marble statue and refused to come. I went to her and tried to win her confidence, but could not get her to enter into conversation. At first I tried her with Acts xiii. 38, 39, which had been so blessed to others, but I felt it was useless, and the sentiment in the word of Jesus being present in my mind, "This kind goeth not out but by prayer and fasting," I looked for direction, and turned to 1 Peter ii. 24. She felt for her handkerchief. I looked in her face and saw a tear trickling down her cheek, and at length I heard her speak. "What are you saying?" I inquired. "That was my father's text," and she wiped away her tears and told me how her father had died prepared for heaven eight months ago, and this was the text he had rested on. "And you believe your father is in heaven?" "Yes." "And you, too, can be prepared, now, just where you sit, to be with your father in heaven, and with the Lord Jesus, just by believing your father's text." I felt deeply interested in her case, and by entering sympathetically into her great sorrow, gained her attention to the gospel, and she left professing faith in her father's text and her father's God. He knows her heart. May He finish the work He has begun!

Next day, as I was talking to a Free

Church minister, and telling him of this interesting case, and the direction I got to the right word, he told me it was all true, for he was the minister who visited the dying sailor, and that he had given him the text, "Who His ownself bare our sins in His own body on the tree;" and he continued, "I had a hope of him (he took his word back, and said), I should not say *hope*, but more than hope; the man died a believer in Christ." He said he would call for her and look after her. Do not those leadings look remarkably like as if she were a sheep of Christ's fold, and that the Good Shepherd is raising up one means after another to get her laid upon His shoulder? One under-shepherd is sent to call her by the gospel, another to shepherd her in the right ways of the Lord.

After this, when standing near the door, the lady who had got relief regarding sinning away the Holy Ghost, came up to me and said, "I wanted to see you, to tell you how astonished I was, on going home and seeing your name on the book given me, to find that you were the author of 'The Blood of Jesus,' a book given me in Australia, on my marriage, by my husband's aunt; and it was the first religious book I ever read with any interest. It struck me much that after traveling round the world, I should come into contact with the author of that book, to be set at liberty, that I might rejoice in God's salvation. The circle is now completed, and I am saved." Her husband is in America, and she is here under an eminent physician. Her soul is free. May the Lord bless and keep her through faith unto salvation!

At the noon-day prayer-meeting to-day, one who spoke said, "When I was a young man, people thought that I was going into consumption, and I consulted a physician, who used only a very few of the most powerful medicines. He gave me a dose that made me sleep for two days, and I could not be awakened; and when I came out of it, I had a sense of being poisoned, and a most unquenchable thirst; but from that day I grew better. He nearly killed me, but he cured me. So when I get a very powerful spiritual remedy in the form of a text, I keep to it in conversing with anxious souls: and one I have seen very much used is, 'Be it known unto you therefore, men and brethren, that through this man is preached unto you the forgiveness of sins: And by Him all that believe are justified from all things, from which ye could not be justified by the law of Moses.' I generally take them on to that kindred passage, Rom. iii. 24–26, which explains it. And if that does not effect my purpose, I lead them on to Rom. x. 9, 'That if thou shalt confess with thy mouth the Lord Jesus, and shalt believe in thine heart that God hath raised Him from the dead, thou shalt be saved.' The great thing for solidity and liberty is to get them to rest on the Word of God, and have a true knowledge of the work of Christ."

The person in charge at the place of meeting, who was anxious to have the place cleared in an orderly way at a certain hour, and the meeting solemnly dismissed with prayer, met the objection raised against it by telling us a good story. He said, "I heard a minister telling how we were not to be saved, and then he said he would tell us how to be saved. But, he said, I had only a few minutes to stay, and I looked at my watch as he spoke of the ways in which we could not be saved, and I longed for him to come to the other side and tell us how we could be saved. But before he came to that, I had to leave; and I did so, and what happened? Before I had proceeded many steps on my way, it came into my mind that *Christ is the way*—'I am the way, the truth, and the life'—and I saw it all. That night a friend said, 'How could you see the way when you left before the minister came to tell the way to be saved?' I replied that I was obliged to leave; but I did not need him to show me the way—the Lord revealed Himself to me. So if there is confidence in the God of all grace, the meeting can be closed at a seasonable hour, and servants and young people sent away in proper time, so as not to be out too late, interfering with family arrangements, and exasperating those not in sympathy with the work, and giving occasion to the enemy to speak reproachfully."

X.

FREE CHURCH ASSEMBLY HALL

Dec. 9th.—The Hall was filled to overflowing. Mr. Moody began by praying for power, and that many to-night should be saved. Mr. Sankey sang, "That will be Heaven for me!" Rev. Mr. Wilson read out the requests for, and led in prayer. Mr. Moody asked prayer for a lady in great distress of mind, and for another present to-night, given to drink. He prayed earnestly for both, especially that the latter should be saved from a drunkard's death, a drunkard's grave, and a drunkard's hell.

Mr. Moody asked Mr. Sankey to sing a hymn composed on the dying words of a saint in Philadelphia, a few month ago, "I am sweeping through the gates, washed in the blood of the Lamb," which was sung with deep feeling, and prepared all for Mr. Moody, who announced that his address was for the inquiring, as well as the careless.

XI.

UNITED PRAYER FOR SCOTLAND.

The following paper has just been issued, and sent to every minister of every denomination in Scotland. Let our readers put it into operation wherever they are:

"Edinburgh is now enjoying signal manifestations of grace. Many of the Lord's people are not surprised at this. In October and November last, they met from time to time to pray for it. They hoped that they might have a visit from Messrs. Moody and Sankey of America, but they very earnestly besought the Lord that He would deliver them from depending upon them, or on any instrumentality, and that He himself would come with them, or come before them. He has graciously answered that prayer, and His own presence is now wonderfully manifested, and is felt to be among them. God is so affecting the hearts of men, that the Free Church Assembly Hall, the largest public building in Edinburgh, is crowded every day at noon with a meeting for prayer; and that building, along with the Established Church Assembly Hall, overflows every evening, when the gospel is preached. But the numbers that attend are not the most remarkable feature. It is the presence and the power of the Holy Ghost, the solemn awe, the prayerful, believing, expectant spirit, the anxious inquiry of unsaved souls, and the longing of believers to grow more like Christ,—their hungering and thirsting after holiness. The hall of the Tolbooth Church, and the Free High Church, are nightly attended by anxious inquirers. All denominational and social distinctions are entirely merged All this is of the God of Grace.

"Another proof of the Holy Spirit's presence is, that a desire has been felt and expressed in these meetings, that all Scotland should share the blessing that the capital is now enjoying.

"It is impossible that our beloved friends from America should visit every place, or even all those to which they have been urged to go. But this is not necessary. The Lord is willing Himself to go wherever He is truly invited. He is waiting. The Lord's people in Edinburgh therefore, would affectionately entreat all their brethren throughout the land to be importunate in invoking Him to come to them, and to dismiss all doubt as to His being willing to do so.

The week of prayer, from 4th to 11th January next, affords a favorable opportunity for combined action. In every town and hamlet let there be a daily meeting for prayer during that week, and also as often as may be before it. In Edinburgh the hour is from 12 to 1, and where the same hour suits other places, it would be pleasing to meet together in faith at the throne of grace. But let the prayers not be formal, unbelieving, unexpecting, but short, fervent, earnest entreaties, mingled with abounding praise and frequent short exhortations; and let them embrace the whole world, that God's way may be known upon earth, His saving health among all nations. If the country will thus fall on their knees, the God who has filled our national history with the wonders of His love, will come again and surprise even the strongest believers by the unprecedented tokens of His grace. "Call unto me and I will answer thee, and show thee great and mighty things which thou knowest not."

W. G. Blaikie, D.D., Professor, New College.
Charles J. Brown, D.D., Free North Church.
James Balfour, 13 Eton Terrace.
H. Calderwood, Professor of Moral Philosophy
Lawrence G. Carter, Charlotte St. Bap. Chapel.
A. W. Charteris, D.D., Prof. of Bib. Criticism
John Cooper, late of Fala, U.P.
G. D. Cullen, Royal Terrace.
Cavan, 12 Lennox Street.
Alexander Duff, D.D.
William Dickson, 38 York Place.
David Dickson, Merchiston.
F. Brown Douglas, 21 Moray Place.
William Grant, Bristo Place Baptist Chapel.
William Hanna, D.D., 16 Magdala Crescent
John Kelman, Free St. John's, Leith.
Robert Macdonald, D.D., Free North Leith.
James Macgregor, D.D., Prof., New College.

John Macmurtree, St. Bernard's Church.
John Millar, 26 York Place.
W. Scott Moncrieff, St. Thomas' Episcopal Ch.
John Morgan, Viewforth Free Church.
David M'Laren, Redfern House.
Duncan M'Laren, jun., Newington House.
Samuel Newnam, Baptist Church, Dublin St.
Maxwell Nicholson, D.D., St. Stephen's Ch.
Polwarth, Mertoun House.
Robert Rainy, D.D., Professor, New College.
James Robertson, U.P., Newington.
Moody Stuart, Free St. Luke's
E. Erskine Scott, 25 Melville Street.
Andrew Thompson, D.D., Broughton Place Ch.
John Wemyss, Richmond Place Congregational Church.
Alexander Whyte, St. George's Free Church.
Ninian Wight, Congregational Church.
George Wilson, Tolbooth Parish Church.
J. H. Wilson, Barclay Free Church.
John Young, U.P., Newington.

XII.

"BRING HIM TO ME."

FREE ASSEMBLY HALL.

Dec. 11*th.*—Mr. Moody spoke from Mark ix. 14-30. Ver. 19—"Bring him to me." Some complain that their prayers are not answered, but that is no reason for being weary or waxing faint. The thing is to inquire the reason why God keeps back answers to them.

A lady came to me to-day and said that she feared her two sons were not going to be saved, but they *will* if she continues to pray for them. Ver. 20—Never did a sinner come to Christ yet, that the devil did not throw him down and try to prevent him. Ver. 21—" From a child."

This was a hard case; he had inherited it. Ver. 23—" To him that believeth all things are possible." You cannot believe, mother of these two sons; if you did, you would have the conversion of your sons. Oh, how easy it is for God to take the accursed appetite out of the most abandoned drunkard, and restore him to a right mind; as easy for Him to save as for me to turn my hand round!

"I charge thee, come out of him." A little time of praying and fasting, of being alone with God, of inquiry of Him in what way we hinder His blessing us, that is the thing which we want.

During the American war, when husbands, fathers, and brothers were away on the battle-fields, their wives, daughters, and mothers learnt to pray, and many an hour was spent by them in their closets alone with God. The results were marvelous, and that, too, in the case of the wickedest and most depraved men in the army.

One day at Nashville a great, strong, wicked-looking soldier came to me trembling. He said he had got this letter from his sister, six hundred miles away, and she said that she prayed to God, night after night, that he should be saved, and he said he could not stand to hear that, and he had come to give himself to Christ; and there and then we knelt down together in prayer to God, he crushed and broken in heart.

Oh, what a privilege we have in coming to God in prayer about our friends! Our prayers may not be answered to-day; we may be in our graves before they are but assuredly they will be answered sometime.

Another soldier came to me and said he had got a letter from his mother, saying that she prayed morning, noon, and night for his conversion; that this letter might be the last he would ever get from her, as he might be killed in battle. "I said when I got it, that I would wait till the war was over, and I would go home and settle down and be a Christian: but I hear to-day that mother is dead, that that letter was the last she ever wrote, so I have come to give myself to my mother's God," which he did. Both these men found peace in Jesus, and became bright and shining lights in the army.

XIII.

CHILDREN'S MEETING.

Saturday, Dec. 13*th.*—There were about two thousand persons present at the noon-day meeting to-day, about one-half children and young people It was a time of much blessing, both in prayer, singing, and speaking the word of grace.

Mr. Moody spoke very appropriately from the words, "*I will hold thee by thy right hand.*" He showed that we need Jesus to save us, and then to take us by the right hand and lead us. He told a story of an ill-used dog having been thrust by a bad man through the gratings into the lion's den, and how that lion became its protector, and applied it to illustrate the truth that the Lion of Judah becomes our protector.

XIV.

TWO SOLDIERS.

We saw some precious cases of resting on the Saviour in the Assembly Hall of the Church of Scotland. Among others, we were deeply interested in two soldiers. One of them said they had many warnings: a soldier who had been boasting the one day that he had not read the Bible since he was a boy, was drowned the next. I began there and laid before him from the Word of God the way to be ready for death in any form. Christ was pressed on him, and he professed his faith in Him. The other one was trembling all over with anxiety about his soul. I tried to anchor him on the Word of God's grace, and got a text fixed in his mind, on which he professed to rest. But I was not entirely satisfied that either of them saw the way of life clearly, though I considered them truly awakened. The next night, as I was speaking to an inquiring child, the soldier who had trembled with conviction the night before, came up to me with a look of assured peace and told me that the Lord had had mercy on him and saved him.

"When was it, do you think?" I inquired.

"It was when I was lying awake at four in the morning thinking over the text that you gave me, that Jesus came into my heart."

"The Lord be praised! But now that you are pardoned, you will need to trust in Christ and abide in Him, and be strong in the Lord and the power of His might, and fight the good fight of faith and lay hold on eternal life. 'In the Lord have I righteousness and strength.' Ask God that, by His Holy Spirit, He may enable you to bear all opposition with firmness and meekness, witness a good confession, and seek to bring your comrades to Him."

We heard of the other soldier that he, too, had come to Christ, and become one of His soldiers.

XV.

MEETINGS IN THE NORTH OF EDINBURGH.

Dec. 20th.—There have been meetings held this week in the north of Edinburgh, in St. Stephen's, St. Bernard's, and Free St. Bernard's churches. Fully 2,000 have been out nightly, and have heard the gospel, both in preaching and song, in such a way as to bring home conviction to many consciences. Numbers have waited, and come to the second meeting for personal conversation, and some have found peace, as on the former evenings; but we have a theory that only one in ten of the anxious wait: nine leave to weep, pray, and read in secret. The gospel is the power of God unto salvation to every one that believeth, and thousands are hearing it in all its naked simplicity, and preached with the power of the Holy Ghost. For this we praise God! Why should it ever cease? Why not go round the city and rouse up every slumbering church, and get souls saved in thousands?

We met with one specially interesting case in St. Stephen's. It was that of a respectable married woman, who had felt as if there were something calling her to stay, and she tarried and received blessing. Satan was very busy to get me to leave the place, but he did not succeed. The gentleman who was conversing with her came up to me, and asked me to speak a little to her, which I did, and in course of conversation she said—"I heard you preach at Auchincairn about ten years ago; and my husband wrote to you, and has a tune of his in your "*Praise-Book.*"

This led to my asking her to tell him to come and see me some morning at nine o'clock. He came the next morning; and after we had looked over some music together, I asked him how his wife was feeling now? He told me that she was very anxious to be sure she was right; and he said, "I wish, Mr. Reid, you would speak to me too."

"Then take that Testament, and let us read a little together." We read Rom. iii. 24-26, and 1 John iv. 9-19, and other passages, "giving the sense," and urging him to make a personal application of it to himself, and he would experience life and liberty. He did so, and went away, thanking me for my explanation of the Word, looking as if he had experienced the liberty of grace. The friend who introduced this woman to me saw them both, farther on in the afternoon, and reported to me that he had had as happy a time as he had ever had, talking with them over the Scriptures; and he believes they both have got into the liberty wherewith Christ makes His people free.

A Bible-woman came up to me the

other night, and said, "It was your book, '*The Spirit of Jesus*,' that gave me the impetus that sent me out in this work. I got a great blessing in reading it."

"Where were you at that time?"

"In Unst, one of the Shetland islands." We were glad to hear of this little bit of fruit after many days; for it is ten years since we sent some hundreds of books to be distributed in Shetland, and have heard very little about the spiritual results. But everything the Spirit stirs up to do for Christ will be used by Him, whether we hear of it now or not (Eccles. xi. 6).

XVI.
BIBLE LECTURES IN FREE ST. GEORGE'S AND FREE ST. LUKE'S.

THERE have been two Bible readings in Free St. George's Church, on Dec. 16th and 19th, on *the Holy Spirit* and *Assurance;* and another in Free St. Luke's Church, on Dec. 23, on *the Blood.*

Mr. Moody gives what is commonly known as a Bible reading—only he himself does all the reading and speaking. About 1,500 were present, chiefly of the educated class. Many of the ministers of the various churches were present, and the effect on all these occasions was marked and marvelous. Christians received a great blessing.

XVII.
SPECIAL MEETINGS.

MR. MOODY has addressed some special meetings in Free Assembly Hall.

On Sabbath morning, December 14, he addressed the young men of the Sabbath-morning Fellowship Union.

On Friday, December 19, he preached to young men on being born again.

On Sabbath morning, December 21, he addressed Sabbath-school teachers.

The same evening he preached to the students of Edinburgh University and the New College, on "There is no difference." This was one of the most magnificent sights I have ever witnessed. On the platform with him were numbers of professors of both colleges, and, I believe, the majority of the students. The hall was densely crowded, and I question whether he ever addressed a more intelligent audience, or one that gave him more profound and riveted attention. Had they not had confidence in him, and felt his power, and, we trust, the higher power of God's Spirit and truth, they would not have sat for more than two hours with such quietness. He commanded that immense meeting, of about two thousand men, as no man on that platform, save Dr. Duff, could have done. The living power of God's Holy Spirit was felt giving the word, and laying conviction on the conscience. The gospel given at the end was most touchingly illustrated, and the very appropriate hymn sung by Mr. Sankey, "I am sweeping through the gates," gave a spiritual finish to the whole that had been spoken. It was an opportunity such as no man ever before enjoyed here; and we cannot doubt but that God has given and used it for the conversion of souls and the glory of Christ.

THE ALL-DAY'S MEETING.

This was held on December 17, the first meeting lasting six hours. Subjects—1. Praise and joy. 2. Promises. 3. Prayer. 4. Christian work; and 5. Heaven.

XVIII.
MEN'S MEETING AT THE CORN EXCHANGE.

AT half-past eight o'clock, December 29, there was a meeting held in the Corn Exchange, Grassmarket, which was attended by about 3,000 persons belonging to the poorer classes. The Rev. Mr. Morgan opened this meeting with prayer.

Mr. Moody began his address by telling the well-known story about Rowland Hill and Lady Erskine. Her ladyship was driving past a crowd of people to whom Hill was preaching. She asked who the preacher was, and, on being informed, told her coachman to drive nearer. Rowland Hill, seeing her approach, asked who she was, and when he was told, he said there was a soul there for sale. Who would bid, he asked, for Lady Erskine's soul? There was Satan's offer. He would give pleasure, honor, position, and, in fact, the whole world. There was also, he said, the offer of the Lord Jesus, who would give pardon, peace, joy, rest, and at last heaven and glory. He then asked Lady Erskine which of these bids she would accept. Ordering her coachman to open her carriage-door, she pressed her way through the crowd to where the preacher was, and said, "Lord Jesus, I give my soul to Thee; accept of it."

Mr. Moody went on to urge on his hearers to give themselves there and then to the same Saviour who was that day preached in the hearing of Lady Erskine, and accepted by her. He brought out the freeness of the gospel offer, and the importance of immediately closing with it. He mentioned several instances of conversion—one of them concerning a soldier, who had been at the meeting of the previous night in that same hall, and who had afterwards gone up to the Assembly Hall, had received Christ there, and was now professing himself a Christian man.

Mr. Sankey sang several of his hymns —"The Lifeboat," "Jesus of Nazareth passeth by," and "The Prodigal Child" being amongst them.

The meeting on Sunday night (Dec. 28) seems to have been the most extraordinary of all these meetings. Though there were about 5,000 persons present, the most perfect order was observed, and the deepest interest manifested in the proceedings. After this meeting was over, hundreds pressed up to the Free Assembly Hall, and when the question was put if there were any there anxious about their souls and desiring to be saved, the whole body rose to their feet in answer to the question. The interest shown was such as many of those present had never before seen in the course of a long ministry amongst the people. Mr. Moody expressed himself as more impressed by it than he had been by anything he had ever before seen.

XIX.

SATURDAY AND SABBATH MEETINGS.

On Saturday, at noon, the usual union prayer-meeting was held in the Free Assembly Hall. Besides Mr. Moody, Mr. Robertson, Mr. Cooper, Mr. W. Dickson, Mr. Daniel, and Mr. John Wilson, took each an active part in it. Mr. Sankey was present during half of the hour, spending the remainder in the Established Assembly Hall, where another large meeting was held, under the presidency of the Rev. Mr. Wilson (Jan. 3d.)

On Sabbath morning Mr. Moody delivered an address to a crowded congregation in the Free Assembly Hall on "Daniel," with reference specially to the lessons to be learned from his life. Mr. Sankey sang appropriate hymns.

Mr. Moody preached on Sabbath forenoon in Free St. George's Church, his subject being, "What Christ has done for man." This he treated very generally. In the afternoon Messrs. Moody and Sankey conducted evangelistic services in the Free Assembly Hall and the Free High Church at five o'clock, and in the Established Assembly Hall and Free St. John's Church at six o'clock—these meetings being for females only. The Jubilee Singers sang at each of these meetings. There was an immense meeting in the Corn Exchange, Grassmarket, at seven o'clock. The great hall was filled with people, who stood closely packed together in every part of it. There must have been between 6,000 and 7,000 persons present. Short addresses were delivered by several ministers and laymen, frequent prayer engaged in, and a great number of hymns sung by Mr. Sankey and the Jubilee Singers. These hymns had each of them a bearing on the thoughts or sentiments that formed the themes of the addresses by which they were preceded. There was the most perfect quiet observed by the vast assemblage, and both addresses and hymns were listened to with the utmost attention.

In his address Mr. Moody pointed out that though it was because of Adam's sin man was condemned, it was not because of it that any one would be lost, but because they neglected to lay hold of the remedy.

Mr. Moody preached to about *fifteen thousand* this first Lord's-day of 1874, at seven different times. His passion for saving souls is self-consuming. Let all Christians pray that he may be upheld by God, in body and soul, and blessed more and more.

XX.

EVANGELISTIC MEETINGS IN BERWICK-ON-TWEED.

Tuesday, January 13th, was a memorable day in the religious history of Berwick. It having been announced that Messrs. Moody and Sankey were to visit the town on that day and hold meetings, large numbers of people were brought into the town by the several lines of railway, from distances of twenty and thirty miles. The forenoon trains down the vale of the Tweed were extremely crowded; but additional carriages were provided for the return journey. In one carriage, an interesting account of the Edinburgh meetings of the previous day was read from a daily paper; in another carriage we heard the singing of hymns; and in all, the one subject of conversation was the meetings to which most of the passengers seemed to be on their way. The readiness with which people were allowed to enter at the various stations into compartments where there was only standing-room, was something new in the experience of railway traveling. The first meeting began at noon, being the mid-day prayer-meeting, which is held daily in the Rev. James Stevens' church for one hour. The Rev. Mr. Chedburn presided, and short prayers were offered by a number of ministers and laymen, between which hymns were sung with much spirit. Messrs. Moody and Sankey arrived from Edinburgh shortly after the meeting commenced, and both delivered short addresses. The under part of the church was filled, many strangers being present. Two meetings were held in the Corn Exchange; the former beginning at two o'clock P.M. The great hall of the Exchange was filled, and the passages occupied, though not so closely packed as in the evening. Mr. Sankey sang, and Mr. Moody preached from Rom. iii. 22, "There is no difference." The acoustic principles on which the hall is constructed are not good, and Mr. Moody was imperfectly heard in many parts; but Mr. Sankey's fine voice was heard in every corner. Mr. Moody closed his discourse with the touching narrative of the return of a prodigal; and Mr. Sankey immediately sang with thrilling effect his Christian song, "The Prodigal Child," beginning, "Come home, come home, thou art weary at heart," etc. It seemed to take the vast congregation by surprise, and was the first thing that powerfully affected them. It was most aptly chosen, and gave a very favorable illustration of what is called "singing the gospel." A number of anxious inquirers waited, and were conversed with after the meeting was closed.

The next meeting was held in Wallace Green Church at six o'clock. The large church was well filled in the lower part, with a few people in the galleries; but the great body of the people had gone to the Corn Exchange to wait till seven o'clock, the hour announced for commencing the service there. When Messrs. Moody and Sankey met with the ministers in Dr. Cairns' room at Wallace Green Church, a message was brought that the great hall of the Exchange was already filled in every part. Two ministers were then appointed to address the vast assemblage while Messrs. Moody and Sankey were engaged in Wallace Green. The Rev. Messrs. Mearns, of Coldstream, and Leitch, of Newcastle, and afterwards Mr. Moody, addressed the audience in the Exchange. The male part of the audience seemed to preponderate. It was a vast mass of earnest listeners. We observed ministers of all denominations present from the towns and villages of the neighborhood, extending over a wide district, many of whom remained for the evening meeting. Reference was made in one of the addresses in the Exchange to the case of a mother who, by believing, had entered into peace in the afternoon of that day, and requested thanks to be returned for the blessing she had received, and prayer to be offered, for the recovery of her prodigal son. This was mentioned as the first convert of the day, and the fact was received as a proof of the presence of the Holy Spirit in answer to the prayers which had been offered for a great blessing to accompany the services of that day. This circumstance seemed to make a deep impression on the audience, and enabled them to realize the fact that the Spirit of God was indeed among them in answer to prayer. It was afterwards found in the inquirers' meeting that many had been so deeply impressed in the Exchange that they felt constrained to come among the anxious, asking to be directed to the Saviour. Mr. Sankey's singing excited wonderful interest. "Sweeping through the gates," "Jesus of Nazareth passeth by," and others, seemed to produce a deep impression.

When Messrs. Moody and Sankey left Wallace Green for the Exchange, the meeting was continued in the former place, and addresses by the Revs. Dr. Cairns, R. Scott, of Berwick, and P. Mearns, of Coldstream. The audience gradually increased, till near the close of the third address, such a crowd rushed into the spacious church as to fill every passage above and below. It was soon explained that this was the second meeting, which had been adjourned from the

Exchange to the church, where there were rooms for conversing with the anxious in a more private manner. After the protracted services of the day it might have been expected that all the people would have gone home, as it was now half-past eight o'clock, and many had been occupied with a succession of services from noon. The second meeting, too, possessed no peculiar attraction, consisting only of short addresses with praise and prayer. But the people were evidently moved by an influence which all could feel, but not fully explain. After two days, Dr. Cairns thus wrote of it to the *Daily Review:* "I cannot attempt to describe the appearance of Wallace Green Church at the evening meeting on Tuesday, when the overwhelming meeting in the Corn Exchange was dismissed, and those who gathered for prayer, with the anxious inquirers, crowded in to fill every corner of the spacious church. The shadow of eternity seemed cast over the great congregation. Many were observed to be in tears; and as the inquirers, with hurried and trembling step, passed into the vestry (though others found a more private entrance), the deepest awe and sympathy pervaded the meeting. This continued for a full hour, and such a gathering I hardly ever expect again to see in this world."

While short addresses were being delivered by Mr. Moody and others, an impulse seemed to fall on individuals, one by one, which powerfully drew them into the anxious meeting, where ministers and laymen were appointed to converse with them.

Altogether, Tuesday was a memorable day in Berwick; the like of it, as Dr. Cairns remarked, had never before been seen in the memory of its inhabitants.

Meetings were held on the evenings of Wednesday, Thursday, and Friday. A large number of additional inquirers waited for conversation. Dr. Cairns expresses a hope, in which all must cordially unite, when he says, "I feel constrained to add my testimony to the profound impression which has, by the blessing of God, been made on the town. I trust it will be as solid and permanent as it is at present visible."

On Sabbath evening the Exchange was crowded, when addresses were delivered, and a large number of additional inquirers waited for conversation. The meetings were continued during the week, the ministers of the town being assisted by friends from Edinburgh."

XXI.

SUMMARY OF THE AWAKENING IN EDINBURGH.

Jan. 8th.—During the last two weeks much progress has been made by Messrs. Moody and Sankey in reaching the masses of the population of Edinburgh with the gospel of their salvation.

Tens of thousands of men, women, and children of all classes of the community have crowded the halls and churches where they have preached and sung of Christ and the gospel.

Multitudes of men assembled in the Corn Exchange, and multitudes of women in the Assembly Halls and adjoining churches on the Lord's day to hear words whereby they might be saved; and on the week days the daily prayer-meeting, noon and night, was crowded with eager, anxious throngs of Christians or anxious ones; while in the Newington U.P. Church and the Canongate Parish Church, fully three thousand came together nightly to listen to the singing and preaching of the glorious gospel of Christ.

Bible lectures have been held in the Free Assembly Hall, Viewforth Church, West Coates Church, and Free St. Mary's, and thereby many have received clearer light on the gospel, more stable standing on the sure foundation, and blessed freedom from bondage.

Mr. Moody's excellent plan of making the Bible speak for itself by quoting text after text and commenting on them, and enforcing them by striking illustrations, has been of eminent use among Christians who had life but no liberty. Christ has said through him to many a bound and groaning one, "Loose him, and let him go."

Mr. Moody's clear preaching of grace reigning through righteousness and salvation by grace without the works of the law, and the believer's place in Christ where there is now no condemnation, and sin shall not have dominion over us, because we are not under law but under grace, is fitted to give immediate relief to burdened, unclear, and legal Christians, of whom we have crowds.

ently been in contact
eachings, such as one
our day: for he has
words of grace and
rystal river of divine
: from the muddy
ieology; and if we,
re still to get a hear-
who have hung as if
nistry of Mr. Moody,
e same simple, scrip-
ct manner. He has
and glorified Christ,
host by believing in
e and grace, and his
: the power of God
umbered souls. We
ıny as 30,000 have
ning voice.
: is no doubt deep,
:raordinary, as com-
of things spiritually
ing of those earnest
y the ordinary and
iyer and preaching,
: Acts of the Apostles
ing when all the dis-
:ontinuing with one
supplications, and in
Ioly Ghost are bend-
: to the one work of
God magnified by the
ng souls. When we
at bulk of the minis-
eople of Edinburgh
iost nothing else for
t giving themselves
d to co-operate with
to make the gospel
·; and when we con-
been this concerted,
itrated effort towards
iave hardly seen so
ght reasonably have
'e very sure if there
grieving and quench-
t of God amongst us
s work, both secretly
d have wrought with
the harvest of souls
h more abundant.
een, as in other days,
ly smitten simultane-
ıs arrested as in the
id of the Most High
ide to stand still and
od. Might the Lord
power as would have
of tens, anxiously in-

quiring what must we do, if there had been an entire exclusion of "the flesh" and a total self-surrender on the part of Christians, more regard for the glory of Christ, less grieving and quenching and more honoring of the Holy Ghost?

We do not quite sympathize with some of the things which have been said about Mr. Moody's preaching, and especially that he is not eloquent. What, we would ask, makes the meetings flat when he is absent but the want of a quality he possesses? and what makes them full of life and spiritual emotion when he is present, but just the superior divine eloquence which flows in his burning words, as if an electric current were passing through every heart?

He is the most powerful speaker—the most eloquent preacher—who most fully carries an audience with him and produces the greatest results; and if Mr. Moody is judged by such a rule, he is one of the most eloquent of living men, for none of us here who are ministers feel the least desire to speak if he is present, for with all our university training we acknowledge his superior power as a heaven-commissioned evangelist. He has the all-powerful eloquence of a man full of the Holy Ghost and of faith, and fired with indomitable zeal for the glory of Christ and the salvation of souls. He may be devoid of rhetoric (and that, we suppose, is meant), and he may use his freedom in extemporizing grammar to suit himself, but withal Moody is the most eloquent, as he is the most successful preacher amongst us. The Lord be praised for giving such gifts to men, and for the thousands of souls He has converted by him in this city, or set into the liberty of grace by a fuller knowledge of Christ and His finished work.

What masses of young people from the schools crowded the meetings during the holidays! And so great has been the attraction of the singing of the one and the eloquence of the other, that hundreds of young persons, especially of the higher classes, who were formerly accustomed to go to the theatre, opera, and pantomime, gave them up deliberately, and from choice and the force of conviction attended the gospel and prayer-meetings. Men who can draw away our educated children by the hundred in this city that boasts of its education, from these haunts of pleasure and amusement, to hear of Christ in preaching and song, and embrace Him as their Saviour, and cling to them as their friends, have that spiritual education which ennobles the character, implants delicate feelings, generous sentiments, tender emotions, and gracious affections, which the young very quickly discover and reciprocate.

But we have no doubt that a very great part of Mr. Moody's superiority over most ministers as a preacher of the gospel, arises from his superior knowledge and grasp of the Holy Scriptures.

Messrs. Moody and Sankey's principle for gospel work is the recognition of the divine unity of the one body of Christ; and accordingly wherever they go they say, in effect, A truce to all sectarianism that the Lord alone may be exalted: let all denominations for the time being be obliterated and forgotten, and let us bring our united Christian effort to bear upon the one great work of saving perishing souls. It is a charming sight to look back over the past eight weeks and think of men who, it appeared, were for all time to come in religious antagonism because of their controversial differences on the Union question, sitting side by side on the same platform lovingly co-operating with those American brethren and with one another for the conversion of souls. All old things seemed to have passed away, and all things had become new, and all rejoiced together in the blessing which has been so richly vouchsafed by the God of all grace.

There has been such a commingling of ministers and Christians of all the churches—all sectarian thoughts and feelings being buried—as has never been witnessed in this city since the first breaking up of the Church of Scotland, more than 140 years ago. What all the ministers and people of Scotland were unable to achieve —a union of Christians on a doctrinal basis—God has effected, as it were, at once on the basis of the inner life by the singing of a few simple hymns and the simple preaching of the gospel:—for as the unity of the nation was secured by the one purpose to make David king over all Israel: "All these men of war that could keep rank came with a perfect heart to Hebron, to make David king over all Israel; and all the rest also of Israel were of one heart to make David king, and there was great joy in Israel" (1 Chron. xii. 38); so the one purpose to have the Lord Jesus exalted and made supreme, and His glory in the triumph of His gospel and the salvation of sinners made manifest, has united the ministers and Christian people of every name in the metropolis of Scotland: "and there was great joy in that city" (Acts viii. 8). "Be it known unto you all, that by the name of Jesus of Nazareth, whom ye crucified, whom God raised from the dead, even by Him" hath been "shed forth this which ye now see and hear." "This was the stone which was set at nought of you builders, which is become the Head of the Corner. Neither is there salvation in any other, for there is none other name under heaven given among men whereby we must be saved" (Acts iv. 10–12). "This is the Lord's doing; it is marvelous in our eyes. THIS IS THE DAY WHICH THE LORD HATH MADE. We will be glad and rejoice in it" (Ps. viii. 23, 24).

Mr. Moody is overpoweringly in earnest, and he brings in the direct, straightforward, decided methods of a thorough-going, energetic man of business into his addresses, in conducting meetings, and his dealing with souls, and, as a preacher generally stamps his own image upon his converts, we may hope to see a brood of decided Christian witnesses and testifiers arising out of this time of awakening, that will let it be known that the glory of the Lord Jesus is the uppermost purpose in their hearts.

This witness-bearing has already begun in colleges and schools, in families and workrooms, in drawing rooms and kitchens. There are discussions going on everywhere regarding both the men and the movement. In ladies' schools there are young converts testifying for Jesus, and boldly confessing Him as their Saviour; evening parties, through the influence of the young believers in the household, are being converted into Christian assemblies to talk over the preaching of Mr. Moody, and to sing in concerted worship the hymns and solos which have been introduced by the inimitable singing of Mr. Sankey.

These two quiet and humble Americans have all but turned society in Edinburgh upside down, and, by the grace of God, have given its citizens the merriest Christmas and the happiest New Year that they have ever enjoyed, by gathering them around the Lord Jesus. It seems as if a voice from heaven had been heard saying, "O clap your hands, all ye people: shout unto God with the voice of triumph. God is gone up with a shout, the Lord with the sound of a trumpet. Sing praises to

God, sing praises; sing praises to our King, sing praises; sing ye praises with understanding" (Ps. xlvii. 5–7). W. R.

XXII.

WHAT GOOD HAVE MESSRS. MOODY AND SANKEY DONE IN EDINBURGH?

This is a question which, in its inward aspect, can be answered only by Him who knows the hearts of men; but that which is visible and apparent can be set down in writing.

For one thing, Mr. Moody has given the Bible its due place of prominence, and has made it to be looked upon as the most interesting book in the world. This is honoring the Holy Ghost more than all the prayers for His outpouring that have been offered; for it is getting into the mind of God as the Psalmist got, when he said, "Thou hast magnified Thy WORD above all Thy name." His addresses on such themes as "How to study the Holy Scriptures," and "The Scriptures cannot be broken;" his own Bible lectures, which were so full of Scripture, and helpful to hundreds of Christians; his constant reference to the Bible, and quotation from it in his preaching; his moving about amongst the anxious with the open Bible in his hands, that he might get them to rest their souls on the "true sayings of God;" and his earnest exhortations to young Christians to read the Word, and to older and well-taught Christians to get up "Bible readings," and invite young Christians to come to them, that they might be made acquainted with the mind of Christ, all showed how much in earnest he is to give due prominence to the Holy Scriptures.

Mr. Moody has also given us a thorough specimen of good gospel preaching, both as to matter and manner of communication. It is not a mixture of law and gospel: his gospel is "the gospel of the grace of God," "without the works of the law," "the gospel of God" coming in righteously and saving the lost, not by a mere judicial manipulation and theoretically, but by grace, power, and life coming in when men were dead, so that we have not only sins blotted out by the blood of Christ, but deliverance from sin in the nature by death and resurrection, and life beyond death, so that a risen Christ is before us, and we in Him, when it is said, "There is, therefore, now no condemnation to them which are in Christ Jesus." There is "justification *of life*" in his preaching, immediately that we are "justified *by His blood*."

He has also distinguished with much decision and precision between the Adam-nature and the new creation in Christ, and made it as clear as noon-day that salvation is not the mere setting right of man's existing faculties, but the impartation of new life in Christ, a new nature, a new creation, so that there exists two utterly opposed natures in the one responsible Christian man, and that "these are contrary the one to the other:" and the knowledge of this gives young Christians immense relief, and a solid foundation for holiness at the very commencement of their Christian course. New creation in Christ—not the mending of the old creation—is Mr. Moody's idea of Christianity: and it is the divine reality which many are now enjoying.

This also leads to the Pauline theory of holiness, as preached by him. He has imbibed very fully the theology of the Epistle to the Romans on this point, and insisted with much earnestness that Scripture has it that Christians are not under the law in any shape or form, and that this is essential to holiness:—"For sin shall not have dominion over you; for ye are not under law, but under grace" (Rom. vi. 14); "But now we are delivered from the law, that we should serve in newness of spirit" (Rom. vii.) His doctrine is that the law never made a bad man good or a good man better, and that we are under grace for sanctification as well as for justification; and yet the righteousness of the law is fulfilled in us who walk not after the flesh (that is, under law) "but after the Spirit" (Rom. viii. 4.) His clearness in distinguishing between law and grace has been the lever of life to many souls.

Our American brethren have also been of great use in showing us what may be accomplished in the conversion of souls, if the heart is only fully set upon it, and there is a determination to have it. They came to us with that distinct aim and object in view; and the Lord gave them the desires of their hearts; and, as the result hundreds of souls have professed salvation. They gave themselves to "this one thing," and they stuck to it, brushing aside all other things: even the conventional courtesies of life were made short work of by Mr. Moody if he spied an anxious soul likely to escape. His friends might introduce some notable stranger at the close of a meeting, and feel rather annoyed that, instead of conversing with him or her, he darted off in a moment to awakened souls; but he made that his work, and everything else had to be subordinate to it. "This one thing I do," seems to be his life-motto; and in sticking to this all-absorbing object, he has read us a noble lesson of holy resoluteness and decision. If we who are ministers have similar faith and expectancy, and work like our American friends for the conversion of souls, the conversion of souls we shall have. Our Lord said to those who were to be the first preachers of His gospel, "I have chosen you, and ordained you, that ye should go and bring forth fruit, and that your fruit should remain" (John xv. 16); and when they were endued with the Holy Ghost and with power, they did "bring forth fruit" in the conversion of souls (Acts ii. 41; iv. 4); and their fruit remained (Acts ii. 42), and has done so, in the millions of souls saved in all ages down to the present day.

Our friends have been the means of rescuing hundreds of souls in this city from impending and everlasting damnation. Their labors have been especially fruitful in the conversion of young women and girls, who in course of time will be in the important position of wives and mothers; and if the thousand of them that appeared at the young converts' meeting, to receive Mr. Moody's farewell address, should all hold out, it will be an unspeakable blessing that has been conferred by God on this community through their instrumentality.

Persons at a distance have wondered at us having so many ladies among the anxious, and the question has repeatedly come to us, "Where are the men? Your anxious inquirers are nearly all women, as we read of them in your reports." If such persons had been present on Friday, Jan. 16, and run their eyes over the young converts in the Free Assembly Hall, between eight and nine o'clock, and counted, as was done, the 1,150 that were present, and failed to find 150 of them men, they would no longer have been at a loss to see why the greater proportion of the cases of awakening mentioned are women.

But we believe also that any one who would affirm, from the excessive preponderance of women over men on Friday at the young converts' meeting, that the movement had only laid hold of women, would be very wide of the truth; for although the meetings went on for three weeks almost without men, towards the close there were many young men who were brought under the power of the truth. It is, however, well known that most young Scotsmen, from a variety of influences and motives, even though converted, would rather be excluded from the meeting than face the ordeal through which those had to pass who received tickets; and had there been a converts' meeting for men to come to without any examination or receiving of tickets, hundreds would have attended it.

In a time of awakening it is also well known that women who are religiously impressed will go through fire and water to comply with the wishes of those who have been made useful to their souls. They will do anything they are asked to do; hence the mass-meeting of women on Friday, the 16th. But not so with men—especially Scotsmen—hence their absence, notwithstanding that many are known to have been converted.

Before that meeting was held, we had given it as our calculation, based on the facts that had come under our own observation daily in the inquiry-meeting, that there might be 1,500 souls converted, or who had professed to be converted, believing themselves to be so. We are still of the same judgment, and that very many more of them are men than that converts' farewell meeting revealed. Twelve hundred women and three hundred men and boys seem to be the proportion and sum total who have professed conversion. Hundreds of them may go on flourishingly, and bring forth thirty, sixty, and a hundred fold. Hundreds may go back, die out, or be choked with the world, and many who have divine life in their souls may collapse, and the work may have to be done over again, and they revived and set at liberty, because of the lack of teaching. This has been our observation of the results of past revivals, having been in nearly all that have taken place in this country for the last six-and-twenty years. But we see no necessity for this sad outcome of a blessed work of grace, if the professed converts were fully taught in all the precious truth of God with regard to their place in a risen and glorified Christ, as Romans, Ephesians, and Colossians, spiritually and competently expounded, would teach them. Good milk, and plenty of it, makes an infant thrive and grow. "As new-born babes, desire the sincere milk of the word, that ye may grow thereby. Grow in grace and in the knowledge of our Lord and Saviour Jesus Christ." The only way not to fall is to grow, and growth and strength are by the truth.

One night it was publicly reported (and we were confined to the house at that time with a heavy cold, and could not be out to verify it for ourselves) that 600 men had come up from the Corn Exchange, and fallen on their knees on the floor of the Assembly Hall, professing themselves willing to give themselves to Christ. What a pity that Christians should exaggerate like that, and give the enemy cause to ask incredulously, Where were your 600 Corn Exchange converts when the converts' farewell meeting was held? Have they gone back from Christ already?

A similar band of men, 400 strong, came up from the Corn Exchange on the subsequent Sunday evening, and filled the body of the Assembly Hall; and to an outsider and onlooker they would have appeared to be 400 anxious inquirers; but on being tested at the close (as was done), they were found to be mostly Christian men—many of them helpers in the work; and it turned out that there was not a score of anxious souls amongst them. There had been hundreds of men more or less impressed that night in the Corn Exchange; but it was preposterous to suppose that West Port, Grassmarket, and Cowgate men could be got to any extent, to make a long and difficult pilgrimage, up infinite steps of stairs, to some terra incognita in the regions beyond, in order to be conversed with about their souls!

There is nothing more disliked by Mr. Moody than exaggerated representations as to the numbers at his meetings, and as nothing is more hurtful to the solid progress of the work than romancing about numbers, it should be carefully avoided. There is nothing we have paid more attention to than strict and sober accuracy in all our reports, and we have thereby sought to secure and retain the confidence of our readers.

How we should praise God that there are hundreds who have been made Christians in our city, and that thousands of lips are filled with the melody of joy and praise, as the result of the visit of our beloved brethren, Messrs. Moody and Sankey!

THE WORK IN DUNDEE.

I.

Messrs. Moody and Sankey began their work at Dundee on Wednesday, the 21st of January, 1874. A united prayer-meeting was held in the Steeple Church. Over 2,000 tickets were issued for the meeting, and long before the hour announced for the commencement of the proceedings the church was crowded.

On the following day a united prayer-meeting was held at noon in Free St. Andrew's Church, Mr. Moody presiding. This was continued from day to day. Children's meetings were also held on Saturday, and Sunday meetings followed in various parts of the town. The work was terminated with an "All-day meeting" on the 6th of February. At the close Mr. Moody addressed a young converts' meeting in the evening, when about four hundred were present, professing to have been saved through the labors of Messrs. Moody and Sankey during their visit at Dundee.

The Rev. Mr. Sharp writes as follows of the movement:

II.

MESSRS. MOODY AND SANKEY IN DUNDEE.

I am glad to say the amount of blessing that has fallen upon Edinburgh seems to be imported to Dundee. From the very first all the meetings have been very largely attended, and the whole town seems to be moved. It would take up too much of your space to give even an outline of the glorious results arising out of the visit to this town of these two honored servants of God. Hundreds of anxious souls wait every night to be spoken with, as well as many at the close of the mid-day meetings. Day after day the interest has been increasing. He would be a bold man, or even minister, who would dare to dispute the good that many have received to their own souls. We have had personal experience, day after day and night after night, of hearing from the lips of persons themselves who have professed to have found peace to their own souls in believing in Jesus as their Saviour since these meetings began. I do not believe the people of Dundee ever witnessed such a sight as was seen here last Sunday.

Mr. Moody gave an address to workers in the Kinnaird Hall in the morning. The admission was by ticket; the place was filled. He also preached in other places through the day. But what I refer to principally is the evening services—a meeting at half-past five, and again at half-past seven, were held in the Kinnaird Hall, which holds about 2,000. No one can form the least idea of the scene in Bank street, where the hall is situated, even after the hall was filled; the street seemed filled from end to end with the crowd, eager to gain access, but could not for want of room. Many were awakened by the impressive addresses of Mr. Moody, as well as the beautiful hymns sung by Mr. Sankey. His melodious voice, giving such charm to the soul-stirring words, seemed to produce a most powerful effect upon the large audience, and hundreds remained at the close to be spoken with, and many gave evidence of having received much blessing.

I have no time to enter into the full particulars, but allow me first to say, the whole of the meetings are largely attended both by males and females, both by young and old, and what seems to be so pleasing, a most harmonious feeling appearing to pervade the whole town amongst all classes and all denominations—ministers and people all seem to rejoice together. I am glad to say, many who have hitherto been living without God have been brought to peace in believing in Jesus.

Mr. Moody's address at the Bible meeting yesterday seemed to make a deep impression on all present. His subject had reference principally to searching and studying the Word of God, and the good to be derived to our own souls by so doing. The large audience that had assembled in Dr. Wilson's church,—which was crammed, and many could not get in at three in the afternoon,—appeared to be loath to leave, and looked as if they could have listened another hour to such profitable instruction, and such glorious truths that fell from the lips of the speaker. Even the very youngest in the meeting seemed to listen with most intense interest.

God seems to be working powerfully in Dundee through the instrumentality of these two God-honored servants of His. What to myself is very encouraging is the *want* of that opposition which is so common, and which is so often raised by the wicked one.

No one can fail to see the happy and cordial feeling all over the town; everybody seems to be pleased with one another, and however much some people may object to such gatherings, no one who has the least spark of the milk of human kindness flowing in his veins could fail to be pleased as well as delighted to see the happy, cheerful, and friendly smile to be seen in the faces of each and all as they greet one another in the street; and to myself, it seems a little heaven below to see how happy every one looks as they leave the meetings, and more especially is it to be admired and soul-cheering to see how the young children, along with their parents, seem to enjoy the meetings; and, glory be to God, many a parent's heart has been made to sing for joy to see so many of their children giving their young hearts to Jesus. And oh, what a glorious sight to see and to hear parents and children now singing together with one heart and one voice, " I am so glad Jesus loves me," and I pray God they may be enabled to hold on and hold out to the end. Yes, to hold the fort and wave the answer back to heaven, " By Thy grace we will."

Having attended many of these meetings in Edinburgh for weeks together, as well as many of the meetings here in Dundee, and from all I have seen and got to know from personal experience, I am satisfied God has been working mightily with them here, and the power of the Spirit of God has been felt in the conversion of many souls. I am, sir, yours, etc.,

ALEX. SHARP.

DUNDEE, Feb. 4, 1874.

THE WORK IN GLASGOW.

I.

On Sunday, Feb. 8, 1874, the Evangelists began their work at Glasgow. The following letters from the Rev. Dr. Andrew A. Bonar, slightly abridged, vividly present the development and progress of the work there.

MESSRS. MOODY AND SANKEY'S MEETINGS IN GLASGOW.

FIRST LETTER.

Dear Brethren,—You wish to know something of the work of God in this city. The rumor of what God was working elsewhere, especially as the cloud of blessing seemed to come nearer us, had prepared the way for our American brethren's visit; indeed, there were cases here and there where persons were awakened to conversion by the single rumor of others being so blessed. Let me give you notes of what has been passing here during these few days, with all the freedom of one writing a letter to a friend.

Messrs. Moody and Sankey began their labors in Glasgow on Sabbath morning, the 8th. At nine o'clock in the City Hall, a delightful and most stirring meeting of Sabbath-school teachers, numbering about three thousand, was held. Mr. Moody took this way of engaging the prayer and sympathy of 3,000 workers for Christ in the beginning of his labors. Some of the ministers in this city were in a certain way witnesses of the effect produced, teacher after teacher coming into church just as the bells ceased, with happy, thoughtful, solemn faces. The evening's teaching could not fail to feel the influence of that morning. Half-past six was the hour for the evening evangelistic services, but more than an hour before the time the City Hall was crowded in every corner, and the immense multitude outside were drafted off to the three nearest churches, which were soon filled. Mr. Moody's subject was "The Gospel" (referring to 1 Cor. xv. 1-4), illustrated and enforced in his usual style, downright, earnest, and powerful. Mr. Sankey's singing at both meetings began at once to be felt as indeed "the gospel" preached by singing, impressive and melting, as well as most attractive. Is it another of the Lord's many new ways, in these last days, of graciously compelling men to come in, like the Grecian mother's intense agony of desire expressing itself in the song that lured her wayward child back from the precipice to safety?

The daily prayer-meeting at twelve o'clock was begun on the Monday following—held in the United Presbyterian Church, Wellington Street, which accommodates 1,500 persons. Mr. Moody, after the many requests for prayer had been taken up, started with the passage in 2 Chron. xx., which records Jehoshaphat's prayer, especially dwelling on verse 12: "Our eyes are upon Thee, for we know not what to do." The church was full. Mr. Sankey's singing is aided by a voluntary choir of male and female voices, every one of the number throwing their heart into this work as a means of winning souls; and altogether there is a liveliness and interest, as well as a solemnity, in the crowded meeting such as has seldom been witnessed. Christians and Christian ministers of all denominations, from the country as well as town, come to this meeting; it is a meeting that sends us back to Apostolic days, when the multitude were of one heart and of one soul" (Acts iv. 32), praying "with one accord" for the setting forth of the power of the Holy Ghost in the city. "It is not preaching that Scotland needs," said one brother, "it is prayer and power." Our brethren reckon this hour of prayer to be the most important of all the meetings, since it is here that believers are to be filled with the Spirit to overflowing, and then go forth to the unsaved.

As I do not promise to write to you chronologically, let me give one sample of our meetings. On Wednesday the chairman read Luke v. 17-31, with many racy remarks and pointed appeals; and when the meeting was thrown open, five or six persons in turn spoke briefly. A minister told the anecdote of a Highland chieftain, who used to say that it was not right to ask "blessing" merely; God wished us to ask "showers of blessing" (Ezek. xxxiv. 26). A friend from Edinburgh stated that there was no symptom of decline—every day the prayer-meeting in the Assembly Hall thronged, and every evening some cases of blessing at the evangelistic meetings. He told also of drops falling in a district in Dumfriesshire. One of the ministers of the city drew attention to our Lord's conversations with souls, urging on all this means of laboring for the Lord, and stating his conviction, from what he had come in contact with, that hundreds of souls were ready to speak their mind to any who would approach them. Prayer was offered, and part of the hymn, "Jesus the water of life will give freely, freely, freely," was sung. A minister from Edinburgh confirmed by some further facts what had been stated in regard to the blessing there.

An elder pressed the privilege of taking part in the making known salvation to those around us. A minister from the country gave interesting details of awakening begun in his congregation and neighborhood since the Week of Prayer, and spoke of the holy solemnity resting on all who came together, night after night, for prayer, so that the very walls of the place seemed consecrated. But let me tell of the Evening Meetings this week. The three first evenings Mr. Moody and Mr. Sankey were in the north-east part of the city, in the Barony (Established) Church for an hour and a half, and then in the Free Barony Church. Both places have been filled to the door, night after night, and many inquirers have remained at the close. Mr. Moody's address on "There is no difference," and on "The Son of Man is come to seek and save that which was lost," were both awful and most melting, and full of saving truth. The singing of "There is life for a look at the Crucified One," evidently moved many. In all the meetings it is quite common to see tears trickling down the faces of men when "Jesus of Nazareth passeth by" is sung. A young woman was awakened on Sabbath morning by the hymn sung by Mr. Sankey, "I am so glad that Jesus loves me."

Perhaps I should mention here that one of your London ministers was present on Thursday, and candidly said to the meeting that he had come the day before full of prejudice against these gatherings; but that all his prejudice was gone. He urged upon all present (referring to Mr. Moody's subject that morning, 2 Kings, iv. 1–6) to come, bringing not only empty vessels, but vessels large and deep.

Thursday evening's meeting was in the City Hall, and consisted wholly of men, invited by ticket. The very look of the meeting was solemnizing, such a sea of faces, every face looking at the speaker with fixed and intense earnestness. "Except a man be born again" was the subject; there had been much prayer offered in prospect of this gathering of men, and it was answered. Mr. Moody was enabled to speak in marvelous power, and the Spirit assuredly was working, so that from time to time the whole mass of souls seemed moved, and bent down under the truth. The hymns sung, too, appeared to have a wonderful power on that audience of men. When at the close those were invited to remain longer who were on the Lord's side or wished to be, above a thousand kept their places; and when, after four brief prayers had been offered in succession, they were let go, a large number of anxious souls remained. Many of these last were very deeply concerned. In short, it was one of those meetings that can never be forgotten. "The power of the Lord was present to heal."

Surely the Lord is gathering in His elect in haste before the great and notable day of the Lord. And as in the days of the Forerunner, He made men willing to go out in thronging multitudes to the desert, seeking out the preacher, the preacher not needing to seek out them; so it is now. "The kingdom of God is preached, and every man presseth into it." Applications for visits of our two brethren come in from all the region round: Greenock, Dumbarton, Paisley, Rothesay, Hamilton, Millport, Saltcoats, Bothwell, Barrhead, Rutherglen, and other localities.

SECOND LETTER.

Perhaps we in Glasgow are at that stage of the movement described in Acts ii. 42, "Having power with all the people; and the Lord added to the Church daily such as should be saved."

At the daily prayer-meeting on Friday, 13th, thanks were given for the most interesting meeting of 4,000 men in the City

Hall on the preceding evening. Mr. Moody's helpful word that day was in regard to the three classes of believers we meet with everywhere—those who have got to the length of John iii. 15; another class, who know by experience John iv. 14—they have the living water springing up in them; and a third, and best, who answer to the description in John vii. 38, 39—true believers, and pouring out on others "rivers of living water."

The evening meetings were held again in the two Barony Churches, Established and Free, and many anxious remained behind to converse.

On Saturday the meeting (as usual on that day) was specially for children—a lively and impressive meeting. The church was filled with young people, and there have been decided conversions in connection with these gatherings for the young. At the same hour the usual prayer-meeting was carried on in Ewing Place Chapel, close by, and the place was filled.

On Sabbath morning there was another gathering of Sabbath-school teachers, at nine o'clock. These were not the same company as last week, but from another part of the city. The City Hall was the place, but it could not contain all who sought admission. The address by Mr. Moody, on Matt. xx. 1–15, made the privilege of working for the Lord appear so honorable and so pleasant—especially when he called on the laborers among us to leave the Householder to give whatsoever He might think right, and not "bargain for a penny a day"—that many felt truly humbled, and all were fired with new desire to win souls, a work and privilege which angels almost envy us. On coming out, it was interesting to notice that a row of outside listeners had stationed themselves close to the building, eager to catch at least the songs of praise.

None of the evangelistic services are held at the usual hours of church service, but much prayer went up for these meetings from many congregations throughout the day.

At five o'clock the City Hall was filled with *females only;* and so deep was the impression, that about a hundred inquirers remained to be conversed with, some of whom were led into light and liberty.

At eight o'clock there was a vast assembly, of *men only,* in the City Hall. They were packed into every corner; and outside were nearly as many in vain seeking entrance. It was, like Thursday evening, a memorable time. Mr. Moody's subject was "Whosoever,"—salvation absolutely free, all gift; nothing between a sinner and eternal life but his unbroken will. The mass of men listened with intense interest; now and then you could see a tear, or the head bent in deep emotion. When Mr. Sankey sang the hymn, "I am Coming to the Cross," nothing could exceed the rapt, silent attention. When he came to the verse—

"In the promises I trust,
   Now I feel the blood applied:
I am prostrate in the dust;
   I with Christ am crucified,"—

not a head in the vast multitude moved, every face expressed deep feeling. This verse was repeated amid still deeper silence and emotion. At the close, when an invitation was given to those who minded to remain for twenty minutes simply for prayer, above a thousand remained, and thereafter a large number waited for conversation, though the hour was late.

In the daily prayer-meeting, one of our brethren undertakes to arrange beforehand the requests for prayer—a most important service, for they mount up from 150 to 200 every day. On Monday, the 16th, several brethren stated what they knew or had heard of the progress of the work in our city and elsewhere. From Dundee it was reported that real Sabbath-school work had got an impulse; and one school was mentioned where a great awakening seemed begun. The evening of this day had been appointed as a time when all inquirers awakened during last week should meet for counsel and conversation with Mr. Moody and Mr. Sankey, assisted by Christian workers, in the hall of the Free Barony Church. Above two hundred came, and of these a hundred were men, all willing, as far as they could, to tell "what's the trouble?" It was felt by all who took part to be a time of singular solemnity—reapers gathering up sheaves for the Lord's garner. This was, indeed, an encouraging result of one week's prayer, preaching, and singing.

I do not attempt to give an account of every meeting from day to day, though there has been no day without its incidents worth preserving. There is over the city a breathing of the quickening Spirit Christian workers find it easy to approach men on the matter of salvation.

Mr. Moody began his Bible-readings in the afternoon. They are held for the present in the Free College Church (Dr. Buchanan's). That day Dr. Fairbairn, Principal of the Free College, presided. Many ministers of all denominations were present. Mr. Moody's subject was, "The Blood of Atonement," "the scarlet line that binds together every leaf of the Bible." The audience that crowded the church was of a more educated and fashionable kind than some in the evenings, and they heard a most clear, powerful, decided statement of saving truth, illustrated by many touching incidents; and well worth remembering was the preacher's testimony that, wherever he had been, in any part of the world, he had found that those ministers who preached the blood were the men who were winners of souls. We called to mind, as he said this, the counsel of a much-blessed Methodist to those around him, "*Live in the Sacrifice! Live in the Sacrifice!*" These afternoon Bible-readings are thronged, and this is itself an important fact; for attention to the Word of God, and the true sense of it (Mr. Moody remarked the other day at a prayer-meeting), was the prominent characteristic of revival in the days of Nehemiah (see chap. viii. 1–8, and ix. 3). Indeed, it is a question how far any revival is likely to yield much permanent fruit where a real hunger for the Word does not characterize it. In Josh. i. 8 and Psa. i. 2, 3, constant meditation on the Lord's law is enjoined as the grand secret of spiritual freshness and growth.

That same evening the evangelistic meeting was held (as it has been all this week) in John Street United Presbyterian Church. The crowds increase instead of diminishing. The Wesleyan Chapel, on the opposite side of the street, opened to receive the overflow, has been filled, and no night passes without fresh cases of anxiety.

At nine o'clock the Christian young men of the city met in Ewing Place Chapel, to consult with Mr. Moody as to what they might do to forward the Lord's work. Mr. Sankey gave tone to the meeting, singing—

" Oh, what are you going to do, brother ?
   Say, what are you going to do?
You have thought of some useful labor," etc.

When Mr. Moody had made some suggestions and got many of those present to tell their own views, he asked if as many as agreed to enter on such a plan as was suggested would rise to their feet. The whole number (there must have been 700 present) at once rose; and already these young men have begun to meet every night, between nine and ten, to stimulate each other to watch for souls, and to lay hold on other young men. The hour is fixed thus late in order that there may be no interference with other meetings, and in order to give opportunity to those who are kept late by business. What a field Glasgow presents, may be inferred from the fact that the young men of the city between the ages of fifteen and twenty-five number 70,000.

At Wednesday's prayer-meeting, Neh. viii. 1–12 was the passage read. The passage led to the subject of "joy," which not only the president but the successive speakers took up. Then was read a letter containing brotherly salutations from the Edinburgh Daily Prayer-meeting; and when it was mentioned that at that very hour they would be praying for us, the meeting engaged in silent prayer for their brethren there.

I think it was on Thursday that thanksgiving was requested by two students brought to Christ in the meeting of the preceding evening. Another thanksgiving came from a Sabbath-school teacher, for an awakening begun in her class in answer to prayer offered for it a week ago. We heard in private many interesting cases that cannot be published, but which will soon be felt. One, however, I may mention. A man went to pay a debt long due; surprise was expressed at his coming on that errand, but he explained all by saying, "*I was brought to Christ last week.*" Is not this a brother of Zaccheus? (Luke xix. 8.) We shall see speedily much more of such fruits in all classes of society, to the glory of Him who gave Himself for us "that He might redeem us from all iniquity, and purify unto Himself a peculiar people, zealous of good works."

There is blessed work going on in other parts of Glasgow and in the neighborhood. Our beloved brethren were sent by the Master to fire the train laid in many places before they came. Laborers, who only sowed before, are now reaping. But beyond all this, in connection with their preaching and singing, the Holy Spirit is quickening souls that were never touched before. It is a day of the Lord's power.

Worldly men are "doubting whereunto this will grow" (Acts v. 24), and God's people are rejoicing that He is "making ready a people prepared for the Lord."

THIRD LETTER.

The work goes on. The Lord Jesus, sitting at the right hand of God, is "confirming the word by signs following." We hear of conversions coming under the notice of workers for the Lord in all parts of the town. Last Friday the subject of the noon prayer-meeting was Jas. i. 8, and in the Bible-reading in the Free College Church (Dr. Robert Buchanan presiding), the topic was "Heaven." It had been "Grace" the day preceding. At the close of the evening meetings in John Street United Presbyterian Church and the Wesleyan Methodist Church, the number of inquirers was large; but so well was all arranged for conversing with them, that though there must have been above a hundred at one time, yet all was order and stillness, the different workers in separate seats, and some in separate rooms, dealing with the anxious. No idlers were permitted to look on, and there was time for helping individuals to get at a clear understanding of their own difficulties and hindrances, which in many cases is the main thing needed, in order to their right apprehension of the gospel.

We pass on to the Sabbath morning meeting in the City Hall. It was one of Christian workers, 3,000 of whom filled the place. Mr. Moody spoke on Isaiah vi. 8, "Send me." At five o'clock the hall was filled with females only, and at eight with men only. As other neighboring churches were open at the same time for the overflow, at least 10,000 persons that night heard the gospel in a special manner. From half-past six onward there was an important gathering from all the meetings in the old College Church, of all who were anxious, and above 200 souls were there in the course of the evening. It was a busy scene, the workers dealing with individuals, or occasionally in little groups. It was the business of heaven that was carried on. The wares of God's market were exhibited and pressed on the acceptance of sinners, without money and without price. I could not help going back to the memories of other days, for in this old College Church, in Whitefield's time, there ministered one whose whole heart was in revival work. This was Dr. John Gillies, who not only laboriously compiled the "Historical Collections," one relating to remarkable periods of the success of the gospel (a book of deep interest to all who care for the winning of souls), but used also to send forth to his parishioners a short weekly paper—a rare thing in these days—giving them information about the work of God in America, Holland, Germany, as well as at home. What a joyful sight would Sabbath night's meeting of inquirers have been to him, gathered on the spot where he so often and so pathetically called on his people to cry for the outpouring of the Holy Spirit on Glasgow and Scotland! He prefaced his first "Exhortation to the hearers in the College Kirk" by such burning words as these—"Such multitudes of my own kin, my brothers and my sisters, going to hell, never to get out again? Break, break, hard heart! You who read these lines, think not my words strange, but weep with me, if you are men, and not stones. O Thou who didst weep over Jerusalem, Thou alone canst give us comfort in this overwhelming calamity! Heavenly Father, for Thy Son's sake, be pleased to stir up many diligently to preach the Kingdom, when such multitudes of souls are in danger of perishing!" This was in 1751. Who can tell but that that man of God may, even at this hour, have been made glad by hearing the tidings told "in the presence of the angels of God," that many sinners are repenting in Glasgow, and that there have been many who looked to Calvary from the spot where he used so to yearn over souls.

The Monday prayer-meeting is a time for reports of the work, and not less for thanksgiving. Luke xvii. 12–19 was the Scripture read. The inquiry-meeting in the evening in John Street Church was very large, and full of interest. While it was going on, there was held in the City Hall, at seven o'clock, an evangelistic meeting, which had some peculiar features. The place was full. Mr. Moody merely opened with a few remarks on the object of the meeting after prayer and praise, and then addresses were given, brief and pointed, by a great variety of speakers. The attention of the large assembly throughout did not flag, and some were deeply impressed. Mr. Sankey began by singing—

"Not ajar the gates of glory—
They are open day and night."

Prayer was offered, and then Dr. Marshall Lang, of the Barony Church, spoke earnestly on the ark, showing its door open for invitation to men, as is Christ's pierced side, and showing the ruin of all outside the ark. Mr. Moody related the first incident that impressed him, and awakened in him concern, viz., the deep feeling manifested for him by his Sabbath-school teacher. The hymn was then sung by Mr. Sankey and the choir, "Light in the valley." Dr. David M'Ewan, of John Street United Presbyterian Church, then for some minutes proclaimed the way of life through the Lord Jesus, illustrating his statements by the experience of John Wesley, and another hymn was sung, "The Lifeboat." Mr. Bonar, Finnieston Free Church, gave an incident that recently occurred near Dundee, in connection with these lines of that hymn—

"Safe in the Lifeboat, sailor, cling to self no more,
Leave the poor old stranded wreck, and pull for the shore,"

and then proclaimed the way of salvation from John vi. 40. He was followed by Dr. Wallace, of Campbell Street United Presbyterian Church, who set forth the sinner's Saviour by an illustration taken from what he had witnessed a few days ago in the Highlands, the rescue of a perishing sheep. Ps. xxiii. was sung, and Mr. Pirrett, Burnbank United Presbyterian Church, gave a warning word from the case of one he had known who wrested to his undoing the truth, "Once in Christ, always in Christ." Mr. Sankey and the choir sang "Depth of Mercy," and as they came to the chorus from time to time, "God is love, I know, I feel; Jesus lives," etc., the deepest stillness pervaded the meeting. Mr. Campbell, of the Wynd Free Church, spoke to those present who might be anxious, and Mr. Howie dropped a closing word, urging to immediate application. The meeting lasted an hour and forty minutes. Inquirers were invited on leaving the hall to join those met in John Street Church.

Wednesday, Feb. 25th.—Yesterday's noon meeting was full of interest. Dr. Cairns, from Berwick, presided. He read Isa. xii. previous to telling us some of the "excellent things" which the Lord has been doing in Berwick. After "declaring His doings," he earnestly and lovingly urged all ministers and believers to use this present tide-time, by going to individuals and bringing before them the matters of salvation, not waiting till they came to ask. Mr. Moody afterwards took up this remark, enforcing it, and pressing it on all who would win souls; for when the Son of Man came to "save," He first of all went to "*seek*" the lost. Mr. Goldie, from Stirling, Mr. Bogue, from Stockton-on-Tees, and Mr. M'Nab, from Ardrossan (the last of whom had been taking part in the London work), each gave details of what the Lord was doing. The City Hall evening meeting had been specially kept for friends from Greenock, Port-Glasgow, Johnstone, and other places on the same line of railway, who had intimated their desire to come, but there were fully as many present from the city. The most memorable meeting that evening was that of the young men in Ewing Place Chapel at nine o'clock. A deputation of four young men from Edinburgh began the evening's work, each in turn speaking with persuasive earnestness and affection. Then Dr. Cairns, with freshness and fervor, as if he had returned to the days of his youth, addressed them, and was followed by Rev. J. H. Wilson, from Edinburgh, who brought them to the point, "Why not to-night?" When Mr. Moody came in, he saw the impression resting on the meeting, and proposed that at once, on the spot, opportunity should be given of dealing with all *who desired to take Christ as theirs.* The three front seats of the church were cleared, and an invitation given to those young men present who professed to have this desire to come forward. The seats were immediately filled, and when three other seats had been cleared as before, another stream of young men poured in. Some one present counted in all 101, and there were others who felt as those who came forward, though they kept their seats. During prayer intense emotion was manifested by many and an awful solemnity was felt to pervade the place. It was a night not to be forgotten. Christian friends remained conversing with the anxious till about twelve o'clock. And I may add that next night the chapel was filled with 900 or 1,000 young men, who were again addressed by the deputation from Edinburgh and others. When Mr. Sankey had sung "Almost Persuaded," Mr. Moody asked all to retire except those who wished to converse about their souls' salvation. There must have been not fewer than 140 who remained, and Christian workers were occupied with them till near twelve o'clock. It was a glorious harvest field.

The Wednesday noon meeting, I must not forget to say, was as crowded as ever. The Rev. J. H. Wilson read Acts ii. 42-47, as descriptive, in its main features, of the work of God in Edinburgh, and most urgently did he press all Christians to say, every man to his neighbor, and every man to his brother, "Know the Lord," giving illustrative incidents occurring in all ranks of society. Dr. William Wilson, from Dundee, followed, and bore testimony to the blessing in that town, and its continuance. If the meetings were not so crowded as before, there was nevertheless evidence, day by day, that the Spirit was still working in the midst of them. Mr. Taylor, of Kelvinside Free Church, gave a word of exhortation, and Dr. Wallace, of Campbell Street United Presbyterian Church, called attention to the young women of Glasgow, of whom there must be far more than 70,000.

Thursday, Feb. 26th.—To-day's noon meeting was lively and warm. After singing and presenting to the Lord in silent and public prayer the numerous requests, Matt. vii. 7-11 was read, in connection with which Mr. Moody detailed several most interesting answers to prayer he had just heard of in letters from America. One of the cases was the following:—He had once, after most urgent solicitation, preached in a small church in the prairies, where one Christian woman continued praying day and night for the pleasure-loving young people, whose only enjoyment seemed to be the song and the dance. A letter received that morning brought the cheering tidings that in that same spot thirty-two young men were now on the Lord's side, and working for Him. Mr. Keay, of Free Trinity Church, spoke of last night as the most fruitful in his ministry. He had been detained in his vestry for five hours, partly conversing with inquirers, and partly hearing the story of deliverance given to awakened ones. Dr. Buchanan, Free College Church, read a letter from all the evangelical ministers of Kirkwall, and thirty elders, inviting our beloved and honored American brethren to come and help them. He then asked all to unite in prayer for these brethren in the Orkney Isles. A brother related a remarkable blessing following on his personal dealing with a sea captain, afterwards lost in the Black Sea, who, saved himself, became a blessing to his house. Mr. Gebbie, minister of the Established Church, Dunlop, pressed the duty of personal dealing. One of the deputation from Edinburgh suggested that those who dealt with souls should be very cautious of mentioning in public any of their conversations with them. Dr. Marshall Lang referred to the remark made yesterday by Dr. Wallace, regarding the young women of Glasgow, for whom more might be done by Christian ladies. After a few minutes of silent prayer, some notices were given, and the meeting closed.

The Bible-reading yesterday was the closing part of Mr. Moody's subject, "Heaven, its treasures, etc.," intensely interesting and useful. To-day's subject was, "The Love of God," repeated in John Street Church, in the evening, to a densely-crowded audience. At the close of the address in the evening, not less than two hundred remained for conversation. There are new inquirers every day, and throughout the city ministers are continually discovering traces of the Lord's footsteps, and are themselves stirred up to more earnest, prayerful, expectant ministrations.

Thanks have this week been offered for at least three students brought to Christ; and thanks for the conversion of two youths who, on Sabbath last, came to hear simply by way of amusement, and who came into the inquiry-meeting for no other end. The Lord "was found of them that sought Him not."

On the memorable Tuesday night of the young men's meeting, a youth from Ireland was led, on his way to the theatre, to turn aside into Ewing Place Chapel, and it became his birth-place for eternity. He had come to town with a full purpose of finding out its gaieties and pleasures, and of returning home to tell his companions what he had enjoyed. That night he was "apprehended" by the Holy Spirit, under the solemn, stirring appeals of the deputation from Edinburgh. When opportunity to come forward for special prayer was given to all who would fain be decided for the Lord, he pressed forward at once; all the time that prayer was offered by various pleaders, his one cry from the heart was, "Lord, have mercy on me, a sinner!" He saw and understood the way of salvation, but one difficulty remained, viz., "What would he feel or do on the morrow?" This difficulty, however, was removed by the remark of one with whom he conversed, who said, "Surely, if you trust Jesus to *save* you, you may trust Him to *keep* you ever after." He returned to his room that night only to tell his astonished friend that now he was a new creature. Every thought about theatre, opera, amusement, had given place to the one absorbing thought of salvation, and a Saviour found!

FOURTH LETTER.

"There is something far better than gold. God thinks so little of gold that in the New Jerusalem it is used for paving the streets." In one of his Bible-readings Mr. Moody made the above remark, and every Christian man will understand it. Even the joy of winning souls, and of seeing souls won to Christ, is better joy than the world's best. And the Lord is giving largely of His gladness to His own at this time.

The "thanksgivings" at the various meetings are worthy of notice, were it only in the way of reminding our readers that, like frankincense put to the meat offering (Lev. ii. 2), this grace must be in actual exercise wherever true prayer is going up. "In everything by prayer and supplication, with thanksgiving" (Phil. iv. 6), is the divine rule. The leper who has been blessed must return to give glory to God.

There is a children's meeting held every Saturday; it is crowded. Other persons are not admitted unless they are bringing some children with them. As yet, there is no very general movement among the young, though there are many cases of decided conversion in several of our Sabbath schools. When Mr. E. P. Hammond was here among us, six years ago, the great blessing came on the young, and only drops fell on the older ones. The Lord is sovereign as to times and ways of working.

The meetings this week are as interesting as ever. We have had help from brethren who sympathize in the work, such as Mr. George Wilson, of the Tolbooth Church, Edinburgh; Mr. Maclaren, from Manchester, so well known in the churches; and Mr. Arnot, from Edinburgh. At the daily prayer-meeting, it is now an everyday thing to see the pulpit-stairs, and the three front seats, filled with ministers from town and country. There are about a hundred ministers oftentimes, and most truly was it said the other day, "Every minister is worth a hundred other persons," in reference to the vast influence they may exert on their people when they are themselves filled with the Holy Ghost. Whitefield used to say, in regard to this, that "every minister's name was *Legion*." Will the Lord's people cry mightily to God in their behalf at this season? If we ministers get the fullness of the blessing, if we get a fresh and full anointing from the Holy One, thousands throughout the land will soon know it.

But let me turn to the young men's meetings. On Sabbath morning, at nine o'clock (March 1), there was a gathering in the City Hall of the young men of the Glasgow Christian Associations. With these 3,000 before him, Mr. Moody spoke from the words, "Run, speak to this young man" (Zech. ii. 4), Mr. Sankey and his choir singing such hymns as "Jesus, the water of life, will give, freely, freely, freely;" "What are you going to do, brother?" ending with "Hold the fort, for I am coming!"

On Tuesday evening, Mr. Maclaren, of Manchester, addressed a vast assembly of the same class in the United Presbyterian Church, Wellington Street. He spoke from John iii. 16, "God so loved the world," etc. After setting before them in this glorious passage the manifestation and outpouring of the love of God, he called attention to this feature of the verse—viz., that it put the negative side ("not perish") first, and the positive side ("have everlasting life") afterwards. The reason of this is, that every man *who does not believe perishes*. Notwithstanding that great love of God, there is certain destruction to every soul that closes itself against Him. "He that believeth not is condemned already;" and this "is the condemnation that men *love the darkness*."

Mr. Arnot, of the Free High Church, Edinburgh, addressed a similar gathering on Wednesday night, taking 2 Pet. i. 3-7 as his subject, impressing on them the two subjects of faith and holiness, and the connection between these two graces.

Every night at nine o'clock the young men meet, and the work among them makes steady progress. On Monday evening Mr. Moody, as usual, came in near the close of the hour, read Rom. x. 10, and made some happy remarks on the importance of "*confessing Christ with the mouth*." Through neglect of this, many are left in something like darkness, and have llitte joy. We ought to speak for Christ; at the same time we must beware of spiritual pride. Heart utterances are what we want, not flowing eloquence. Whenever the devil whispers, "That was a good address," you are in danger.

After a hymn had been sung, an opportunity was given to those recently brought to Christ to tell "how great things the Lord had done for them, and how He had had compassion on them" (see Mark v. 19). The first young man who spoke began by saying, "I was one of the 101." He meant by this expression to refer to the memorable Tuesday night last week, when a breathing of the Holy Spirit passed through the assembly, and 101 young men came to the front seats, asking to be prayed for, and guided into the truth. The speaker added, "I have been wishing to be saved for many years. When those who were *sure* that they were Christians were asked to stand up, I felt that I could not honestly do so, though I was a member of the Church, a Sabbath-school teacher, and was one of the ten who had sent in a request for our warehouse for prayer on our behalf. I kept my seat. Mr. Moody then asked all that were Christians to leave the three front pews. I occupied one of these, and when the others went out I kept my place. Thinking that I had, perhaps, misunderstood him, Mr. Moody kindly said to me, 'Are you not a Christian?' I said, 'I am not.' But that very night I found Christ."

A young student next spoke. "I also was one of the 101 that night. Though taking part in Christian work, I felt my need of what I had not found. That night, at the meeting for conversation, five of the young men in succession spoke to me; and, unknown to the other, quoted to me John v. 24, 'Verily, verily I say unto you, He that heareth my word, and believeth on Him that sent me, *hath* everlasting life.' I was at length enabled to apprehend the truth, and I now thank the Lord for saving me, and pray that all here may be brought to Christ."

Another spoke. "I had been seeking Christ a long time. That night, when I was going away without relief, Mr. Moody came up, and took me kindly by the hand. He looked at me—I might say he put his two eyes right through mine—and asked me if I would take Christ now. I could not speak, but my heart said, Yes."

The above may give some idea of the intensely interesting scene. A dozen more declared what God had done for their souls. Not only that night, but on some of the after evenings, a similar scene has been witnessed. Last night, one very intelligent young man told briefly, but very clearly, what his state had been till he was awakened on Sabbath evening last, and how miserable he had been on the following days, not being able to see that salvation was for him. Getting a ray of light, he went home, read John iii. 36, "He that believeth on the Son hath everlasting life," and sought on his knees to be led into the truth. "And," he added, "God heard me. I believed then; I believe now; I am a ransomed soul." And that the Lord is in these meetings who could doubt who saw, two nights ago, sixty or seventy of these young men, when the invitation was given, rise up from their seats, as an intimation that they desired to be prayed for, and be led to Christ. Our Scottish youth are not at all demonstrative in regard to their feelings; it is not usual for them to tell out what is passing in their hearts on the subject of their state towards God; and this fact makes the present movement all the more remarkable. Till they have got *faith* they are slow to speak about their *feelings*.

On Wednesday, at mid-day, there was a prayer-meeting of fathers for their children, and on Thursday of mothers. To-day a meeting of sisters was held in Ewing Place Chapel, to which so many came that the place was filled. I understand that there was deep solemnity in the meeting, and much impression. Mr. Moody stepped in to speak an encouraging and guiding word; and it may be this meeting of sisters may yet become a mighty power among the young women of Glasgow.

If time and your space permitted, I have at hand more facts, which would all be interesting. Above 200 persons came on Monday evening to the United Presbyterian Church, John street, as persons who had recently been awakened. Of these, seventy-five professed to have found Christ. Last night there was a large number of inquirers in the Free College Church. This week the evangelistic meetings are in the Free College Church, the overflow passing into the Park Established Church.

I am writing you about Glasgow specially; but you may like to get notice of God's work in less known localities. Requests for prayer come to hand from all quarters—*e. g.*, one came to me, asking my congregation to pray for a work of God in the district of John O'Groats' House; and another from Christian friends who live near Cape Wrath. Preaching on Wednesday at Auchterarder, I found unmistakable traces of God's doings in that quarter; and passing on in the evening to Dollar, found an assembly of above 1,000 souls, eager to hear the Word; and at the close, besides others, about fifty of the boys and young of the Dollar Institute waited for conversation and inquiry. At the Stirling noon-day prayer-meeting, next day, there was a large attendance. There have been not a few awakened there of late, and the interest is deepening. The ministers of all denominations take part most cordially. There, too, I heard of work going on not only in such places as Alva and Dunfermline, but in obscure parishes. Souls are coming from great distances to ask the way of life at the lips of those who can tell it, and these souls awakened to this concern by no direct means, but evidently by the Holy Spirit, who is breathing over the land. It is such a time as we never had in Scotland before. The same old gospel is preached to all men as aforetime; Christ, who was made sin for us, Christ the Substitute, Christ's blood, Christ's righteousness, Christ crucified, the power of God and the wisdom of God unto salvation; but now the gospel is preached "with the Holy Ghost sent down from heaven." And amid all this the enemy is restrained, so that we are solemnly reminded of Rev. vii. 1–3, the time before the coming of the Lord, when the four angels are charged to let no storm burst, not to allow the wind even to ruffle the sea's smooth surface, or move a leaf of any tree, till the seal of the living God has been put on His elect. Is not this sealing going on daily among us? Are not the four angels looking on? Surely it is time to seek the Lord, that He may rain righteousness upon us.

FIFTH LETTER.

"The kingdom of Heaven suffereth violence, and the violent take it by force" (Matt. xi. 12), was said of John the Baptist's days. In answer to such prayers as he sent up during his thirty years in the deserts of Judea, the Holy Ghost was at work, and everywhere were found men in right and real earnest about salvation. You might have seen them thronging the road to Jericho and the wilderness, leaving home, comforts, business, friends, intent on the one great matter that filled their minds day and night. "The kingdom of heaven suffereth violence"—men pressed into it with all the eager determination with which soldiers press into an assailed city (like Coomassie)—"and the violent took it by force." Those who were thus intensely earnest snatched, as the word means, the kingdom at once, as the robber does the purse he covets, seizing his opportunity. All this we see before our eyes in the present time of revival; men are truly in earnest, and they catch the gift of God at once, while the cold formalists wonder and dispute against sudden conversions, "not knowing the Scriptures, nor the power of God."

When I closed last week's letter, I mentioned various places in Scotland where God was working. There are many other districts equally interesting. At Aberuthven, near Auchterarder, almost every house in the village has some one under its roof awakened by the Spirit. In Dumfriesshire, at Lockerbie and at Moffat, not less than seventy in each place have been awakened. Near Glasgow, not Chryston only, but other places, such as Kirkintilloch, are shaken. At the daily prayer-meeting last Monday, it was stated that there had been not less than 300 inquirers and converts in the inquiry meeting in Free St. David's on Sabbath evening. A friend mentioned that at Dalmellington, in Ayrshire, a work had begun; seven had been lately converted. Dr. Black, of United Presbyterian Wellington Church, gave extracts from a letter from England, showing a work begun in a district where there had been no special means. Mr. Moody read, from letters just received, accounts of friends brought to Christ. Mr. Wells, of Free Barony Church, stated that he had a list of seventy persons in his congregation who had received blessing during the meetings. Mr. Barlas, of United Presbyterian Church, Belgrave street, stated cases occurring in his district. Mr. Taylor, of Free Church, Kelvinside, spoke of this last week as the happiest in his ministry since he came to Glasgow. He had seen abundant proof that the Spirit of God was at work in the midst of the city. His visits as a pastor brought to light most interesting cases, in all grades of society. All sorts of instrumentality also seemed to be employed. He had been told of one awakened by the singing of the hymn where these words occur—

"Let some droppings fall on *me*—even *me*."

One day thanks were given for a person who had been blessed while the hymn, "Jesus of Nazareth passeth by," was being sung; and several other cases were reported in which the same hymn had been blessed. Last night I met a Christian working-man, who joyfully informed me that "in the building-yard where he worked, this week there had been two boys and three men brought to Christ." "I give thanks for six," was on a paper handed in at the prayer-meeting; while a disciple, who had for many years been pleading for the conversion of near and dear relatives, asked the meeting to join him in thanksgiving for a daughter saved, a nephew and several nieces. A letter said: "We cannot leave Glasgow without telling you that the brother whom we told you of as having come here to attend the meetings, left for London this evening, we firmly believe, resting in Jesus."

A lady asked prayer for her own conversion, stating, "I have come from Switzerland on purpose to be present at the meetings. I have every reason to believe in the power of prayer, having been cured through prayer, at a small village in Switzerland, after having been dangerously ill for thirteen years. I should be extremely sorry to leave Glasgow without receiving what I came for. I have been well brought up, but am not a Christian." A case like this reminds us of Acts v. 16: "Then came a multitude out of the cities round about, bringing their sick folks, and them that were vexed with unclean spirits."

And yet more, this other, from a person about twenty miles out of town: "Dear sir,—Would you kindly forward four tickets to admit to the morning meeting on Sabbath first to the City Hall. I have never had the pleasure of being present at any of these precious meetings that have been held in Glasgow, though a constant reader of the reports given in the various newspapers; but I will be in Glasgow on Sabbath first along with three friends. Going in the spirit of anxious inquirers, we pray God that it may be our blessed privilege to come home having found that Christ is indeed precious to each of us."

Another day, at noon, four young men, from a mining district in Ayrshire, were found waiting at the close of the meeting to speak to Mr. Moody, if possible. He had gone out; but they sat down in the inquiry-room with one of the ministers who was still there. "Are you all of one mind? are you all in Christ?" was the question put to them. "Three of us are Christ's, but our friend here (pointing to the fourth) is not." The minister entered into conversation with the unsaved but anxious one, and found out his state of mind. He showed him that Christ was offering to be his substitute, and to appear in the presence of God for him, and asked, "Will you believe in Him as He so offers Himself to you?" In a moment the lad's countenance changed, and, half springing from his seat, he struck the Bible with his hand, exclaiming, "I see it all!" The scale had fallen from his eyes, and he, with his three friends, who had been to him like the friends of the palsied man, left the room to return home by the train, rejoicing.

One other case. A young man attracted Mr. Moody's attention at an inquiry-meeting, an intelligent young man who had long been anxious. Mr. Moody discovered that one thing had hindered his full decision, viz., want of courage to tell his wife all that was passing through his mind. But last Sabbath afternoon he was enabled to go home and frankly tell all he felt. It turned out that she too was in deep anxiety, only waiting to have the ice broken. The result has been complete deliverance of soul to that young man, who is now able to help others in the way. A similar case to the above is the one brought out in the following letter from one in Edinburgh, which Mr. Sankey read yesterday: "I have such good news to tell you. When you were here, you wanted me to write to my sister about Jesus and coming to Him; but my old sinful heart went dead against it. Dr. S., however, began to tell me that my health was very precarious, and all your advice came back to me. I *did* write to my sister, a girl about sixteen. My want of faith has been reproved; for I had such a letter from her, telling me she had felt sure all this winter that there had been a change in me, and why I had not written to her before; and she ended by confessing that she could resist no longer, but had taken Christ, and, God helping her, would live for Him. Please pray for her, and encourage all young converts to write to their friends. Another thing you wanted me to do went fearfully against the grain, and that was to hold meetings. Had anybody told me last year that I'd ever come to do such a thing, I'd have scoffed at them. So you see it's nothing of myself, but something that makes me, in spite of myself, long to work for Jesus. Will you pray for my Canongate meeting? I've got such bad characters; oh, if I could only reach them! drunkards, and profane people who don't believe in hell—my heart just yearns over them. It was a fearful effort at first to speak for Christ, but now 'I love to tell the story,' for 'All to Christ I owe.' As I came home last night I heard such beautiful singing at the head of one of the lowest streets here. Coming up I found some young men were singing 'Depths of Mercy,' in parts, and whenever they had gathered a crowd, invited them to accompany them to the meeting. A great many followed them. I know it will cheer Mr. Moody to hear of the hint he threw out being thus taken up by these young Christians. The work here goes on wonderfully; it is too great to be spoken of.—Your loving friend in Christ."

I am scarcely leaving myself space to speak of other parts of the work. The evangelistic meetings have been held this week again in the Free College Church. The subjects have been, "Where art thou?" "How long halt ye between two opinions?" and "Sir, remember." The last of these addresses was awfully solemn. Mr. Moody related, as an illustration of memory being ready to yield back all the past at God's touch, how he himself in early days was nearly drowned, sinking twice, and caught the third time he came to the surface. During the time he was under water, all that was buried in his memory came up before him. And so the memory of Abel's blood flowing from the deadly wound is ever before Cain, and so with all the sins of sinners. This makes hell terrible beyond measure, and there is no sleep there. "If I did not believe in hell for ever, would I (said he) come here to preach night after night? If I did not believe in that hell, I would be off to my home by the first boat that sails from the Clyde."

The Bible-readings have been in the Park Church (Established), and the subjects this week have been "The Holy Ghost," "Jacob," "Daniel." All the meetings are crowded to the door more than ever, and there is daily fruit.

On Sabbath morning, the members of the Glasgow Young Men's Society for Religious Improvement, filled the City Hall at nine o'clock A. M. There were young men present who had walked in that morning from Englesham, Kilbride, and other places. We do thank God that Mr. Moody's "hands are made strong by the mighty God of Jacob," so that he is able to work night and day, and certainly he never wearies in spirit. He read Luke xix. 1–13, and spoke briefly to the saved on "Occupy till I come," urging them this week to resolve by God's grace, every one of them, to speak to and seek the conversion of at least one soul each. A large number, when opportunity was given, rose in response to this appeal, and during the week it was evident that the resolution was not forgotten. One petition came in to the noon-day prayer-meeting " from a young man who promised to seek to bring a soul to Jesus; pray that he may have grace to be faithful in dealing with two young men in his warehouse who are anxious to find peace in Jesus." The main part of the address was to the unsaved, and when, in his closing prayer, Mr. Moody stopped for half a minute, there was profound silence over that assembly of 3,000 young men, broken in upon by the yearning, urgent pleading, "O Lord, speak to them! speak to them Thyself!"

At the Young Men's meeting in Ewing Place Chapel on Tuesday evening, Mr. John Burns, of Castle Wemyss, presided, and gave a most hearty address. He read from and held up the character of Nehemiah. "We business men in this great city are exposed to many temptations, and are often in great perplexity. Let us, like Nehemiah, 'pray to the God of heaven.' Speaking from experience, I can testify to the value of prayer in the case of business men. I have great faith in prayer, silent and instant prayer. We have not time during the day, in the midst of business, to go to our knees; but let our hearts go up." He then spoke of the work now going on, as a work of God, from which the best fruits might be expected. He was followed, in a few words, by Archdeacon McLean, who accompanied him. As usual, a large number of the young men waited for inquiry. One of themselves, on Wednesday evening, spoke as follows; and his words were felt by all present: "I would like to say a word as to the power of prayer. Seven years ago, about a stone-throw from where we now stand, a young sneering infidel retired to his bed on a Sabbath evening. About three hours after, that same youth rose and cried to God to have mercy on his soul. Some of you may say, 'Oh, that's a story made up, and far-fetched.' No, it is not. I was that youth. When I retired to bed, three of my young friends were assembled in another room, wrestling with God for my conversion. I could not sleep. I arose and went into them, and asked them to pray for my soul. I found the Saviour; and, blessed be His name, I have followed Him ever since. My companions scoffed, and said that it would soon pass away; but I have been kept. I have tasted all the pleasures of life in other days, but I am here to testify that the love of Jesus is sweeter than all. Young men, don't be deceived; the pleasures and the philosophies of this world pass away. Take Christ, and He will satisfy the longing soul." Night after night, there are not only such addresses, and many inquirers, but also many conversions.

I had intended to tell you a little about the ministers' meeting for prayer and conference, on Wednesday, at which about 200 were present, of all denominations; and many from the country. But this must be reserved; and meanwhile let me entreat every reader to pray for a baptism of fire, a gift of " Power" (Acts i. 8) to every minister of Christ who carries Christ's message to the Churches.

II.

DR. BONAR'S letters bring down the record to March 13th. The noon-day meetings and those for the children continued to be maintained with unflagging interest. On the 13th of March the requests for prayer were as many as 150 in number. Sunday morning, the 15th inst., Mr. Moody delivered his lecture on Daniel to 3,000 young men assembled in the City Hall; and Mr. Sankey at the close of the service sang "Daniel's Band" and "Hold the Fort," with the strong choruses, in which the vast assemblage joined. "Evangelist Meetings" were begun in the eastern district of the city on Sabbath evening, the 15th inst. On Monday there was another service, when it was determined to continue the meetings in the same church during the remaining evenings of that week. At the same time there were four evening meetings for young men, under the charge of other workers, going on in different parts of the city. There were also ministers meetings, and meetings of parents, with meetings of the Good Templars and meetings held by the Evangelists in surrounding districts. On the 30th of March they were present at the daily prayer-meeting held in the ship-building yard of Alex. Stefters & Sons, where there was a large attendance of the workingmen, the number being variously estimated from 1,500 to 2,000. All the partners of the firm were present. On the 5th of April a praise meeting was held in the City Hall. On the 8th and 9th of April Mr. Sankey and Mr. Moody attended the noon prayer-meeting held in the Town Hall at Greenock; and on the 16th there was a church convention on the revival movement, which was largely attended.

III.

THE *Daily Mail* thus sketches the work, of the revival in Glasgow, its agencies, and results :

During the six days beginning with Tuesday of last week the suburb of Hillhead was nine times flooded with crowds hurrying to the Crystal Palace. This unique glass house is the largest place of public assembly in Scotland, and can seat about four thousand, while a thousand or two more may be crowded into it. Tuesday evening was for the young women. Hundreds appealed in vain for tickets after 7,500 had been distributed, and hundreds who had them struggled in vain for admission. The building was crowded up to the fainting point, and the meeting was partly spoiled by its numerical success. On Wednesday the young men who were ticket-holders. darkened the Great Western Road more than an hour before the time of meeting. All comers were welcome on Thursday, so long as there was any room. In spite of the rain the Palace was filled by seven o'clock, and about one-half of the audience seemed to be young men of the middle classes. On Friday the noon prayer-meeting was transferred to the Palace, which was comfortably filled with the better, or better-off classes. Friday evening's meeting was the most significant of the series. Tickets for it were given only to those who, on applying for them in person, declared that they believed themselves to have been converted since January 1st, and gave their names addresses, and church connection, which information, we are told, is to be forwarded to their several pastors. It was publicly stated that about 3,500 received tickets on these conditions. As the Americans did not arrive till six weeks after the New Year, and as the tickets were not exclusively for the frequenters of their meetings, it was hardly fair in one of our contemporaries to insinuate that the object was to number and ticket Moody's converts. The children had their turn on Saturday at noon, and the working people at night. On Sunday morning the young women were admitted by ticket, and at six o'clock P. M. the Palace was filled both inside and outside, as an Irishman would say. While several ministers, along with Mr. Sankey, conducted the service inside, Mr. Moody addressed a crowd in the open air that filled the whole space between the Palace and the gate of the Botanic Gardens. Many hundreds did not even get the length of the garden gate. The estimates of the vast throng—mere guess-work, of course—range from 15,000 to 30,000. A month ago, in the same place and under the same auspices, another meeting was held for six and a-half hours. We refer to the "Christian Convention," which Dr. Cairns declared to be " unparalleled in the history of the Scotch, perhaps of British, Christianity." It was reported that about 5,000 were present, of whom some 2,000 were ministers and office-bearers from Scotland and the north of England.

Now these are conspicuous facts, and challenge the respectful attention and sympathy of all, whatever their religious views may be, were it on no higher principle than that of the ancient poet, "I am a man, and deem nothing human uninteresting to me." Some have already photographed the humorous side of these religious assemblies, and proved what, we dare say, nobody will deny, that some blemishes cleave to them. We are persuaded, however, that many of our readers will not be disinclined to look at the higher aspects of "these wondrous gatherings day by day." For we are not aware that so many large and representative meetings have been drawn together in Glasgow within six days by any cause or interest whatever during the present century. Here is a novel addition to " the May meetings "— a new General Assembly, with representatives of almost every class of society and every Protestant Church in the land.

The religious movement, of which these meetings are the most outstanding manifestation, dates, so far as it met the public eye, from "the week of prayer" in the beginning of January. The ministers and office-bearers of almost all the churches then met, and formed a committee to arrange for united prayer-meetings, and also for the expected visit of the American Evangelists. The record of what was going on in the north of England, and especially in Edinburgh, had previously inflamed, as well as informed, many of the more receptive and sympathetic souls. Tokens of growing interest had also been appearing in many quarters, and evangelistic services, such as those conducted by Mr. Brownlow North and others, had indicated that the spiritual thermometer was steadily rising. During the first week of January St. George's Church was crowded at noon, while the overflow was accommodated in Hope Street Free Gaelic Church. After the first fortnight Wellington Street U. P. Church was made the centre, where, on an average, about a thousand met daily for prayer. In the second week of February Messrs. Moody and Sankey began their work among us; and for the last three months they have conducted meetings every day, with a few exceptions. The mind experiences a sense of fatigue in detailing their efforts. They certainly have not spared themselves. Here is something like an average week-day's work— 12 to 1 o'clock, prayer-meeting; 1 to 2 o'clock, conversation with individuals; 4 to 5 o'clock, Bible lecture, attended by some 1,200 or 1,500; 7 to 8:30 o'clock, evangelistic meeting, with inquiry meeting at close; 9 to 10 o'clock, young men's meeting. The tale of some Sabbath-day's work is even heavier: 9 to 10 o'clock, City Hall; 11 to 12:30 o'clock, a church service; 5 to 7 o'clock, women's; 7 to 9 o'clock, men's meetings in City Hall. Very few men possess, or at least exercise such powers of service; though, in addition to the aid from the realm beyond on which true workers rely, we doubt not that congenial and successful Christian work may sustain a man beyond any other form of human effort. Admission to these meetings was usually by ticket—a necessary precaution against perilous overcrowding. The animated scenes of last week in the

Botanic Gardens prove that the interest has not waned, even after three months use and wont had worn off the edge of novelty.

Accepting this as a genuine Christian work, it may be worth while to fix attention on some of its leading characteristics and results. We would say here, in passing, that we cannot well understand why some educated minds, without granting a hearing, condemn religious revivals out and out on philosophical grounds. Viewed on the human side, the philosophy of revivals, as they term it, is just a department of the philosophy of history. In no region has progress been uniformly steady and gradual; but it has been now and then by great strides, by fits and starts, and such events as the Germans call epoch-making. In all the affairs of men there have been tides with full floods. Every channel along which human energies pour themselves has had its "freshets." We are all familiar with revivals in trade, science, literature, arts, and politics. Times of refreshing and visitation are not much more frequent in sacred than in secular history; and they indicate the most interesting and fruitful periods in both.

To say that the work betrays some imperfections, and that there have been many objectors, is only to say what has been justly said of every great enterprise, civil and religious. But this revival seems to be distinguished from all previous revivals by the circumstance that it has been endorsed by something like the Catholic consent of the churches. From the outset, nearly all our leading ministers, and not a few of our foremost laymen, identified themselves with it. They sat and sang together on the pulpit stairs and platform at the daily prayer-meeting. A Highland member of the Free Church Presbytery lately protested against some of the accompaniments; and in a court that numbers about 150 members, there was not one to second his lament. One of our most conservative churches—the Reformed Presbyterian—gave its unanimous and cordial approval the other day at its Synod.

The unfriendly letter-writers fall into two classes. Some sign themselves clergymen, and are much exorcised about their clerical status. If any in these days will make it their chief concern to stand upon their official dignity they shall find by-and-by that they have not much ground to stand upon. No evangelists, however, have come among us who have more respected the position and influence of the ministers. Mr. Moody's first statement at his first meeting in the City Hall was, that he met with the Sabbath-school teachers first, because he knew that no class would welcome him more heartily, with the single exception of the ministers, and that it would be presumption in him to lecture them. The other class of unfriendly critics write in the interest of intellectualism and culture in its "broadest" sense. We suspect that the "sages," whose profession is, as one of themselves has said, that they are neither great sinners nor great saints, are the enemies of revivals only because they are the enemies of the things revived. Would they object, for instance, to a revival that gave body and popular attractions to the worn-out ideas which they commend as the *ne plus ultra* of attainable truth? At all events, it will not do for them to say that only the women and the children have been attracted, for there has been nightly a most imposing muster of the vigorous manhood of our city, and the City Hall has been often found too small to accommodate the men who flocked to some of the special services.

Mr. Moody is very fortunate in having such a colleague as Mr. Sankey. He has enriched evangelistic work by something approaching the discovery of a new power. He spoils the Egyptians of their finest music, and consecrates it to the service of the Tabernacle. Music in his hands is, more than it has yet been, the handmaid of the gospel, and the voice of the heart. We have seen many stirred and melted by his singing before a word had been spoken. Indeed, his singing is just a powerful, distinct, and heart-toned way of speaking, that seems often to reach the heart by a short cut, when mere speaking might lose the road. Most people admit that the work has been conducted in a very calm and soberminded fashion. Mr. Moody is credited with a large share of shrewdness and common sense. He has not yielded to the temptations that powerfully assail his class. He does not give himself out to be coddled and petted by well-meaning, but injudicious admirers. We have not noticed in him that offensive affectation of superior piety that provoked a sarcastic acquaintance of ours to say that some revivalists seemed to begin their story as Virgil makes Æneas begin his —"I am the pious Æneas." He keeps close to the essentials, and is free from such crotchets as often narrow the sphere and destroy the influence of evangelists. It is not irritation but balm that he tries to bring to our religious divisions. It must be owned that a premium has not been set on the hysterical, the convulsive, and the sensational forms of religious excitement. The proverbial weakness for numbers has been more apparent in some of his sympathizers than in himself. Nor does he make himself responsible for the reality of every apparent conversion. He has set his face sternly against the religious dissipation in which some of his most indefatigable hearers rejoice. Novelty-hunters and marvel-mongers have not been gratified. Sight-seers have been usually excluded from the meetings for inquirers, and only "workers" have been admitted. That there has been nothing necessarily repellant to thoughtful and educated people is proved by the number of middle-class young men in sympathy, and by the fair proportion of them at the "Converts' Meeting," and also by the crowds of genteel people at the quiet afternoon Bible lectures. Though he has introduced some novel methods, he has stuck to the simple old truths, and his convictions are in entire accord with Scottish orthodoxy. His straightforward, business-like, slap-dash style gives a fascinating air of reality to all he says, while his humor, capital hits, vivid and homely illustrations, and now and again his deep feeling, seldom fail to rivet the attention of his hearers. He has not a round-about and far-off way of handling divine things, and hence many accuse him of abruptness, brusqueness, and undue familiarity. The Christian life he commends is manly and genial, intense, and yet not strained or twisted. These features go far to explain what would be called in America his personal magnetism. Many ask, "But will it last? What is to come out of all this?" In Edinburgh, they say that since the Americans left the impression has been steadily increasing, and that it has entered influential spheres almost untouched before. The summer-scatterings will severely test the reality of the movement, but perhaps they may also scatter a share of the stimulus along both sides of the Clyde. The avowed end from the first has been that the ordinary congregational channels might be flushed and flooded with fresh energy. Such extraordinary efforts are most successful, though their success is less apparent when they add new power to ordinary agencies. If this be the result, the friends of the movement will have no cause for disappointment, while its enemies will point to the absence of demonstrative accompaniments as a proof that it has entirely collapsed.

We may expect that something will be gained from the experience of the past months. New methods of conducting meetings are already finding favor. Some may be in danger of surrendering hastily their individuality, and adopting modes of speech and action foreign to them. We may easily ascribe too much to the new methods of the American Evangelists. Their success is due largely to the fact that they approached the Scottish churches on the side on which they are weakest. It would seem that Scottish styles are about as popular in America as American styles have proved in Scotland, and for the very same reason. At the Evangelical Alliance in New York the speakers from our country were most appreciated, because they were strong where Americans felt themselves to be weak. The career of Dr. Hall in New York is also a notable case in point. By all means let us have more elasticity, and a greater readiness to adopt and adapt whatever is serviceable. But after all, new methods will not help the churches a great deal. The surprise and force of contrast soon wear off; and if men go too far for a little in any direction, they take their revenge in abandoning what formerly they over-praised. Age and repetition by-and-by make the most skillful methods dull and conventional. The grand need is far deeper—an inward vitality that makes men and churches fresh, various, and fruitful. If, as we are told, multitudes in all the churches have been recently quickened, new bottles should be made, as well as borrowed for the new wine.

Some confidently expect a more general coöperation of Christians than has hitherto prevailed. Dr. James Hamilton's quaint illustration has been so far verified. When the tide is out, each shrimp has a little pool of salt water, which is to him all the ocean for the time being. But when the rising ocean begins to lip over the margin of his lurking place, one pool joins another, their various tenants meet and mingle, and soon they have ocean's boundless fields to roam in. It will be a pity if an ebbing tide carries each back to his little narrow pool.

The relation of this work to the masses has been much discussed. Those who blame Mr. Moody for not working among them should remember that the tickets for all the meetings were distributed by the ministers of each district, and that in some cases the non-church-going had the preference. Recent speeches in Presbyteries and Synods show that many are anxious to give a home mission direction to the movement. Quickened life in presence of neglected multitudes must approve its sincerity by zealous mission work. We hear that the committee have already purchased a monster tent, capable of holding 2,000, and that it will soon be one of their chief rallying points. The young ladies of the choir, who give invaluable aid, are likely, it is said, to continue at their post. This would be a very graceful and telling way of bringing together the East and West End. Hundreds of young ladies with splendid voices and an expensive musical education might thus find a grateful relief from ennui, and a healthful substitute for other excitements. The work among the masses gives them a fine opportunity of gaining a recompense for all the trouble and cost by which they have become gifted musicians. It will be a new power to them, and to many preachers who can appreciate such coöperation. Even if we accept the estimates of the most sanguine—if we admit that thousands have been deeply stirred—if we grant that many of the best fruits of this work would not care to ask tickets for Friday night's meeting, and if we thus double or quadruple the numbers present, we shall then have only one or two per cent. of our population who are reported to have come decidedly under the influence of the movement, while the lowest classes have hardly been touched by it. Now, authorized reports speak of some outlandish parishes where one-tenth, and even as many as one-third, of the inhabitants have been deeply impressed. That is to say, in some out-of-the-way places the apparent results have been ten or twenty times greater than in Glasgow. The friends of this work have, therefore, reason for hoping that—to use the words of one of their favorite hymns—"still there's more to follow."

IV.

THE FAREWELL MEETINGS IN GLASGOW.

THERE is no doubt that the instrumentality of Mr. Moody has been greatly blessed to many in Glasgow during the past months. God often chooses for His purposes the weak things of the earth, but He as often takes men of great natural gifts, and uses them powerfully to impress others. Mr. Moody belongs to the latter class; he is possessed of great ability and tact, and of a most remarkable capacity for work. He has also great knowledge of the Scriptures, and is so evidently in earnest, believing he has God's commission to speak, that his power is above that of most men. Mr. Sankey has the same genuine ring, and by the judicious use of music has attracted many; but the power to keep them has mainly been given to his fellow-worker.

Many who have remarked on the subject of the revival have directed their observations towards our American brethren alone. They have doubtless been as the mainspring of the movement, used by God for that purpose, but all the other parts have also been in operation, many clergymen and laymen carrying on simultaneous work.

The last week has been a most impressive one; the interest may be said to have culminated in the assemblages in the Kibble Palace. On Tuesday evening, the 12th current, a great meeting of *women* took place there; all classes and ages were largely represented, and the bearing of the majority was most devout. Although upwards of five thousand were accommodated within the building, the issue of tickets had been so liberal that nearly two thousand more could not gain admittance, and were addressed on the green outside by various clergymen. Even among those who could hear nothing the greatest good humor prevailed. It was noticeable that many stood round the closed doors, where they could neither hear nor see, preferring not to listen to the other preachers, as they had come in their thousands, attracted by the fame of "Moody and Sankey."

On Wednesday night, upwards of seven thousand men managed to find sitting and standing room within the Palace, packing exceedingly close. The vast assemblage was most decorous, and obeyed orders implicitly. The full, strong singing of the hymns was a sound to be remembered. Numbers of inquirers gathered afterwards in the opposite church, and many could state that there was a good result of that night's work.

Admission to these two meetings had been exclusively by ticket; but on Thursday night the Palace was open to all, so that the doors had to be shut, leaving large numbers outside. The meetings throughout were conducted in the usual vigorous style, Mr. Moody being president, and delivering pointed addresses—clergymen and laymen relating their own experiences, and bringing forward instances and reports of the work elsewhere.

On Friday evening came the meeting for those who professed to have been converted during the last few months. The tickets for this had only been given to those who placed their names and addresses, and the name of their minister, on a Register opened for the purpose; and of which they were informed, extracts would be forwarded to the clergymen to whose churches they belonged, thus to prevent thoughtless application for converts' tickets, an ultimate check being established. The Palace was comfortably filled, and the utmost order prevailed. In looking over the assemblage, it was apparent that the great proportion consisted of young people, probably under twenty-five years of age. There was a preponderance of females, about in the proportion of three to two. All were well-dressed, clear-eyed people, in the ring of whose voices, when singing the opening hymn of praise, more than the common sound was heard. It was a glorious sight. Some, no doubt, may have joined that throng without due warrant; but with such exceptions, each individual had found his title clearly written in the Word of his Lord and Master. One instance came within our knowledge of two ladies who, receiving tickets under a misapprehension, personally came to deliver them up. This was the more honorable, as many would have entered by any means in their power, had the *Register* not stood in the way. An instance, indeed, occurred of one old lady, who, having made up her mind to enter, would take no denial, and indignantly gave the peculiar reason for insistence, that she was well known to the best people of the West End, and to the magistrates of the city and county.

On Saturday came the meeting for children, and in the evening another for grown people; but the greatest gathering of the week took place on the Sabbath, when, after a meeting in the morning for women who had not obtained admittance on the previous Tuesday, the gates were thrown open in the afternoon to all comers. Such a crowd came as had never been seen in these parts before. Many a time during the week the Great Western Road had been darkened for an hour and a half with the living stream, but that night for three hours the stream was incessant. Vehicles were not easily to be had on Sunday, so nearly all came on foot—all classes, "gentle and simple," young and old, blind and lame. The Palace was immediately filled, but the afternoon sun was so hot there that soon the whole had to turn out on the green; there a crowd, variously estimated at from twenty to thirty thousand, was soon gathered. Some apprehension was entertained that mishaps might ensue inside or outside the building; but by the exercise of considerable firmness, and compliance with orders on the part of the people, the danger was happily averted. Mr. Moody addressed the crowd standing on the box of a private carriage, and by those within comfortable earshot he was considered to have surpassed himself in earnestness and force. The singing of the sweet hymns by such a strength of voices, sounding upon that quiet sunny Sabbath evening from amid the fresh foilage of the gardens, was deeply impressive. Such a sight, too, had probably never before been seen within the limits of the land. While the main body dispersed, filling all the approaches and the public roads, about five thousand Christians, and those professedly anxious about their own state, gathered inside the Palace, and for the last time heard the voice of the man for whom such an affection has sprung up in the hearts of many. The scene was impressive when an English speaker, with rapid and energetic utterance, reminded the assemblage, many of whom had the greatest cause to thank God for all He had recently done for and by them, that that man (Mr. Moody) ought to be constantly remembered in the prayers of all to whom he had proved an instrument of grace. Many were much moved. Mr. Moody then took a farewell of the people, most of whom he could never hope to see again in the body, and as a final message declared that many Christian friends in that place and elsewhere had agreed to unite in prayer that night for those then gathered together who might be anxious about their own state, conscious they were not saved. The twilight was rapidly deepening when he asked those in such a case to rise to their feet in sign of their desire. The solemnity of feeling was indeed deep when from four to five hundred persons quietly rose all over the house, and as quietly resumed their places, actuated, evidently, by something outside their ordinary lives.

Was not that something like the Spirit of God? We shall, perhaps, never know here; but when from among the ranks of those who have felt His power we find men and women quietly affirming to friends and strangers the inner change which has been wrought in them, and then going out to work for Him; when we find this widespread, and representatives of all classes among the believers, have we any right further to question that God has been working, and will continue to work, in men's hearts powerfully? Be the instruments who they may, are they not of God's choosing? Some of his own servants may have been kept from joining in the work of promoting the awakening; but now that the first instruments of this general awakening have left us, it must be, it is, the sacred duty of all the stated ministers in the field, to take up the work where it now stands, and to carry it on; thankful and joyful in the fresh vigor infused into the spiritual life of many, and jealous only for the extension of the Master's kingdom. It is our duty now to consider whether, in much of our gospel teaching, we have not, by our own default, been beating the air, preaching above and beyond our hearers, clothing the simple lessons of God's Word in such conventional language as to be unintelligible even to the educated; how much more so to the illiterate. Laymen have become alive to the power they possess when filled with love to the Master, to tell in plain language of a risen Redeemer, and to impress and convince their fellows, so that many have been forced to exclaim in wonderment when the central truth of the gospel has burst upon them, "Is this all? is this the divine but simple truth I have missed for so many years? Our teachers might have made this plain long ago."

Blessed time when the mists are breaking,
When men are awakening! Ours be it to
Help, not to hinder, God's harvest.

A. S. D. C.

THE WORK IN PAISLEY AND GREENOCK.

MESSRS. MOODY AND SANKEY IN PAISLEY.

THE visit of these American evangelists to Paisley, long looked for, took place in April, and has been attended with great success. It was arranged that mid-day services should take place each day in the Oakshaw Street U. P. Church, and at the service there on Wednesday evening Mr. Moody was present. The church was crowded to welcome him. Mr. Moody presided, and was supported by the Rev. Dr. Lees, of the Abbey, the Rev. Mr. France, the Rev. Dr. Thomson, and nearly every other minister in Paisley. After one of Mr. Sankey's hymns had been sung, prayer was offered up by Mr. Moody, who afterwards read the 5th chapter of St. Luke, and eloquently expounded the gospel therein contained. Thereafter addresses were delivered on certain portions of the Scriptures, and prayers were offered up by the Rev. Dr. Thomson, and the Rev. Messrs. France, Hutton, Dodds, Crouch, and Sturrock. Other hymns from the collection used by Messrs. Moody and Sankey were sung, and the service closed with the benediction. In the afternoon, the venerable Abbey was filled with a highly respectable audience to hear a Bible lecture from Mr. Moody, and his devout manner of elucidating certain passages of Scripture seemed to impress his hearers in no ordinary manner. In the evening Mr. Moody made a third appearance at the Free High Church, the largest church in Paisley, and there could not be fewer than two thousand people present. At the conclusion of the services, which were of a most impressive character, there remained about one hundred persons, who expressed their anxiety to inquire further into the Scriptures.

On Thursday at mid-day Mr. Sankey was present at the Oakshaw Street U. P. Church, which was, if possible, even more crowded than on the previous day, the staircases and passages being filled. The Rev. Mr. Clazy presided, and was largely supported by ministers of various denominations. Mr. Sankey presided at the organ, and during a very beautiful service sang a number of his hymns, which were listened to with great attentiveness and reverence, and a large number of people appeared considerably moved by the pathetic rendering of such choice words. In the evening Mr. Sankey was at the Free High Church, which was again crowded long before the service. It was then announced that another service would be held simultaneously in Free St. George's Church, and as soon as the church could be lighted and prepared, it was filled in every available space. Mr. Sankey divided his attention between the two services, and sang several of his hymns to the great delight of the separate congregations.

An esteemed brother, writing from Paisley on Tuesday, 14th inst., says: "For two months past, a daily prayer-meeting, twice a week evening evangelistic services, and a series of Sabbath evening services, have been held in Paisley, under the conduct of a committee of the ministers of various denominations, embracing the great majority of the ministers in town. The Sabbath evening joint services have been crowded by earnest audiences, and many of the other meetings have been largely attended. The visit of Messrs. Moody and Sankey, for three days last week, was the occasion of overflowing meetings and general stir in the community. On the night of Mr. Sankey's attendance, a second church was opened, which he visited, and a third might have been filled with the dissapointed crowds. The addresses of Mr. Moody at the noon meetings, Bible lectures, evening services, and in the inquiry rooms, as well as Mr. Sankey's evangelistic singing and simple, heartfelt utterances, were attended by a large amount of visible impression. The inquiry, or after meetings, were scenes of great interest. On the three several nights, especially on the last of Mr. Moody's striking and earnest appeals, large numbers remained behind, many under strong emotion, desiring to be conversed with by Mr. Moody, the ministers, and other approved friends who stayed for this purpose. Several hundreds probably were conversed with from first to last individually, or in small private groups, over the great area of the Free High Church, where the meetings were held. Many of these professed, and seemed to have received light and quickening. In private life, and within the circle of private ministerial observation, not a few of these have been noted as the apparent subjects of important change. The interest seems to widen, and measures are being taken by the ministers to cope with the growing spirit of inquiry. One pleasing feature of the case is, the impression abroad among the young. Naturally there are criticisms and hesitancies with some as to the conduct and tendency of the movement, but these are neither numerous nor formidable. The marked approval of the fathers and leading ministers of the various denominations, and their personal superintendence, and happy joint action at all meetings, combined with the unassuming manner and quiet earnestness of the American strangers, has prepossessed and assured the mind of the religious public.

A new visit of Messrs. Moody and Sankey this week is likely to give fresh impetus to a good and great work, which, it must be the prayer of all who have closely witnessed it, nothing may arise to mar.

Wednesday, April 15.—The former visit of Messrs. Moody and Sankey last week to Paisley, produced such a deep impression that the evangelists were induced to return this week. On Tuesday Mr. Moody was present at the afternoon service in the Free High Church, and from the crowded attendance it was evident that the interest had not in the least abated. As on the previous occasions, Mr. Moody was supported by a large number of the ministers of various denominations in town, several of whom also took part in the service. In the evening another service was held in the same church, at which Mr. Moody was also present, and delivered, if possible, even a more forcible address than any of those he had previously given. The whole service seemed to result in awakening an interest so manifest, that at its close there could not be fewer than between 400 and 500 persons remaining to enter into conversation and inquiry with the evangelists. The good that has been done is apparent in many instances, and not a few who were prejudiced before to the movement have readily come forward, and now support it with all their power. It has been resolved to hold meetings each evening this week in Paisley, in addition to the mid-day services. On Sabbath Mr. Moody will return, and will devote nearly his whole time at services to be held in the town. It has been arranged that there shall be three special services. At nine o'clock in the morning he will meet with a number of young men in the Free High Church, and at other times in the day he will be present at Free St. George's Church, and at Oakshaw Street U. P. Church.

Following up the return visit of Mr. Moody, Mr. Sankey was present at several services in Paisley on Wednesday. In the evening a service was held in the Free High Church, and overflow meetings in Free St. George's Church and Oakshaw Street U. P. Church. Mr. Sankey was present at each, and there were crowded congregations. There is no doubt that the movement has made considerable progress in Paisley, and has awakened many to religion and study of the Scriptures who hitherto paid little regard for either. At the close of the meetings, on Wednesday, 15th, large numbers of inquirers remained for conversation with the evangelist, and were in conference till a very late hour. A wish has been expressed that both Mr. Moody and Mr. Sankey might visit Paisley together, and it is believed that the result would be an ample reward.

VISIT OF MESSRS. MOODY AND SANKEY TO GREENOCK.

ASTONISHING results have attended the visit of Messrs. Moody and Sankey to Greenock. At the usual mid-day prayer-meeting, the Town Hall has been quite crowded; while at the evening evangelistic services, several churches have had to be opened to accommodate the overflow. At the close of all the services, numbers of anxious inquirers have remained to be spoken to, and a special corps of ministers has been told off for this branch of work. One night last week, no fewer than five hundred anxious ones remained. They were of all ages and both sexes, and the scene which was presented was affecting in the extreme. In a short time many were announcing that they had found peace, and others that they had received assurance of their conversion. One of the more noteworthy occasions on which Mr. Moody has spoken in Greenock was a meeting for the study of the Bible. The audience consisted entirely of elders, Sabbath-school teachers, and Christian workers. Mr. Moody took for his subject, " Heaven;" and, in the course of his remarks, he said it would be impious to doubt that all things were possible to God. God could convert souls to Himself in a moment; and once a man was converted, his name was written in the Lamb's book of life, and could never be effaced. At the same meeting, it was stated by a gentleman from Edinburgh that the work in that city was spreading to all classes, and that special prayer-meetings are now being held by the school-boys of Edinburgh. On Sabbath, 12th inst., Messrs. Moody and Sankey were engaged at meetings all day. At nine in the morning they held a conference with Sabbath-school teachers; immediately afterwards, the members of the Working Boys' and Girls' Religious Society, with their teachers, were addressed; at eleven Mr. Moody gave a Bible reading in the Rev. Mr. Macrae's U. P. Church, Greenock; and in the evening, both Messrs. Moody and Sankey were present at a meeting in the Town Hall.

Another valued correspondent writes: Messrs. Moody and Sankey's labors in Greenock have excited an amount of interest deeper and wider than any similar services have done in former years. Indeed, it may be said that no similar services have ever been held here. There is a specialty and uniqueness about these which distinguish them from all preceding efforts in the same direction. It is very observable that it is the regular church-going population which has been mainly, though not exclusively, affected. The class of anxious inquirers who have remained to be personally dealt with after the public meetings, has included many who have maintained a Christian profession—in some cases for years—but who sorrowfully admit that they never till now realized the power of divine grace in their souls. Large numbers of all ages and attainments have professed anxiety, of whom the greater proportion belong to this class. Many of the younger members of Christian families have been quickened to decision, and the additions to the formal membership of all the churches are numerous in consequence. In one or two instances ministers have found among the anxious inquirers members of their own flock, whom they had regarded as well established in Christ. Several very young children have been wonderfully moved; and among the Working Boys' and Girls' Society instances of apparent conversion are numerous. Ministers and members of all denominations have wrought, on the whole, harmoniously; but it is to be regretted that here and there a spirit of exclusiveness has shown itself where it could hardly have been expected, and which, if it were encouraged, would do much to hinder the work, if not to arrest it altogether.

Wednesday, April 15th.—At the earnest request of the committee for united prayer Messrs. Moody and Sankey made arrangements for revisiting Greenock this week. Mr. Sankey has been present at all the noon-day prayer-meetings in the Town Hall, and the attendance, though not nearly so large as during the first week of his visit, has been most encouraging. At the evening meetings in the same place the audiences have all along been very large, and great numbers have remained till the second meetings to be spoken to. On Wednesday, at the "sweet hour of prayer," Mr. Sankey said that on the previous evening the second meeting had been the most interesting of any that he had seen yet in this country. The large galleries of the Town Hall were filled with anxious ones; and, although they were eager for spiritual conversation, a solemn stillness was preserved along with the earnestness. As the work of the ministers progressed, numbers of inquirers became convicted, and a great many announced with holy joy that they had found peace and been converted. Mr. Sankey, noticing that some of the audience were in mourning, read a touching letter from the father of a young woman who had, only a short time previous, accepted Christ. During the reading of the letter not a few were affected to tears. On Tuesday evening the Rev. Mr. Stewart, Ewing Place Church, Glasgow, presided, and conducted the services. He spoke of the work going on in Glasgow, and narrated several instances of conversion that had come under his own notice. A second meeting was held, to which about 1,200 persons remained.

EDINBURGH.

I.

RETURN OF MESSRS. MOODY AND SANKEY TO EDINBURGH.

MESSRS. MOODY and SANKEY paid a visit to Edinburgh in May, and continued for three days. On Tuesday Mr. Moody appeared at the daily prayer-meeting, but as he was not generally expected, there were not above two hundred more than the ordinary attendance present, and the Free Assembly Hall was not above one-half filled. But in the evening the hall was crowded to excess in every part. The Moderator's gallery was reserved for ministers, of whom a large number were present. There were amongst these many who had come from remote districts of the country to be present at the Assembly meetings, and whose keen and special interest in the proceedings was very evident. While the congregation awaited the arrival of Messrs. Moody and Sankey, devotional exercises were conducted by several of the ministers of the city. Mr. Moody, on entering the hall, reminded the meeting that, when he left Edinburgh two months ago, he requested the prayers of the converts here that a blessing might rest on the work they were going to engage in at Dundee, Glasgow, and other towns; and as these prayers had been abundantly answered, he asked them to join with him in thanksgiving. Thanks for this were accordingly offered up, and a hymn, "He leads us," having been sung, Mr. Moody delivered an address on the words of encouragement frequently found in the Scriptures addressed to God's people—"Fear not." The lecture was enforced by frequent reference to Scriptures. The impression was also deepened by hymns sung by Mr. Sankey.

At the close, Dr. Andrew Thomson, the Moderator of the United Presbyterian Synod, made an earnest appeal in the name of the churches, the missionary societies, and the perishing heathen, to the young men present to recruit the ranks of the ministry at home and abroad. They would never, he said, rue the day they laid themselves on Christ's altar. He spoke to them in the name of ministers of every denomination seated round the platform, and also in the name of the honored evangelist who presided. They had found it, he assured the meeting, a blessed thing to preach the gospel. None of them would like to change places with a king, a peer, or a judge on the bench, or to give up their ministry for all the world. No true convert, who had ever put his hand to the plough, had desired to look back. Theirs was the noblest, the happiest, the most blessed life that a man could spend on earth.

Mr. Moody advocated the adoption of a shorter course of study for young converts who were willing to devote themselves to evangelistic and missionary labor. He believed hundreds and thousands of young men and women in America and this country would come forward to work for the Lord if they were not kept back by the eight or ten years of study required. It was intimated that all the young men disposed to respond to the appeal made by Dr. Thomson should, after time for prayer and consideration, have an opportunity next Monday evening of offering themselves for the work of the ministry or of missions. The meeting was then closed.

Wednesday, May 20th. — The daily prayer-meeting was densely crowed to-day. The body of the hall was reserved for ministers till eleven o'clock, but the most of those who had been in town had left and others had not come, for very few appeared; and it was marvelous with what a rush the body of the hall was taken possession of at eleven o'clock, when the people who were waiting outside were allowed to come in. Shortly after eleven Mr. Sankey appeared, and led the congregation in the singing of some new hymns. The requests were then read, and the Rev. Knox Talon prayed. "Safe in the arms of Jesus" was then sung; and Mr. Moody called the attention of the people to the 22d, 23d, and 24th Psalms, reading portions of them.

This meeting ended at one o'clock, many of the audience having been in the hall since about ten o'clock; and they seemed to adjourn nearly *en masse*, and take possession of Free St. Luke's, where Mr. Moody was announced to lecture at three o'clock, for before two o'clock that large church was crowded, even to the third gallery, with an audience of not much fewer than 2,000. The singing of favorite hymns went on for an hour, then Mr. Sankey sang the following hymn as a solo:

Nothing but leaves! The Spirit grieves
  Over a wasted life:
O'er sins indulged while conscience slept
O'er vows and promises unkept:
  And reaps from years of strife—
Nothing but leaves! Nothing but leaves!

Nothing but leaves! No gathered sheaves
  Of life's fair ripening grain:
We sow our seeds; lo, tares and weeds,
Words, *idle* words for earnest deeds,
  We reap with toil and pain,—
Nothing but leaves! Nothing but leaves!

Nothing but leaves! Sad memory weaves
  No veil to hide the past:
And as we trace our weary way,
Counting each lost and misspent day
  Sadly we find at last—
Nothing but leaves! Nothing but leaves!

Ah, who shall thus the Master meet,
  Bearing but withered leaves?
Ah, who shall at the Saviour's feet,
Before the awful judgment-seat,
  Lay down for golden sheaves,
Nothing but leaves! Nothing but leaves!

Mr. Moody announced as his subject *The Two Adams*, and showed that we are either in the first or last Adam. The address was an exceedingly thoughtful and comprehensive one.

II.

THE PRAISE MEETING.

The Assembly Hall seemed to be taken possession of by much the same audience as soon as the doors were opened, and hundreds were disappointed of getting in, and were addressed in the College Quadrangle and the Free High Church. We do not think we have ever seen the Free Assembly Hall so densely crowded. The meeting, which was a short one, was commenced by Mr. Moody reading portions of the last few Psalms from the 145-50, and giving brief and appropriate comments on them. He also read 2 Chron. v.; Ezek. v. 11; 2 Chron. xx. 21; Acts xvi. 25; Jer. xv. 9.

Dr. Bonar quoted 1 Peter i. 3, "Blessed be the God and Father," etc. He said the very essence of praise was, as the word *blessed* literally meant, speaking well of God; and the best way to praise Him was to speak well of Christ, to testify the Gospel of the grace of God.

Rev. Robert Howie, of Glasgow, then addressed the meeting, and showed that we have to praise God for what He has given, and for what He is; but that it is a higher thing to praise God for what He is than for His gifts; and if we saw more of God we should praise Him more. If I may be permitted to speak in the name of my brethren, I would say we owe more to God than you do here. We rejoiced to hear of the work here, and longed that we might have similar blessings, but we have had more than we could have thought of. There were 3,500 converts at the farewell meeting, but that does not represent above one-third of those we know have been converted. And on the last Sabbath, about 20,000 assembled and were addressed in the open air, and four or five thousand went into the Crystal Palace, and about two thousand of them rose up, asking to be prayed for—seeking the Saviour. I have to give special thanks—first, for a praying mother; then my own conversion, and for being in the ministry in times like these. We have had a great work of grace.

There have been great meetings in shipbuilding yards, containing thousands of men. Messrs. Moody and Sankey went and had a meeting, and 1,000 men came. We believe that 10,000 have been converted in Glasgow since the year began; but what are these among so many, when our population consists of 600,000? There was one remark Mr. Moody made that he never allowed a day to pass without speaking to some one about the soul's salvation. If each one of the thousands of saved ones would do this, how many would be saved! Let this be the continual expression of our praise.

Rev. Mr. Mair, of Morningside, said he had been fifteen years a minister, and he had to praise God that this past blessed Winter and Spring had been the best time in the course of his ministry. If he had had £1,000 given him for his missions and church-work, he would have thanked the donor, and thought much of the gift; how much more thankful should we be to God, who had, week after week, been giving precious souls? Last communion was a time when the new power was experienced, when from sixty to one hundred were added to the Church. It was a real communion, for souls were feeling really joined to the Lord.

A minister from the country gave thanks for blessing to himself and the district where he labored. He said we had no idea of the depth and extent of the work in the country.

Rev. John Duke, of Dundee, thanked God for a plenteous rain in Dundee in connection with the visit of the American brethren. About 400 had been converted, and they were going on well. They were working also in giving tracts, teaching in Sunday-schools, helping evangelists by singing in the streets. He had had a young communicants' class, the like of which he had not had for six years.

Rev. John Morgan praised the Lord for his own conversion; for putting him into the ministry in times like these, and in circumstances such as he had had. During the eight years of his ministry, he had admitted 2,500 to the fellowship of the Church. And more especially would he praise God for the privilege of being in the work going on in this city last Winter and Spring.

Rev. John Kelman praised God for many blessings. He mentioned one cause for praise, that there had been such good weather during the visit of their American brethren to Edinburgh: only four days had been foul.

Rev. James Robertson, of Newington, said: They had truly been getting of late into the rapids of the stream that makes glad the city of God. Often in early days he had, after awakening sermons, watched for his minister at the corners of the streets, eagerly wishing for some opportunity of speaking with him. He believed there were many such in all congregations—their hearts longing even to bursting with concern about salvation. It would be to such like cold water to the thirsty to have special invitation every Sabbath-day to meet with their minister alone. On a Monday morning he had been visiting a dying father in the ministry, who asked, "What were you preaching on yesterday?" "I preached a whole sermon to the unconverted." "Oh," said he, "preach many, many whole sermons to the unconverted. I would often do that if I had my work to begin again. We are far too ready to take for granted that people know the way to be saved." In his last moments, another saint was heard whispering, "Bring, bring." One article was brought after another, but the waving of his hand showed that none of them was what he meant. Then at length, with a great effort, he uttered:

> "*Bring forth the royal diadem,*
> And crown *Him Lord of all*."

Mr. Sankey then sang a hymn of the lost sheep found, and the meeting was closed with the doxology.

III.
NOON PRAYER-MEETING.

Thursday, 21st.—The daily prayer-meeting was held to-day in Free St. John's Church on account of the meeting of the Assembly in the New Assembly Hall. The hour was eleven o'clock. At the hour Mr. Moody entered and asked the meeting to join in singing the 46th Psalm,—

> "God is our refuge and our strength,
> In straits a present aid ;
> Therefore, although the earth remove,
> We will not be afraid."

Mr. Moody then led in prayer, asking for special blessing on Edinburgh, on the Assemblies convening on that day, on deputations of young men going out in the Lord's work, and very specially on the great meeting to be held in the Queen's Park at five o'clock.

The subject for the day was, *The thief on the cross*, Luke xxiii. 39, etc. There is the conversion of all classes of people in the Bible — the rich and poor — the virtuous and the vicious, and the vilest of the vile; this thief was one of this vile sort. There were several thoughts expressed regarding what converted the thief. Mr. Moody said it might be hearing Christ's prayer, "Father, forgive them." Rev. Mr. Morgan thought he may have had a praying mother, whose heart he broke, and she knew only when she went to heaven that her prayers had been answered. Mr. James Balfour said he had heard the Lord said, *It is finished*, for Christ died before the thieves. Mr. Moody drew some lessons from the dying thief, such as, (1.) The strength of his faith; (2.) That Christ is never in a position in which He cannot hear prayer; (3.) That salvation is distinct from all ordinances and works. He could neither be baptized nor have the Lord's Supper; and He had nails through hands and feet, and could neither work for God nor run on His errands to carry blessing to others. He said he was struck with what Dr. A. Bonar had said in Glasgow—that he had asked his Bible-class to find another instance of a death-bed conversion, and they could not find one. We speak of this being the eleventh hour, but perhaps to the thief it was only the *first*, for it was likely the *first* time he had heard Christ. Let us now sing the 18th Hymn, which fits right into this subject,—

> "O bliss of the purified, bliss of the free ;
> I plunged in the crimson tide opened for me.
> O'er sin and uncleanness exulting I stand,
> And point to the print of the nails in His hands.
> O sing of His mighty love, mighty to save."

The Rev. James Robertson referred to having visited in his cell a condemned man, who said, "Ah, sir, when I am gone nobody will *remember* me." Didn't the dying thief speak as if he fully knew that Jesus was going away to be the *Intercessor* for transgressors? He could not say, "I have fought a good fight." Yet who sings more loudly or more gladly, "Salvation to the Lamb." He got not weapons to fight, but wings to fly. The last act of the Redeemer's life was the saving of a soul. No other door into heaven but that which this thief went in by. "I would like to know *you* in heaven," said a friend to old John Newton on his death-bed. He replied, "Well, you will find me at the feet of the thief who was saved upon the cross." It is a shame to be deep in debt on earth, but it will be our glory in heaven to be the deepest drowned in debt to "free grace and undying love."

Rev. W. Bremner, of Glasgow, said the Jews wanted evidently to degrade Christ by having the two malefactors crucified with Him; but Satan was outwitted, for the prey was taken from the mighty, and one of them was taken with Him to heaven, to show His power to save even when on the very brink of hell. Are there not some here on the brink of hell? To-day is your only opportunity as far as you know; be saved now.

Mr. Balfour referred to the ever memorable watch-night, and how we prayed that many might be made "fishers of men" this year, and how wonderfully the Lord has heard our prayers.

IV.

GREAT FAREWELL EVANGELISTIC MEETING IN THE QUEEN'S PARK.

On Thursday, 21st, at 5 o'clock, a great gathering assembled in a natural amphitheatre on the way to St. Anthony's Well, in the Queen's Park, to hear the farewell singing of Mr. Sankey and the farewell preaching of Mr. Moody. It was the largest open-air meeting we have ever seen convened to hear the gospel. We took pains to get a fair estimate of the number present, and we came to the conclusion that there were between TEN and ELEVEN THOUSAND. There were never above ten thousand present at one time, for during the hour when the preaching was going on, hundreds came and hundreds went in the outskirts of the crowd.

The audience was singing "*Jesus loves me*" as we came within hearing of the meeting, and from the number of voices joining in the singing, we concluded that the majority of those present had attended Messrs. Moody and Sankey's meetings. The Hundredth Psalm was then sung, and that was the only thing in which all the audience seemed to join—at least in the first verse—for they knew the tune. Then Rev. Mr. Grant read, at Mr. Moody's request, John iii., and Rev. Wm. Fraser prayed. We then had a hymn, and Mr. Moody preached "the gospel to ever creature," and thoroughly illustrated it by striking anecdotes and incidents. His voice reached us outside, and we could follow him, but with the constant conversations — children playing and shouting, and a number of dog-fights, those not accustomed to him averred that they lost some of his words, and were rather strained in following him. It would have been a better arrangement if Mr. Moody's preaching had come on half an hour earlier. With such an immense audience the effect would have been greater had they been addressed sooner; for the night being rather cold, about a thousand persons left during the time he was preaching—especially of the outsiders: the very persons most needing the gospel. We trust that the Word was with power to many, for the preacher spoke plainly, earnestly, and faithfully, and gave an A B C gospel, so that all might understand. It was a solemn time. There was a mass of men, and women, and children, many of them unsaved, and needing to hear words of salvation, and they heard them. It was an impressive sight to see masses of human beings hanging on or sitting on the shelves, and to all appearance in the clefts of the rocks behind the preacher, for it reminded us of the time when men and women will be crying to the rocks to fall on them and cover them from the face of Him who sitteth on the throne, and from the wrath of the Lamb, and the blessed contrast made us glad, for here was one standing on the rock beseeching sinners in Christ's stead to be reconciled to God. It was the day of grace, and not the day of wrath! It was said in a newspaper that Messrs. Moody and Sankey were hustled and mobbed, at the close, by ill-behaved people. This may have been, but we did not see it: it appeared to us rather that those who pressed upon Mr. Moody were loving friends wishing to bid him farewell, and he had to flee from them, which he did, and escaped to the carriage waiting on the Queen's Drive below, pursued by hundreds, all anxious to shake hands with him. Mr. Sankey, in following, had equal difficulty in getting away from the thousands that wished to have a last shake of his hand. The crowd of appreciating persons whom we saw at the side of the carriage were eager, loving friends: and we saw none of the other sort. There never was such a scene witnessed in Edinburgh, or anywhere else, so far as we have ever heard. We believe it is an index of the feelings entertained for our excellent friends, and the token of genuine regard and Christian love for the blessing the Lord has made their labors in the gospel to many souls.

V.

HOW THE AWAKENING FOUND US.

THE minds of the ministers of Scotland were occupied for ten years with a project for the Union of the Churches. An outsider could see no sufficient reason why the non-established Churches there should remain apart when they were at one in doctrine, polity, and worship; but after spending ten years of precious time in trying to have a union consummated, they failed, and negotiations for it were finally broken off. All this time, their minds being so full of this union work, and of the controversies in connection with the prosecution of it, their proper work of seeking the salvation of the lost, and the growth in grace of the saved, had not been carried forward so vigorously as it might and should have been. The saving of sinners had nearly come to a stand-still; and many were feeling the burden of souls, and imploring the Lord to send a spirit of awakening and revival, when Messrs. Moody and Sankey came to Edinburgh, and the blessing of God seemed to come with them, and to diffuse itself over the city and the country at large. We know that it was immediate, for it came into our own family the first night that Mr. Moody preached: and that it has continued and increased in power and volume, the late farewell meetings in Glasgow and Edinburgh attest. The soil is now productive, and a breath of spring-tide freshness is in the air. Sinners in great numbers are coming to Christ, and associating with His people.

And the Lord Himself has done this great work in such a way as to hide pride from man. Who would ever have framed such a prayer as this, and presented it at the throne of grace: "O Lord, in Thy great mercy send the two laymen called Moody and Sankey from the city of Chicago, to be the instruments in the revival of Thy work in the cities of Edinburgh and Glasgow, and throughout Scotland?" Any person who would have dared to pray in that fashion a year ago would have been deemed a lunatic; for who had heard of such men? And yet they are the men God has chiefly employed to accomplish the great work in which we this day rejoice. We were very much struck one day at hearing a Free Church minister thanking God publicly that He had sent those honored evangelists to do the great work they were doing, and thereby rebuke and humble ministers for not having put themselves into His hands to be used by Him in doing that work, but that strangers should have to be sent by Him to do it. And the union has come, too, in a higher way by the power of the Spirit; but since this was written the Moderator of the Free Church Assembly has delivered his opening address, and he has struck the right key-note. Let us read what he has said thank God, and take courage.

VI.

FREE CHURCH MODERATOR'S OPENING ADDRESS.

THE last third of Dr. Stewart's opening address was occupied with the present awakening. He spoke as follows: When differences of opinion and divisions arise among those who are honestly and earnestly seeking the good of His Church, the Great Head, the Lord Jesus Christ, often heals these divisions in a way they dream not of, and discovers "a more excellent way" for bringing about the end all have in view, viz.: His glory.

We have had a very remarkable example of this in connection with our late troubles. At the very time when the proposal of an incorporating union with brethren of other churches seemed to be relegated to a far distant future—when an answer to the many prayers that "we all might be one, even as the Father and Son are one," seemed to be withheld—when pseudo-philosophers, with profane levity, were proposing a prayer gauge, to test the efficacy of prayer—the Lord manifested Himself as a faithful and a jealous God—jealous for His own glory and faithful to His promises—as the hearer and answerer of prayer, by pouring out a blessed and copious effusion of His Holy Spirit upon our land, whereby many have been converted and saved, and a deep and most solemn impression has been produced upon the minds of men of all ranks and degrees. The result of this blessed visitation has been the healing of breaches among beloved brethren, and the producing such union of heart and co-operation among the godly and earnest-minded laborers in all our churches as warrant the hope of union on a broader basis than we had dreamt of, when "the Spirit of the Lord shall lift up a standard" against Popery and infidelity, "coming in like a flood," or when in some other way "God's set time to favor you" shall arrive. He has promised that His people "shall see eye to eye when He turns again the captivity of His Zion," and meantime, "in brotherly love preferring one another," let us watch and pray for it, "more than they that watch for the morning."

Permit me to say, as a comparative stranger, that of all the business transacted by the last General Assembly, that which affected and refreshed me most was the conferences regarding the state of religion throughout the country in general—the manifestations of spiritual life in the various congregations under your charge, and the measures adopted or recommended for overtaking the spiritual destitution still, alas! too prevalent throughout the land. It was a disappointment that more time could not then be spared for the consideration of subjects of such permanent interest, but I trust this year they may occupy both more time and a more prominent place in our deliberations.

It has pleased God to make use of two strangers from the other side of the Atlantic as the instruments through whom the spiritual awakening which has gladdened, and still is gladdening, many parts of Scotland, broke forth; and readily and heartily, I am sure, we are ready to render all due honor to beloved brethren whom the Lord Himself has honored—but, at the same time, we must not lose sight of the fact, that by these conferences in our Assemblies on the state of religion, by the deputations sent down to visit the various Presbyteries with the same object in view, and by increasing prayer and spiritual effort on the part of ministers, elders, deacons, and other godly laymen, the ground had already been prepared, the good seed had been copiously sown, and all that was wanting was that "God should give the increase." Blessed be the Lord our God, for He hath given the increase, and many of you, beloved brethren, who for many a year "went forth weeping, bearing your precious seed, have at last returned rejoicing, bringing your sheaves with you."

One more remark, and I have done. Rev. fathers and brethren, there is still another aspect in which it seems to me we should regard the blessed work of the Holy Spirit, in awakening and reviving the churches of our land at this time, and that is in the light of "a baptism with fire" ere times of trouble come, ere we are called "to contend earnestly for the faith once delivered to the saints." In looking back upon the history of the Church of Christ, we can trace many instances in which such "times of refreshing from the presence of the Lord" were the preludes to seasons of warfare and distress, of storm and tempest, when men's hearts were ready to fail them through fear, and many made shipwreck of their faith. Surely this is one aspect in which we may regard the outpouring of the Holy Ghost on the day of Pentecost, for scarcely had its blessed effects been felt, than persecution of the infant Church began. The outpouring of the Spirit and remarkable awakening that accompanied the great Reformation of the sixteenth century was at once the prelude to centuries of persecution and martyrdom, and the preparation of God's people for "resisting unto blood, striving against sin."

In the history of the Church of Scotland this has been often verified. The revival of religion in Scotland, so affectingly described by the historian Kirkton, was the means God used to fortify the hearts of His people against the *dragonades* of the bloody Claverhouse. When infidelity, heresy, and moderatism were deluging the Church about the middle of last century, the Lord the Spirit raised up a standard against them, in the remarkable revival at the Kirk of Shotts, and generally throughout the West of Scotland. And in later times, when the "ten years' conflict" had fairly begun, a new baptism of fire was given to our Church in the revivals at Kilsyth, Dundee, and many places in the North, to prepare men's hearts for the Disruption with all its trials and sacrifices, its astounding liberalities, and its ultimate triumph.

If we take into account the state of society in Britain at the present day—the growth of libertinism, communism, and infidelity—the influx of Jesuits into it unopposed—the rapidity with which Popery is again acquiring the mastery in the Church of England under the name of ritualism, and the supineness with which good men within her pale regard the humiliating spectacle—we have not to look far for a field of elements of trouble and danger to all the churches of our land; for these two enemies, Popery and infidelity, however divergent the ends they aim at, will act together as faithful allies in the endeavor to crush out vital religion. With such conflict in prospect—conflict in which, if we would be faithful to our Lord and Master, and keep carefully the sheep of His pasture, we must needs take our part—we ought with gratitude to recognize in the present effusion of the Holy Spirit a similar preparation vouchsafed to us against the day of trouble, and to consider seriously what attitude it becomes us as a Church to assume in defence of the gospel.

"Finally, my brethren, be strong in the Lord, and in the power of His might."

VII.

DR. THOMSON'S CLOSING ADDRESS.

The Rev. Dr. A. Thomson, as Moderator of the United Presbyterian Synod, said in his closing address: "There is no part of the Synod's proceedings that has been so interesting alike to yourselves and to the Christian public as the conference on the great religious revival and on evangelistic work. While some brethren were perplexed by honest difficulties which explanation removed, and while on the part of none of us was there any disposition shown to give a blind and unqualified approval to everything that had been done or spoken, there was soon manifested a universal readiness to acknowledge in the present awakening a blessed reality, and to own in it with adoring gratitude the work of the Holy Spirit. There was a tone of glad and solemn interest, an eagerness for information and for practical suggestions, and an earnest desire that the blessing might spread like a vestal fire over the land. The hearts of the brethren beat warmly and in unison. The extensive and thorough nature of the measures recommended by the Synod in the sending of deputations to Presbyteries, the issuing of a pastoral address, and the exhortation to every minister and session to seek revival in themselves and in their flock, and then to make their churches the centres of an earnest evangelism to the regions around them, proves how much the Synod had become of one heart and one soul in this mighty movement. If the injunctions of the Synod are carried out with prayerful and persevering energy in all our congregations—from Shetland to the Mull of Galloway, and from Berwick to Brighton—it will be a blessed year for our Church, many a full net will be brought to land, and long before another Synod, the cry will have gone up from many a congregation, "And now, O Lord, we thank and praise Thy glorious name."

THE WORK ELSEWHERE IN SCOTLAND.

At Dunbarton, that busy ship-building town, Mr. Moody preached in the South Church twice and gave a Bible lecture at the Free Church; on the 24th of May he and Mr. Sankey were at Kilmarnock, when many from neighboring towns and parishes were drawn together. On that day three meetings were held. The first was specially designed for workers; while in the case of the other two, the numbers proved so great, that after the Low Church was filled with its thousands, it was found necessary to open King Street U. P. Church, which was also speedily filled. The meetings in the Low Church were conducted by Mr. Moody, and those in King Street by the ministers of the town and a few others, while Mr. Sankey went between the places, and took part in all the meetings.

On Monday the evangelists proceeded to Saltcoats. A meeting was held in the afternoon in the Parish Church, which was densely crowded, and crowds again assembled in the evening in it and the North Church. On Tuesday at noon a meeting was held in Irvine Established Church.

Although there was less than a day to make the necessary preparations, yet at the hour the large building was filled. Mr. Moody preached most powerfully and impressively from Mark xvi. 15, 16, and Mr. Sankey sang one of his most eloquent appeals. A well-attended meeting was held in the evening in the Relief Church, and was conducted by the ministers of the town. The brethren could not wait to attend this meeting, as arrangements had been made for their presence in Ayr. Long before four o'clock, the Old Church there was crowded to overflowing. Meetings were also held in the evening, at eight o'clock, in the Old Church, and in Cathcart Street U. P. Church. It will thus be seen that Ayrshire has shared with other places largely, though in a brief time. I am content that results should tell the power, and yet as an eye-witness to the meetings in Kilmarnock, Saltcoats, and Irvine, I cannot refrain from testifying to the expectancy and solemnity that pervaded the large gatherings, and the deep interest and attention manifested.

PERTH.

"Good unto all men is the Lord,
O'er all His works His mercy is."

WE, in Perth, have found it true. It was last Winter when the Christmas fogs were round us that we heard once more "tidings of great joy." The power of the Lord was present in Edinburgh, and the men with whom the right hand of the Lord was working were expected in our city.

Often during the years since the Revival of 1860, the prayer had gone up that the Spirit of quickening might be poured out on Perth, and now we thought the Lord will answer these prayers—our prayers, the prayers of the strangers who have been amongst us, and the prayers of our departed ones. But God's "due season" had not come.

Noon-day prayer-meetings and evangelistic meetings were commenced, and we looked from week to week for the presence of Mr. Moody and Sankey in our midst. For twenty weeks the noon-day meetings were continued, and God gave many drops of blessing during that time. There was a cheering work amongst the children. Professor Martin, from Aberdeen, held meetings for five weeks, and in these meetings many little ones gave their hearts to Jesus.

The Professor left us, and it was resolved to make, for one month, a strenuous effort to win the older people to Christ by holding evangelistic services every night in the week, Saturday excepted. During that month Messrs. Moody and Sankey have come—and we lift our hearts to God and are glad that He has chosen His own time for sending them, and that He has sent His blessing with them now.

Mr. Sankey came amongst us on Friday, the 29th of May. There was a meeting in the evening, at which Dr. Black, of Inverness, and Mr. Robertson addressed the people, and for the first time in Perth, we heard that night Mr. Sankey's new song.

On Sunday morning a meeting was held for "Christian Workers, and those disposed to work," when the City Hall was crowded. Earnestly and thrillingly Mr. Moody spoke, and when Mr. Sankey sang "Go work in my vineyard," that large congregation listened with glistening eyes, and hearts kindled anew in love to Jesus and a desire to work more diligently for Him who suffered so much for us.

In the evening there were meetings on the South Inch (at which between 4,000 and 5,000 people were present), and in the City Hall, with overflow meetings in two churches. If it were the spirit of curiosity which prompted such multitudes to stand about for hours until they could get into a church, we must thank God that He has brought good even out of this evil.

Mr. Moody has held the usual course of meetings every day. Noon-day meeting in the City Hall, Bible-reading in the Free West Church, and evening meeting in the North United Presbyterian Church. The number of inquirers has been very great, and many, very many, have found peace. It has been a quiet, strong tide of blessing; it is as if God had sent His servants to unlock the flood-gates of His grace, and the water of life has swept out in deep and steady currents, leaving no place for the breaking waves of excitement and mere feeling. Especially is this to be noticed in the Bible-readings, when from day to day the large church in which Mr. Moody lectures, is crowded with people reverently and simply studying God's Word.

Besides the above meetings, there has been an overflow meeting in the City Hall each night, at which many have been blessed. And in the Free West Church there is a meeting for children. There is an inquiry-meeting after, from which many little ones go out trusting intelligently and heartily in Jesus, and eager to bring others to Him.

Messrs. Moody and Sankey remained at Perth until the 7th of June, where on the evening of that day Mr. Moody preached to a congregation of not less than 7,000 souls. The following Tuesday the evangelists paid another visit to

DUNDEE.

On Tuesday night two churches were thrown open, but so great were the crowds seeking admission that it was found necessary to adjourn to the Barrack Park, where an immense number of persons of all classes speedily assembled. The sight of so many persons hurrying along the streets from the churches to the park had the effect of arousing the curiosity of many more, who also hastened to the meeting. It was pleasing to see so many in their working clothes, mechanics and others; women carrying their children in their arms—in short, many from the humblest ranks of life; and Mr. Moody preached with his usual pathos and force. At the close of the service, meetings for inquirers were held.

On Wednesday, Thursday, and Friday nights open-air meetings were held in the same place. On each successive night the interest and solemnity seemed to grow more intense. The attendance was very large, the numbers being variously estimated at from 10,000 to 16,000 souls. Nothing could surpass the decorum of the vast assemblage. There was no sensationalism in the service, and no frothy excitement in the audience. One striking feature in the composition of the gathering was the unusually large proportion of men—shrewd, hard-headed, strong-minded men—a class not to be put off their feet by any mere sensationalism. And yet we saw the eyes of hundreds of these horny-handed sons of toil suffused with tears, under the word of God, which was preached with unaffected simplicity. One result of these open-air services has been, that a greatly increased number of men have come forward, asking the question, " What must we do to be saved?"

On the nights of Thursday and Friday, an inquirers' meeting for men only was held in Ward Chapel; and so encouraging has been the immediate outcome in the number of the anxious, that at Mr. Moody's suggestion special evangelistic services for men are to be carried on nightly for the next fortnight. A large staff of male Christian workers have promised to assist in carrying on this special effort.

On the afternoons of Wednesday and Thursday, Mr. Moody held a Bible-reading in Free St. Paul's, which was so crowded on the second day that it was deemed necessary to hold the meeting on Friday in Kinnaird Hall. Long before the hour of meeting this spacious room was crowded even to excess.

On Friday night a meeting was held in Chapelshade of those who have recently professed faith in Christ. There was a large attendance. The young disciples were suitably counseled and affectionately admonished by Mr. Sankey, who presided at the service.

On Saturday Messrs. Moody and Sankey left for Aberdeen. This second visit has been much appreciated, and has given a fresh impulse to the work. It has also served to make it abundantly evident that the work of grace has been going on quietly and steadily in this town and neighborhood during recent months. The ordinary channels of worship and work are full of the river of God—in some cases, indeed, full to overflowing. The pulse of Christian life in this city is beating more strongly and healthfully than it has done for many a day. It is now plain enough that the blessing will be largely permanent and abundantly productive. A thousand earnest souls are longing and praying for greater and still greater things. The impetus given to Christian work in all its departments can scarcely be over-estimated. This is especially true in regard to work among the young.

ABERDEEN.

HERE the work was commenced on Sabbath, the 14th of June, with a nine o'clock meeting for Christian workers, admission by ticket. There were 3,000 issued, and the Music Hall was quite filled, every available place being occupied, either sitting or standing. Mr. Moody, after reading a small portion of the Scriptures, spoke for about three-quarters of an hour from the text, "Here am I; send me." Mr. Sankey, assisted by a most efficient choir of male and female voices, effectively rendered several hymns, among which the principal were, " Hark, the voice of Jesus calling," " Go, work in my vineyard," and " Nothing but leaves." The meeting was a most solemn one, and the audience most attentive.

The evening meeting, at 5 P. M., was on the Links, in the natural amphitheatre of the Broadhill, where a platform had been erected for choir and speakers. It was here that the deep interest in these gentlemen—arising, of course, from mixed feelings of curiosity, or desire to know more of that better way of which they speak—showed itself. One may be allowed to say that the town was moved to come, and see, and hear. Some 10,000 were in position before and around the platform long before the hour of meeting; and yet from before five till past six there were continuous streams of men, women, and children from the city, Footdee, Woodside, Old

Aberdeen, and as far as Dyce, flowing to the one point on the Broadhill. There could not have been fewer than 20,000 to 22,000 on the Links that evening. Mr. Moody spoke from the words, "The wages of sin is death," and was listened to with rapt attention, while the hymns were distinctly heard over the vast crowds in the stillness of a quiet summer evening.

The next meeting was announced for eight, in the Music Hall, but it being filled before seven, Mr. Moody began at that hour, speaking on the subject of the prodigal son. There was much power. The chief hymns were, "Jesus of Nazareth," "Come home," and "Almost persuaded." There were many inquirers. The crowd outside was very great, and Free West Trinity and the Baptist Chapel, Crown Street, had to be opened, and were more or less filled; while several ministers conducted an open-air service in one of the squares. We have never at any time, I may say, seen the city so moved as it was this day.

On Monday a meeting was held in the South Parish Church, with a prayer-meeting at 3 P.M. Amongst the audience there were between twenty and thirty ministers of various denominations. In the evening a meeting was held in the South Parish Church.

Two hours before the time announced for commencing the meeting in the South Parish Church, a crowd had gathered at the door, and no sooner was admission gained than every seat and corner of the large church began to be rapidly filled. It was soon seen that the numbers waiting outside could not gain admission into the church, and provision was immediately made for having an open-air meeting in the quadrangle of Marischal College. Mr. Moody and Mr. Sankey arrived about half-past seven, and prayer having been led, Mr. Sankey sang the already well-known hymn, "Hold the Fort," the choir and the audience joining heartily in the chorus. Mr. Moody read a portion of tenth chapter of Luke, the story of the good Samaritan, and in a few sentences drew a vivid resemblance between it and the mission of Christ to wounded sinners. The reading was followed by the singing of a hymn by Mr. Sankey. "The Lost Sheep" was the subject of the hymn, and it was rendered in such a peculiarly appropriate style that the visible effect on the audience was something remarkable.

A short supplication for a blessing on the meeting was offered by Mr. Moody, who then said he would call their attention for a short time to the text in the second chapter of Luke, "Behold I bring you glad tidings of great joy, which shall be to all people, for unto you is born this day a Saviour."

Mr. Moody only spoke for about twenty minutes, and by this time Mr. Sankey and a portion of the choir had gone to the quadrangle, where there was a considerable gathering. A verse of a psalm was sung in the church, and Mr. Moody proceeded to the open-air meeting, the entire congregation following him. By the time he got on the platform between 4,000 and 5,000 had gathered in the square.

No sooner was the concourse of people comparatively quiet than Mr. Moody wished to hear them all sing the 100th Psalm, after which he began to speak from the text in Mark xvi., "Go ye into all the world and preach the gospel to every creature." The audience before him was of a much more miscellaneous nature than any of his previous ones, a goodly number having been drawn apparently from the Guestrow and Gallowgate, to whom Mr. Moody directed his special attention, addressing them with a ready familiarity. He expressed himself greatly pleased with the character of the meeting; he liked open-air meetings on week days, because all kinds of people could come to them, while no doubt a good many came all eyes and mouths open for curiosity's sake. The text he had chosen was an open-air one, and commanded them to preach the gospel to every creature, and in a few sentences he pointed out how comprehensive was this injunction of the Saviour's. Throughout both his evening discourses, Mr. Moody showed a wonderful fitness for adapting the circumstances around him to illustrate his meaning, thereby giving a kind of personal interest to what he was saying. His address lasted about the same time as the one in church, and at its conclusion he intimated that a prayer-meeting would be held in the Free High Church for about half an hour. While this meeting was going on, those who desired private conversation retired to the hall below.

The prayer-meeting was continued in the church by several clergymen, and did not break up until after ten o'clock; the inquirers' meeting lasted a good time longer.

FORFARSHIRE.

MESSRS. MOODY and SANKEY continued their labors for two weeks in Aberdeen, and from thence paid a flying visit, all they could find time for, to Montrose, Brechin, Forfar, and Arbroath. In each place all their meetings were densely crowded, although they were, for the most part, held in the very large Parish Churches. At Montrose, as in all the other towns, the earnest ministers and Christian people of the place had been making many prayerful efforts to awaken special interest in divine things in this time of blessing.

In Brechin next day they held two meetings, which none of us who had the joy of being present at them can ever forget. Both Mr. Moody and Mr. Sankey seemed to be peculiarly happy and at home in this town, and to speak and sing with even more than their wonted tenderness and power. We had hoped for an open-air meeting, both here and in Arbroath, but Mr. Moody felt unequal to that effort, having hurt his voice in one of our northern mists.

The second meeting was at half-past two, in the Parish Church, which could not nearly hold all who would fain have heard the strangers. The schools of the town had a half-holiday, that masters and pupils might be present, and all the factory workers who chose were also given leave of absence to attend the meeting. Many of these last were present in their working clothes, and bareheaded. Two of the ministers led in prayer. One prayer was specially on behalf of Lord Dalhousie, since deceased, who lay very ill almost under the shadow of the ancient church, where more than 2,000 lifted up their hearts on his behalf.

Mr. Moody preached for an hour with great power on the words, "Ye must be born again;" and after this meeting Messrs. Moody and Sankey hurried off to Forfar, where they addressed another large gathering, called together on a few hours' notice, in the very large Parish Church of that town.

Next day, Thursday, July 2d, our American friends proceeded to Arbroath, with their usual unflagging energy, and I had the privilege of being again with them. They addressed two meetings, which were both held in the Parish Church, as the largest place to be had. Mr. Moody's state of voice prevented him speaking in a third meeting, or in the open air, as had been expected. As usual, very many were disappointed of admission, for want of room; and many more, as I know, from the country district round, did not attempt to be present, knowing that others would be filling the church before they could arrive.

At the evening meeting the church was even more crowded, and the audience included a great many working people. Mr. Moody preached on "The Son of Man is come to seek and to save that which was lost," and the present writer never heard him tell the message of divine love with greater tenderness or power. He afterwards invited inquirers to meet him in a United Presbyterian Church not far off; and about 100, including about 40 children, did so. Both then and since there have been many proofs that the Holy Spirit of the Lord is at work in Arbroath. Evangelistic services have been held every night since Mr. Moody's visit to the town, and a considerable amount of religious interest has been evinced at these. We trust much prayer will be offered up by God's people for yet greater things to be seen among us in this district.

TAIN.

ON Monday, July 13th, Mr. Moody preached to a very large audience in the Free Church at half-past one on Monday. Five o'clock was the hour appointed for the open-air meeting, and this picturesque town presented an aspect never to be forgotten. The special trains have just arrived; the steep way from the station is thronged; vehicles of all descriptions approach by the various avenues into the town; and as we move forward to the Academy Park, the whole population seems astir, moved in one direction, drawn by one impulse. The service proceeds. Mr. Sankey sings the solo, "The Lost Sheep," accompanying himself on the American organ. Every eye is fixed; and as the stirring, earnest statements and appeals of Mr. Moody follow, the gaze of curiosity is changed into the intense earnestness of personal interest. It is the old gospel, yet some there feel it as they never felt it before. It is estimated that from 4,000 to 5,000 were present at this meeting.

At seven the Free Church, capable of containing upwards of 2,000, was densely crowded, many having to leave for want of room. About half-past eight the benediction was pronounced, after a most solemn service, and Mr. Moody requested as many believers, and persons knowing that they were yet unconverted, but who wished to find Christ, as could remain, to do so, while others left.

While a hymn was being sung, those who had to leave did so; others gathered into the area of the church, and the doors were shut. There were some moments of silent prayer, and then, amid deep stillness, Mr. Moody said, "We are all friends here, and I would just request those who believe that they received Christ to-day, and those who desire to receive Him now, to stand up, that we may pray for them." For more than a minute all was still; then Mr. Moody said slowly as one after another rose, "One, two, three, four, five, six, seven;" adding, as a large number now arose, "More than I can count. God be praised!" What a moment was that! "God be praised!" was the language of many a heart. Till eleven o'clock the church was an inquiry-room, Mr. Moody, Mr. Sankey, many ministers, and others being engaged in pointing souls to Christ; and many professed to accept God's gift, and to enter into peace.

Tuesday being wet, meetings were held in the Free Church at twelve and at half-past two. The church was filled to overflowing on both occasions; many remained in the church during the interval, singing hymns, while some ministers were conversing with anxious ones. At the afternoon meeting, Mr. Sankey sang several solos. The breathless stillness—tearful eyes—testified to the power that accompanied these sacred songs. Mr. Moody spoke with a peculiar force and impressiveness, on "I pray thee, have me excused."

After the benediction, very many remained; and when Mr. Moody again asked those who desired to be saved now to stand to be prayed for, about 500 stood up. It is impossible adequately to describe the scene—silence, broken only by that solemn rising. Very many were shedding silent tears—some from a sense of sin and danger, others from joy to see the Lord's work. One minister, who has seen much in connection with this religious movement, lifted his head, which had been bowed in prayer, and seeing these hundreds standing, he utterly broke down, and wept like a child.

Mr. Moody addressed the anxious, and then stated that he must leave, to keep an engagement at Inverness, but would request Mr. Sankey to remain.

Mr. Sankey and many ministers and Christian friends continued in conversation with anxious ones, till nearly six o'clock.

Men and women, the aged and the little child, were there, all with one accord—seeking Christ. Some, in answer to inquiries, stated that to-day, for the first time, they had felt their sin and danger; others had been seeking for twenty years —others for ten years, and various periods.

Those who know the reserve and shyness to mention what is personal in religion, which characterize the people in this quarter, and who consider that many of those who stood for prayer were well known in a small town, will be best able to appreciate the power that could overcome that natural reserve.

HUNTLY.

At Huntley, once famous for its religious gatherings, open-air meetings were held in Castle Park during the first week of July, where, as soon as it became known that the services of the American evangelists had been secured, a largely increased attendance from anything ever seen in Huntly before was everywhere confidently expected. Nor was the expectation disappointed. Some parties actually arrived on Saturday, worshiped in our churches on the Sabbath, and attended the preparatory meetings. At an early hour on Monday the people from neighboring parishes came flocking in from all directions. All sorts of vehicles brought their living freights of both sexes, and the number of pedestrians from neighboring localities was altogether unprecedented. The village of Aberchirder almost emptied itself, and we understand the same may be said of many of the fishing villages along the coast, the exodus from which was so great, that the powers and resources of the "innocent railway" were most severely tasked. The early train from Keith brought 64 passengers to Huntly at 6:40 A.M., the 9:10 train about 2,000, while the train from the south, which arrives here at 8:56, brought fully 3,000 to the meeting.

Immediately after the arrival of these trains, the streets of Huntly presented an appearance such as has never been seen in modern times. A conference was intended to be held in the Congregational Church, but so great was the crowd anxious for admission, that the idea had to be abandoned, as no church in the town would have contained half the number of those wishing to be present, and consequently the forenoon meeting in the Park was commenced at ten o'clock, instead of eleven, as intended.

At this meeting, the lowest estimate we have heard was 10,000, some maintaining there were 12,000 on the ground. In the afternoon, the attendance was much larger, numbers having arrived by the mid-day trains, and also from the country; and when Mr. Moody was addressing the assemblage, it was computed he was speaking to at least 15,000 people, some asserting that the number was little short of 20,000. Notwithstanding the vastness of the crowd, which, by the way, were standing very closely packed together, Mr. Moody was most distinctly heard at its utmost limits.

At the evening meeting, the attendance, though a considerable falling off from what it was in the forenoon, was large. Mr. Moody began by giving some account of his own experience, and proceeded to explain the nature of faith, showing that the reason of men's condemnation was, "that they spurned the remedy." His distinction between "I will not" and "I cannot" was well illustrated, and seldom has it been our fortune to listen to a clearer or simpler exposition.

After Mr. Sankey had led in singing the 40th Psalm, Dr. Black, at the request of Mr. Moody, gave an exceedingly impressive address from Gal. ii. 20. The meeting then adjourned to the Parish Church.

After a short address on the nature and scripturalness of inquiry-meetings, Mr. Moody invited the audience to sing a hymn standing, to give inquirers an opportunity of stepping into the inquiry-room, and a few complied. Mr. Moody, we confess, startled us when he said that the vestry of the Established Church was built for the very purpose, but it was a goodly sight to see it turned to such a use.

NAIRN.

ON Tuesday, July 21st, Messrs. Moody and Sankey were at Nairn. Their visit was preceded by prayer to God for an outpouring of the Spirit, and many of the Christians were looking for much blessing. Mr. Moody presided at twelve o'clock in the U. P. Church. Long before the time announced that the service should begin, the building was crowded. Mr. Moody gave a short address on the three kinds of Christians—Asking, Seeking, and Knocking. Mr. Sankey sang, "Keep praying at the Door."

A Bible-reading took place in the Free Church at three o'clock, and at half-past six Mr. Moody addressed an audience of not less than five thousand, on the Links, on the verse, "Go ye into all the world, and preach the gospel to every creature." At eight o'clock, upwards of one thousand filled the Free Church, where Mr. Moody spoke for half-an-hour, giving a question to each soul, Am I saved, or am I lost? Mr. Moody asked those who wished to be prayed for to stand up, when many did so, and solemn indeed it was to see in the same pew some who stood up to show that they wished to be saved, while those next to them sat still. An inquiry-meeting was held at the close, and about sixty or more were conversed with, while many retired to their homes with an arrow in their hearts. Some professed to close with Jesus, and some left undecided for the Lord. Mr. Moody and his fellow-laborer left for Elgin next morning, while the services are being carried on by the ministers in town and an evangelist. The inquiry-meeting on Wednesday evening was still more interesting, many professing to close with Jesus. The whole town is being moved.

ELGIN.

THE *Elgin Courier* devotes two columns to the two days' visit of Messrs. Moody and Sankey to that ancient town, where meetings were held with results similar to those which have attended them elsewhere.

Last evening, 23d, at seven o'clock, an open-air meeting was held on Ladyhill. The weather was very favorable. Nearly all the shops on the High-street were shut at about seven o'clock. The sun, as he sank to rest in the west, shed his dying glory over the most picturesque scene on the hill-side. It was estimated by some that there were between five and six thousand persons present, it being the largest gathering of the kind we ever remember having seen in Elgin. Tempted by the fine evening, all classes of the people turned out, many arriving from all parts of the surrounding districts. At the foot of the hill a platform was erected, which was occupied by the choir and speakers. The whole hill-side, for a great distance up and round about, was covered with the dense multitude, that presented, with their varied dresses, a most imposing spectacle. On the Market Green there were also a large number of people.

The meeting having been opened with praise and prayer, Mr. Moody spoke for about an hour on the words, "Ye must be born again," with characteristic earnestness and graphic description. Several hymns were then sung, after which the meeting was dismissed, it being intimated that another would be held in the Parish Church, for which there was a great rush. The gates having been opened, the church seats were completely filled in a few minutes. The meeting was devoted to praise and prayer, Mr. Moody leaving to speak with the anxious in the New Evangelistic Hall.

Such a Sabbath-day as the last one we have never seen in Elgin. During the intervals between the different meetings, our streets were thronged with people from all parts of the surrounding districts, of all classes, "set out," of course, in Sunday attire. The number of people from the coast towns—Lossiemouth, Hopeman, Burghead, Garmouth, Buckie, etc., was also (for such an occasion) unprecedently great.

At nine o'clock a meeting of Sabbath-school teachers and mission workers was held in the Parish Church. It was thoroughly representative of nearly all religious workers in the town and district. Most of the clergymen of the town were present.

Mr. Moody's address, specially given to workers in the Christian field, was a most practical one, and was all through powerfully illustrated by most suitable anecdotes, some of which, owing to their rather facetious nature, produced a smile on the faces of those present.

At half-past one o'clock, Mr. Moody preached in the Established Church, which was crowded to the utmost extent.

At five o'clock in the evening the farewell open-air meeting was held on Ladyhill, which was, literally speaking, one huge black mass. For about an hour or so before the time of meeting a perfect stream of people kept pouring onwards up the High-street towards the hill. Ere the hour had arrived, the crowd had grown densely large. There were between 7,000 and 8,000 persons present.

Mr. Moody arrived, with Mr. Sankey, about five o'clock. The first four verses of the 40th Psalm were sung, Mr. Sankey leading. Nothing could have been more beautiful or soul-inspiring than to hear the sound of the fine old tune "Evan," which reverberated from the hill-side. After prayer Mr. Sankey sang "The Lost Sheep." Mr. Moody then spoke from Luke iv. 16. As he concluded, the weather cleared up, and the scene was considerably enlivened and brightened by the rays of the sun. Mr. Sankey sang, "I am coming, Lord," the people joining with him. Prayer was then offered, after which Mr. Moody intimated that meetings would be held in the Parish Church and the Free High Church after the open-air one was dismissed. The crowd then separated.

In a short time both Parish and Free Churches were filled. In the Parish Church an able and appropriate address was given by Mr. Moody on the words, "Son, remember," Mr. Sankey singing a very beautiful hymn. At the after-meeting a large number of anxious inquirers stayed, something like seventy-seven persons standing up, expressing by so doing their wish to become Christians. The meetings in the other churches were equally successful.

A SECOND VISIT TO ELGIN.

AFTER a visit at Banffshire Mr. Moody spent another day at Elgin, and there was great joy on Wednesday afternoon when it was flashed through the country side that on the following evening there was to be another of those great open-air gatherings which every one had enjoyed so much. The meeting is thus described :

It was a strange contrast last Thursday; at five o'clock, in the busy Show at Inverness, at seven in the streets of Elgin, quiet at all times, but that night altogether passengerless and deserted. Surely something unusual was going on—the streets abandoned, the house-doors fast, the shops closed. Through half a mile of the empty streets ours were the only footsteps that echoed on the pavement, and everything was silent and desolate as a plague-stricken city! At last, just on the verge of the town, the stillness was broken by the distant sound of a voice, and the turn of a lane revealed a sight which time can never efface from the memory. There stood the inhabitants, motionless, breathless, plague-stricken indeed—plague-stricken with the plague of sin. The sermon was evidently half over, and the preacher, with folded arms, leaned over the wooden rail of the rude platform. Oh, the sin upon these faces round him! How God was searching the heart that night! I cannot tell you who were there, or how many, or what a good choir there was, or what Mr. Sankey sang, or which dignitary prayed. I cannot tell you how beautifully the sun was setting, or how fresh the background of woods looked, or how azure the sky was. But these old men penitent, these drunkards petrified, these strong men's tears, these drooping heads of women, these groups of gutter children, with their wondering eyes! Oh, that multitude of thirsty ones—what a sight it was! What could the preacher do but preach his best? And long after the time for stopping was it a marvel to hear the persuasive voice still pleading with these Christless thousands?

One often hears doubts as to the possibility of producing an impression in the open air, but there is no mistake this time. No, there is no mistaking these long concentric arcs of wistful faces curving around the speaker, and these reluctant tears, which conscious guilt has wrung from eyes unused to weep. Oh, the power of the living Spirit of God! Oh, the fascination of the gospel of Christ! Oh, the gladness of the old, old story of these men and women hurrying graveward! The hundred-and-one night in Glasgow excepted, never have we seen the Holy Spirit's nearness more keenly realized. These thousands just hung spellbound on the speaker's lips. It seemed as if he daren't stop, so many hungry ones were there to feed. At last he seemed about to close, and the audience strained to catch the last solemn words; when the preacher, casting his eye on a little boy, seemed moved with an overpowering desire to tell the little ones of a children's Christ. Then followed for fifteen minutes more the most beautiful and pathetic children's sermon we have ever heard; and then, turning to the weeping mothers and fathers, concluded with a last tender appeal, which must have sunk far into many a parent's heart.

INQUIRY-MEETINGS.

Long before the close of the address it was evident to all that the Lord of the harvest was going to give us a glorious reaping-time that night. We had not, indeed, been ten minutes on the ground, when a stranger whispered, in the very middle of the address, "Will you come and speak to a woman about her soul?" at the same time pointing out a drooping figure standing near, with face buried in her shawl. We were not surprised, therefore, at the great crowds which entered the inquiry-meetings—in one church for women, another in a large hall for men, while the Christians went apart by themselves to another church to pray. The arrangements connected with these after-meetings were all beautifully managed, and shortly after nine o'clock the whole three were well under way. The women's inquiry-meeting was supplied with relays of workers from the prayer-meeting. The work was on a very large scale, and the workers' report was, that the cases were of a very hopeful character. But the work amongst the men—and this is a splendid testimony to the depth and reality of the impressions—were even on a larger scale still; and the sight in the Evangelistic Hall, where the men's inquiry-meeting was held, is not soon to be forgotten. The whole hall was filled with men, broken up into little groups of twos and threes, talking in hushed yet earnest voices on the great subject of the one thing needful; while behind, in the committee-room, half a hundred young men were gathered in prayer for their groping brothers. Many of these had themselves but newly decided for Christ, and were the fruit of the week's meetings for men, which have been blessed by God far above all expectation.

It is useless to attempt to give even an approximate idea of the extent of the blessing which fell upon Elgin on Thursday night. The whole of Morayshire has shared it, and a powerful hold has been gained in nearly every farmhouse and village throughout the country side.

FAREWELL MEETINGS IN ABERDEEN.

AT the pressing request of a large number of those who had taken part in the evangelistic work set agoing in Aberdeen some months ago, Mr. Moody paid a farewell visit to Aberdeen in August, and addressed several meetings, at the same time taking occasion to urge on to greater zeal those who were engaged in the good work. Mr. Sankey has been obliged to go south to a more genial temperature to recruit his health, but Mr. Moody has wrought on since he left Aberdeen, in different districts in the north, almost without ceasing; the same remarkable results always attending his labors.

At seven o'clock, Mr. Moody met with a large body of young converts in the Free South Church, and addressed to them a few parting words. He spoke on his favorite topic of "Confessing Christ," pleading hard with those who had lately come to Christ to come boldly forward and confess Him.

The Music Hall was crowded to excess long before eight o'clock, the hour at which Mr. Moody was announced to give an address, the passages, orchestra, and galleries being quite packed. "Except a man be born again he cannot enter the kingdom of heaven," was the text on which Mr. Moody based his discourse. Christ did not say these words to a drunkard, to a thief, to a harlot, but to a man who in our days would be made a D.D. or an LL.D.

After referring to the often-doubted possibility of sudden conversions by those who could not understand it, even although there were living evidences of it before them, he bade the meeting farewell, with the hope that they would all meet on the shores of eternity.

Mr. Moody stayed in the hall conversing with anxious inquirers until about ten o'clock, when he drove to a men's meeting in Trinity Free Church, which had gathered at nine o'clock in the expectation that Mr. Moody would give them a farewell address. In the course of the few sentences he spoke to them, Mr. Moody said they could have no idea of the influence the Aberdeen men's meetings had had in other places he had visited. In all of the towns the example of Aberdeen had been followed, and large bands of young men were enlisted in evangelistic work.

A number of the young men then retired with Mr. Moody into an ante-room, to hold private conversation with him, and he continued to converse with them until it was time for him to go and prepare for his journey to Wick by steamer.

MEETING AT CRAIG CASTLE.

On Sunday afternoon, an open air evangelistic service was held on Craig Castle lawn, conducted by Mr. Moody. The weather in the early part of the day was very unpropitious, heavy showers descending, with brief intervals, until four P.M., when the rain ceased, and it continued fair during the evening. The wet detained not a few at their homes, no doubt, but most of those who came seemed to have determined to be present in any case; and by five o'clock a very large company—especially taking into account the thinly-peopled districts from which they had gathered—had assembled on the beautiful lawn in front of the castle. Every valley and hamlet within a radius of ten miles sent its company in gig, cart, or afoot, until at five o'clock about 2,500 people stood on the lawn. The gathering resembled somewhat one of the Covenanter hill-side meetings, save that while the Bibles were still present, the broadswords were altogether absent; and the rendezvous, instead of being a wild, rocky pass, was a hospitable castle, with its fairy dell and leaping linn, celebrated in song, and known as one of the loveliest spots in Scotland.

The beauty of the scene seemed specially to move Mr. Moody, who referred to it again and again in his discourse, which was one of peculiar beauty, power, and pathos. Standing in an open carriage placed near a towering tree, the preacher spoke for nearly an hour from the parable of the marriage feast. A very marked impression was produced, and many retired at the close of the service for conversation with the preacher and other ministers and friends.

The Craig gathering of August, 1874, will, we believe, be ever memorable to not a few as "the beginning of days" to them.

THE LAST WEEK IN SCOTLAND.

During the last days of August, a Farewell Convention was held at Inverness. It was an "all-day meeting," each hour being devoted to a special subject.

After the Convention, Mr. Moody went down the Caledonian Canal to Oban, and there on Friday, the 28th, gave an address, with much apparent blessing, in the U. P. Church. There had been much preparatory work in the town, not only in the open-air meetings, but also in other special services; and in the two preceding months the Rev. H. Bonar and the Rev. A. Bonar had ministered the Word in the Free Church. From Oban Mr. Moody went to Campbeltown, by way of Tarbert, on Loch Fyne, and remained from the 29th till the 3d September, when he left for Rothesay, taking the Tarbert route, and staying at Ballinakill, where many were gathered from various parts of Kintyre to meet him. His work at Campbeltown was deeply interesting, and was crowned with remarkable blessing. He commenced on Sunday, the 30th, by three services; speaking first to workers, then on the blood, and lastly on the grand command, "Go ye into all the world and preach the gospel to every creature." The result after that last address was most striking. Upwards of fifty stood up to ask to be prayed for, and to declare their desire to be Christians. The meeting had been over-crowded, and some went to the Drill Hall, where the gospel was preached by willing helpers; but in the great after-meeting in the church, all were united, and it was felt to be a time of wonderful enlargement and power. On the three following days the interest was deepened at successive meetings; till at the last, on Wednesday night, when Mr. Moody had preached on God's invitation and man's excuses, a very large number were gathered into a hall, either as converts or inquirers; and it was manifest that much fruit had been gathered to life eternal. The work now is laid on the hearts of some who are striving to confirm the souls of the disciples; and, as one means, it has been arranged to have a Converts' Meeting weekly, similar to that in Ewing Place, Glasgow.

The last meeting was at Rothesay, and is thus described:

Meetings for special prayer and evangelistic work have been held here since the middle of October last year. These meetings were held in several of the churches on the Sabbath evenings; in the Victoria Hall, and latterly in the Town Mission Hall on week-day evenings. These services, added to the general interest manifested throughout the country in religious things, led to united meetings for prayer. These daily meetings were brought to a close about the end of May. The meetings in the Town Mission, however, were continued three nights weekly, from the 14th Dec. last, till the present time, and have, we believe, been blessed to not a few. There have been marked cases of interest, and those who took part in the meetings have been greatly refreshed and encouraged, while week by week they were growing more earnest in the work. The prayer-meeting on Saturday evenings has been for some time marked as possessed of more than usual interest.

Several requests from all the ministers and office-bearers in town were sent to Messrs. Moody and Sankey, without success until last week, when, on returning from Campbeltown to Greenock, *en route* for Belfast, Mr. Moody kindly agreed to spend Thursday evening in Rothesay. As soon as the telegram to this effect was received, arrangements were at once made for holding one or more meetings. The news soon spread through the town and island, and it was speedily evident that one building would be insufficient to hold the numbers likely to attend. Accordingly it was arranged to hold a meeting at seven o'clock in the West Free Church, and a second meeting in the East Free Church at half-past eight o'clock. After Mr. Moody's arrival, it was found that only one meeting could be addressed by him, and a change of arrangement had accordingly to be made—a change at first regretted, but which eventually proved to be for the benefit of all. The West Free Church, being the largest building, was accordingly selected, and by seven o'clock was literally packed—passages, pulpit stairs, lobby, etc., being occupied by a dense mass of human beings. Mr. Moody arrived at half-past seven, when Rev. Mr. Thomson took the chair, and gave out the 43d Psalm. Rev. Mr. Ross read several requests for prayer, after which, the Chairman having engaged in prayer, it speedily became manifest that the atmosphere of the church was such as the ordinary means of ventilation could not remedy, so densely was it packed. During the singing of two hymns—"God is Love," and "Jesus paid it all,"—arrangements were being made for conducting the service in the open air.

To the great relief of many in the church, and to the intense delight of hundreds outside, Mr. Moody intimated that the remainder of the service would be conducted by the sea-shore on the Esplanade. Here, in a few minutes, during which the 23d and part of the 17th Psalms were being sung, an immense throng of people, numbering not fewer than 3,000 persons, had assembled round the preacher. After a short prayer, Mr. Moody preached from Mark xvi. 15 and 16. For fully an hour he riveted the attention of his large audience, narrative, metaphor, parable, illustration, and appeal following each other in quick succession and agreeable variety. Towards the close of the service the scene was one never to be forgotten. The firmanent was cloudless and myriads of stars shone brilliantly (for by this time night had fully set in), and were reflected in the Bay, beyond which lay the Cowal Hills dark and massive in the distance. Every now and again the houses in the Gallowgate and the spire of the West Free Church were lit up by flashes of sheet lightning. The Esplanade with its thousands was in front of the preacher. Deeply impressed, evidently, with the position, the scene, and the circumstance that he was addressing probably for the last time a Scottish audience, Mr. Moody concluded a discourse, which for point and power we have not heard on any former occasion surpassed. It was evident the Spirit of the living God owned the truth, for when the intimation was given that a second meeting would be held in the church for prayer and further explanation of the way of life, the building was very speedily well-nigh filled.

An inquiry-meeting was held afterwards in the hall adjoining the church, to which a large number of persons retired, deeply impressed with the concerns of the soul. The night of Thursday, the 3d of September, 1874, will, we believe, be memorable in the history of many a precious soul, and many will take up as their own the words of the Psalmist—"The Lord hath done great things for us, whereof we are glad."

THE WORK IN IRELAND.

BELFAST.

I.

Sept. 6, 1874.—These beloved and honored brethren have commenced their labors of love among us with most marked tokens of God's smile and approval. On Sabbath morning, the 6th inst., the first meeting was held in Dougall Square Chapel, at the early hour of eight. The meeting was exclusively of Christian workers. Long before the hour named, the chapel was crowded. The meeting was conducted in the usual way by Messrs. Moody and Sankey. Mr. Moody struck the key-note of entire devotedness to, and unwearied labor for, the Lord Jesus. All present seemed, in silent prayer, to lay themselves upon the altar afresh, as living sacrifices to the service of God.

The second meeting was advertised to meet at 11:30, in Fisherwick Place spacious church. The desire to hear had crowded the church long before that hour, many going away unable to obtain admission. Mr. Sankey led the praise. Mr. Moody chose as his subject, "Love." The impression upon the minds of multitudes was very deep. Many, we feel persuaded, tasted in fresh power of the love of God, and had their love drawn out to Him who first loved us.

Still a third meeting remained, advertised to be held at 7 P.M., in the largest church in Belfast, capable of holding 2,000. It is considered that not above one-fourth of the people who crowded the streets around the building were able to gain admission. During the service there were visible signs of the presence and power of the Holy Spirit. At the second meeting many anxious sinners remained for conversation. Not a few professed to accept the offered gift of God, from the hand of His promise, even His only-begotten Son.

The daily prayer-meeting was commenced in Dougall Street Chapel, on Monday, at twelve. The chapel was so over-crowded, it was deemed advisable to adjourn next day to a more capacious building, capable of holding 1,400 people. This "sweet hour of prayer" is the centre of the whole movement, and has already proved a blessing to many. On Monday evening the evangelistic meeting was held in Rosemary Street Church. It was, we believe, a most blessed and fruitful one. But the crowd was so great, and causing such inconvenience, as to induce Mr. Moody to alter his plans somewhat, and during the succeeding days of the week he has held a meeting at two P.M. exclusively for women, in Fisherwick Place Church, capable of holding 1,400 people and upwards, and a meeting in the evening in the other church exclusively for men. The Lord has greatly blessed the arrangement. The meeting at two, for women, has been crowded each day. The Spirit of the Lord has been present to heal. Each day increasing numbers remain to be spoken to about their souls, and many profess to have entered into rest through faith in Jesus.

The same report is true of the meeting for men. The Spirit of God is taking of the things of Christ and showing them to very many. The meeting last night was particularly solemn. The work of the Spirit of God seemed more deep and clear, and many professed to accept the offered Saviour.

The Christian community has been deeply stirred; we are filled with expectation; we are looking for great things. The work of the time seems to be fully more extensive and deep than in any place visited by these brethren.

II.

THE work has had a good commencement in Belfast. Numbers thronged and crushed to the churches, so much so that the happy plan was adopted of dividing the meetings, and holding gatherings for women only at two o'clock, and for men only at eight o'clock. Consequently, the large churches are well filled, without any unseemly disorder.

On Friday (11th) Mr. Moody addressed both meetings, taking for his text, "The Son of Man is come to seek and to save that which was lost." With great power and aptitude he proclaimed the Lord Jesus as the "Seeker;" and very touchingly he convinced the people that He was now seeking each individually, seeking to save and to bless them. Mighty FAITH, then, appears to be the secret of Mr. Moody's power. On the hearers he urges DECISION, now to believe, instant salvation on faith in Jesus only. His address was interspersed with telling illustrations, which came right home to every heart. He rapidly referred to the parable of the lost sheep and lost piece of silver, and graphically narrated the sudden conversion of Zaccheus, unmistakably evidenced by the immediate fruit of the Spirit in his change from an extortioner to a restitutor. Mr. Sankey's very sweet solos and touching hymns, accompanied on the American harmonium, seemed to exercise a powerful effect in, as it were, deepening the impression of the Word.

The large church of the Rev. H. M. Williamson, which holds 2,000, was filled with women of all classes; and the one in Rosemary Street, which holds 1,500, had every seat occupied with men. They were mostly shopkeepers and mechanics, and a large proportion such as do not regularly attend churches. After the evening meeting, the Christians were invited to remain and pray for the speakers to the anxious; and the inquirers were directed to side apartments, of which several were filled with those whom the Holy Spirit was convincing of sin, and of the need of the seeking Saviour. Thus, while such a glorious work as has been witnessed in Scotland, has not yet taken place in Belfast, a sweet and encouraging commencement has been made.

III.

THE interest in both meetings is deepening and extending. It is certainly a marvelous sight, "filling the mouth with laughter, and the tongue with singing," to see the crowded meetings of women of all ranks and classes, as they listen with rapt attention to the message of mercy; to mark the manifestations of deep feeling and subdued emotion visible everywhere, and the numbers willing to remain in the inquiry-meeting for conversation and prayer.

The interest in the meeting in the evening is increasing equally. It is dangerous and unprofitable to speak of numbers; but one may say with confidence that from fifty to a hundred, at least, remain each evening, under anxiety of soul, desiring to be pointed to the Lamb of God who taketh away the sin of the world. These are found of all classes, and of all shades of moral and religious character—backsliders, notorious sinners, moral young men, whose consciences are yet tender, and skeptics, whose hearts have been blasted as by an east wind. The majority of the inquirers are young men. This is a special, and I may add a most hopeful, feature of the work. Many seem clearly, in the judgment of man, to have embraced the offered gift, and to be rejoicing in God.

On Saturday, the 12th, there was held one meeting—for children—presided over by Mr. Sankey. The meeting was most interesting, and crowded with earnest young faces.

On Sabbath, the 13th inst., Mr. Moody held a meeting for Christian workers at the early hour of eight, and notwithstanding the hour the place was crowded, so much so that the overflow filled an adjoining room. The address was touching entire consecration to God, and more whole-hearted activity in His service. An open-air meeting was advertised for half-past two o'clock. It was held in an open space, in the midst of the mill-workers of our town. Few, if any, of the thousands who attended that meeting will ever forget it. Very many, I believe, will remember it with joy in the Father's home on high. The attendance was exceedingly great, estimated variously at from ten to twenty thousand! The weather was exceedingly favorable. Mr. Moody's address was founded upon Mark xvi. 15, "Go ye into all the world and preach the gospel to every creature." Mr. Sankey sang "Jesus of Nazareth passeth by." While he did so I could observe in the glistening eye, and the deep sighs of many around where I stood, that it was even so.

In the evening Mr. Moody held a meeting exclusively for inquirers; none else were admitted, the attendance far exceeding our hope—upwards of three hundred. All human computation on this subject must be very indefinite, but when we consider the many who were not present, seeing that the evening service was held at the same hour in all the churches, and add also the numbers at inquiry-meetings held in many of the churches, it will be seen that the shaking among the dry bones has been already very great; in Mr. Moody's judgment, fully greater than during the first week in any other place.

The attendance at the meetings on Monday, 14th, seem to be on the increase. At the meeting for women in Fisherwick Place there were present about fifteen hundred, and at the meeting for inquirers a marked increase; more, indeed, than the Christian workers present could overtake.

As time advanced, this gracious work of God seems to extend and deepen rapidly. On Tuesday the experiment was tried of holding a meeting in the evening exclusively for women, in order to reach the case of workers in mills and warehouses. More than an hour before the time of meeting, the streets around were packed with a dense mass of women; and when the gates were opened the place was filled almost in a moment; and after that, with the overflow, three large churches. In all these meetings, the anxious, willing to be spoken to, were more than could be overtaken. We have reached a blessed difficulty—our inability to find Christian workers in sufficient number, who are able and willing to point the seeking sinner to the Lamb of God.

The number of strangers who from long distances visit Belfast to attend the midday meetings is daily increasing. In this way the work is already extending, and, I trust, will cover the whole island. At its present stage of progress, the most marked features are desire to hear the Word of God, willingness to be spoken to upon the state of the soul, frank confession on the part of many that they do not savingly know Jesus; and, most blessed of all, the equally frank confession on the part of many that they have "found Him of whom Moses in the law and the prophets did write, Jesus of Nazareth."

Last night (Thursday, 17th,) the number waiting to be spoken to was so great that an attempt to speak to each individually was scarcely made. Two or three addresses were given in the way of pointing them to the Lamb of God.

To-day (Friday, 18th,) the mid-day meeting is solely for professing Christians —the subject, "Assurance." In the evening the meeting is intended for such only as are seeking Jesus. Mr. Moody has adopted these expedients because of the want of any hall or building sufficient to contain the crowds seeking admission. Let me venture to suggest to any of your readers who live in cities likely to be visited by Messrs. Moody and Sankey, the wisdom of erecting a temporary structure, if there is no suitable place. It would save the strength of these beloved brethren greatly, and help to concentrate at first the work.

It is a very hopeful feature of the work that it has begun to spread to the adjacent towns. Meetings have been held by others for some four nights in Bangor, ten miles from Belfast; and, considering the size of the town, the work there was equally great. Thus we are looking forward that the work shall extend over the whole province, and over the whole island.

IV.

THE work is taking deep root in Belfast. We may confidently expect "greater things" in Ireland. At the inquiry-meetings there are larger numbers than attended similar gatherings at Edinburgh or Glasgow at the same period. Over 300 anxious inquirers remained to be spoken with at the inquiry-meeting on last Sunday evening. At these meetings there are no addresses; they are announced solely for inquirers—none others are invited to attend. It is not an unwonted feature to find men and women, young and old, voluntarily flocking to meetings for the sole purpose of inquiring the way of salvation and stating their perplexities.

The noon prayer-meeting on Monday was devoted to accounts, from ministers and others, of the results of the work so far.

Intense calm and deep earnestness characterized all the meetings. The Holy Spirit was poured out, not with a rushing wind, but in a still, small voice. An unusual proportion of fine young men waited to be conversed with in the inquiry-rooms. All seemed to feel there are but the two classes, the saved and the lost.

Various were the difficulties felt by inquirers, but all such as anxious souls have expressed from time to time. Some could not understand what "coming to Christ' is; others had previously come, but were staggered because they had not the complete mastery of sin; others, again, had not felt a sufficient sense of danger. Warm-hearted and experienced Christians listened to the difficulties of each and all, and were in most cases enabled by the Holy Spirit to speak the suited word and remove the stumbling-blocks.

The open-air meeting in Agnes Street at 2:30 on Sunday was attended by numbers variously estimated at from 10,000 to 20,000. The fundamental truths of the gospel were forcibly put and ably illustrated. Many were bathed in tears. Multitudes of careless men and women have been awakened.

Singing bears a most important part in the work of God. Deeply effective are Mr. Sankey's solos, not only in touching the heart's affections, but in deepening the impressions made by the Word. The solo, "Too Late," following on Mr. Moody's address on the despair of the lost in hell, had the most solemn effect. The wail, "Oh! let us in, oh! let us in," and the awful response, "Too late! too late! you cannot enter now," are enough to wring the inmost soul of every wavering and undecided sinner.

V.

The second week of the labors of these beloved brethren has closed. As we look back with thanksgiving and wonder, we can only exclaim, "What hath God wrought!" The meeting of Christian workers was held on Sabbath morning (September 20), at eight, in May Street Church. Shortly after seven the place was filled with an audience of, say, fifteen hundred, and the overflow filled also an adjoining building. It is surely a significant fact, and a very blessed sight, to see hundreds of men and women at such an early hour crowding out to call upon God, and to be addressed upon the subject of working for Jesus.

An open-air meeting, similar to the one held on the previous Sabbath, had been arranged for two o'clock. The weather in the morning was very unpropitious; but by noon the day brightened up, and by the hour of meeting it was all that could be desired.

I shall not attempt to estimate the number who heard the gospel on this occasion. I observe a friend from England, who was present upon the previous Sabbath, estimates the number at 30,000. Whatever the number may have been, it was certainly exceeded on last Sabbath. Mr. Moody was greatly helped. The fruit of the seed can only be fully seen when gathered at that day!

A meeting for inquirers only was arranged to be held in the evening of the same day, in the Ulster Hall, the largest public building we have. Admission to this meeting was strictly limited to those professing anxiety to find Jesus. Christian workers were admitted by ticket, a method adopted to avoid the mistake too often made at such meetings of allowing incompetent or improper persons to engage in such work. There is not one Christian in a hundred fitted for this most delicate and difficult service, requiring, as it does, close communion with Jesus, much knowledge of the human heart, and very clear views of gospel truth, and not less a desire to know nothing, and to speak of nothing, save Jesus Christ and Him crucified.

While all this is true, it is equally true that none are oftener blessed in this work than young converts, while their virgin love is yet fresh, and their faith clear and simple. There was at the meeting for inquirers an attendance of some 500—this in addition, it may be noticed, to many meetings of like kind held in various churches at the close of the evening service. It was very touching and stimulating, when an opportunity was given by Mr. Moody at the close of the meeting, to hear many young men read out, in trembling tones, and yet with beaming countenances, some previous promise of the Word of God. It seemed like throwing out a life-buoy to the struggling ones around, who were swimming for life in the waters of death—like the letting down of a cord to the prisoners in the pit in which there is no water. Subsequent information in the young men's meeting proved that these truths were laid hold of savingly by not a few that Sabbath night in the Ulster Hall.

On Sabbath night we had our first meeting for young men, from nine till ten o'clock. To the surprise of all of us, there were about 1,500 present. The beginning is a special work, which, I trust, will spread as in Glasgow. There had been during the Spring a very marked work among the young men in Belfast, in connection with the Young Men's Christian Association. Now it has been deepened and greatly extended.

During the week, the tide of spiritual life seemed to increase each day. The Bible-readings at two o'clock have been full of interest, specially stimulating to many whose spiritual life had hitherto been very dormant. The reading on "Grace" yesterday was, I feel persuaded, made a blessing to many of the children of God, while helping over the threshold many almost saved.

The more special evangelistic services in the evenings have been held this week in St. Enoch's Church. The crowds desiring to hear the gospel have proved inconveniently great, filling the church an hour before the time; while the after-meetings have been so filled, that the work of speaking to the anxious has been very imperfectly done.

Now as to the point reached in the progress of this great work of God, I would not like even unintentionally to mislead any one. The work is great and extensive beyond what I have written, and yet there remains so very much land to be possessed. There are tens of thousands of our population still untouched. Many of the higher classes, church-goers, seem still unmoved. We need continued prayer. We need to realize the "now" of our opportunity—to understand that Jesus is saying, "This thy day."

VI.

THE progress of the work of God in Belfast is still very marked. God continues to own the labors of these dear brethren. The manifestations of the Spirit's presence and power were very marked. In the earlier days of the movement, of the many who were deeply convinced of sin, comparatively few seemed to come to rest and peace and faith in Jesus. It seemed as if a higher tide of the Spirit's power were needed to guide them through the quicksands of difficulty, and over the bar of doubt and distrust, into the haven of rest.

This week, we thank God, it is otherwise. We can say with thanksgiving concerning many, "They which have believed do enter into rest."

The meeting for the young on Saturday (26th) was very striking. Mr. Moody presided. The truth seemed to reach, in the Spirit's power, many young hearts. A meeting for boys under fifteen has been organized. Some of the cases in it are exceedingly touching, affording, I conceive, illustrations of the work of God upon the human heart in its simplest and deepest form. This meeting for boys assembles every evening now at half-past seven.

The open-air meeting on Sabbath (27th) was held in a different part of the town; the multitudes assembling equally great, according to some greater, than on the past Sabbaths. Mr. Moody's address was well calculated to awaken from security and draw sinners to the one Refuge. We have been reaping the fruits in our inquiry-meetings during the week.

On Monday we had no meeting—rather, one of the most remarkable meetings, I shall venture to say, ever held in Belfast. Fisherwick Place Church was open for inquirers from two till ten o'clock. Mr. Moody and other Christian workers were occupied all that time in pointing sinners to the Lamb of God. It is impossible to say how many wounded spirits were conversed with during the day. Many very experienced Christians, who have seen much of the Lord's work in other years, declare they never saw a meeting like it.

Though more privately held, I must not omit to notice a meeting which Mr. Moody had on Sabbath night (27th), with men who profess to have been led to Jesus since these special meetings began, and with others anxiously seeking. I saw many wonderful meetings during the year of grace (1859), but I do not think I was ever so impressed with the glory and beauty of the work of God as when I entered this meeting toward the close of it. It was a sight which would, I think, have drawn tears of joy from any heart to see upwards of 200 young men, the very flower of our youth, one after another acknowledging the yoke of Jesus. Passing just across the street, I entered May Street Church, where more than 1,000 men were assembled to hear the glad tidings of great joy.

In order that as many as possible might have an opportunity of hearing the gospel at these special services, admission on Tuesday night was by tickets, given only to such as had not hitherto heard Mr. Moody. About 3,000 tickets were given on personal application. It was a season to be remembered. The soil, so to speak, was virgin; the attention so marked as to be almost painful in its silence; the presence of God very powerful in the consciousness of every spiritual mind. The inquirers at the close of the meeting were spoken to, as far as they could be overtaken, in adjacent churches, to which the men and women were sent respectively.

The mid-day meeting in Fisherwick Place Church still continues. On Wednesday, Mr. Moody addressed it from Luke ii. 7, " There was no room for them in the inn." Many who are not reached at other places hear the gospel at this hour. It has already proved the meeting-place between Jesus and many a sinner.

A great meeting is to be held in the Botanic Gardens, October 8th. The various railway companies have agreed to run special trains in connection with it. Our desire is that God, who has already done such great things for us, would be pleased to use it to shake mightily Ulster and Ireland. We would wish to assemble 100,000 people to hear the word of God. *We want Ireland for Christ!*

As the week advances, the work of God deepens and extends. The meetings on Wednesday and Thursday, in my judgment, have exceeded anything we have yet experienced. Men and women in great numbers are found crowding the "second meetings," seeking Jesus.

It is, at all events, worthy of remark: the great contrast in outward manifestation between the present work and that of 1859. I have not heard of or noticed any physical excitement—not even an outcry, much less what were then known as "prostrations."

Another asked prayer for a son who has joined the Church of Rome, that he may be led back again to the truth as it is in Jesus.

One of the subjects spoken of was the willingness of our unseen Head in the skies to guide and uphold His members. Do we not take care, going along a difficult road, to keep our foot from slipping? Will He, the Head, take less care to keep His member from falling or going a step wrong, if that member confidingly trusts His guidance? Let us not forget that the guidance He may give may be to go and consult with those of riper judgment and longer experience in His work.

VII.

SABBATH morning (4th) dawned upon us very wet and windy. We had fears that it would be impossible for the masses of the people to meet in the open air; but a little while before the hour of meeting the rain ceased, the sun shone out, and the weather became most auspicious. Here let me say, it has been most noteworthy that, during the last weeks, while we have had most inclement weather, every Sabbath-day, and at the hour of our great gatherings, it has been all that could be desired.

The number in attendance was fully equal to any preceding Sabbath. It may give you some idea of the multitude if I state that the field on which the meeting was held contains about six acres, and that the people stood densely packed from one end to the other. There was profound solemnity. The impression upon the hearts of the people by the truth in the power of the Spirit was very deep, as the sequel will show.

Mr. Moody held his usual meeting on Sabbath evening for those in deep distress about salvation, and for those who had found eternal life during the past weeks through faith in Jesus. The meeting was exclusively for men, and admission solely by ticket. The hall in which it was held was completely filled. Mr. Moody stated in the noon-day prayer-meeting on Monday that, in his judgment, it was *the most remarkable meeting he has had yet in Europe.* To God be all the praise! One after another of these young men—and they comprise the very flower of our youth—rose, and, with clearness and wonderful felicity of expression, in burning words, declared what God had done for his soul. At length, at nine o'clock, the meeting was closed.

Meanwhile another meeting of men was assembling in a church. It was already very nearly filled when we heard the tread of a large company approaching. It was a phalanx of these redeemed youths. They sang the new song. In a spontaneous burst of praise they were telling forth the wonders of redeeming love. No language can describe the scene. The heavenly echoes of that burst of praise, I think, will never be forgotten by any who heard it. The meeting that followed, consisting of some two thousand men, I need not say, was one of profound interest—

Jesus in the midst and the marching glorious.

During each day of this week and at every gathering, more and more of the presence of the God of salvation has been manifested. Let me in a sentence or two describe one, which, in sober language, was most wonderful. Mr. Moody addressed on Monday evening in Fisherwick Place Church, a meeting of men. At the close of his address, all who had recently been found by the Good Shepherd, and also all who were seeking Him, were requested to retire to the adjoining lecture-room. Some six hundred men did so. Mr. Moody again sifted them, by requesting that those only who were deeply anxious to be saved should adjourn to another room. Probably nearly three hundred did so. In breathless stillness Mr. Moody addressed them, very briefly stating that he could do no more for them—that they had heard the gospel, and that it was for themselves to decide. He called upon them to kneel and pray for themselves. They bowed as one man, and now here and now there might be heard the short cry for mercy—a few earnest words of supplication; probably about thirty or forty so cried to God one after the other. Surely the Lord is in this place! was the thought which rose in holy fear in the hearts of all.

After a short prayer by Mr. Moody, he addressed them very faithfully. He again held forth Christ, and invited all to rise who felt that they could there and then accept Jesus. All of that large company, save twenty or thirty, stood up, and solemnly avouched the Lord to be their God. This wonderful sight cannot be described. The glory of it cannot be realized even by those best acquainted with divine things. If there is joy in heaven over one sinner that repenteth, what shall we say of the gladness in the Father's house when the prodigals in companies of some two hundred enter, as it were, at once?

Thursday, October 8, we had fixed for a gathering of the masses in the open-air. Many had fears for the weather, but much prayer in many places over the three kingdoms were offered to God for the success of the meeting. God did for us above what we asked. The weather was splendid; everything as regards order and decorum all that any of us could wish. It was the largest open-air meeting I ever attended. I cannot pretend to fix a limit to the numbers. He who counts the stars knew the history of each present, and what were the dealings of his heart with Christ and the free offer of His salvation. The only regret that seems to be expressed by any was, that the services were so short.

Mr. Moody addressed the vast multitude from the words, "I pray thee, have me excused." With graphic felicity, great clearness, and soul-piercing power, he exposed the miserable pretences by which sinners impose upon themselves in refusing a present offer of present blessedness. The address seemed to strike with convicting power many consciences, and, from many instances coming under my own observation, at the inquiry-meeting in Fisherwick Place Church, I have reason to believe in salvation power.

VIII.

THE great gathering in the Botanic Gardens on October 8th has been our crowning mercy in this season of blessing. We feel as if every prayer had been heard and every heart gratified by our gracious God. As the days pass, and as tidings reach us from the country districts all around, we continually hear of rich blessings bestowed and of precious fruits following. Many carried with them to their homes the spark of renewed life. That spark has, in some cases, already burnt into a blaze. We receive the good news from many places of great readiness to hear the Word of God, and the cry, "Come over and help us," reaches us from many quarters.

Our dear American brethren left us on Saturday for Derry. Tidings have reached us that a great and effectual door was opened unto them in that city. I trust some eye-witness will lay before those interested an account of the doings of the Lord there. Meanwhile, in Belfast, our meetings have gone on as usual. The interest in divine things continues still unabated, many anxiously seeking Jesus, and many finding Him as their Refuge and Portion. In the Young Men's Meeting, held every evening at nine, in Fisherwick Place Church, the work of God makes great progress. Monday night was especially a night of great power.

Messrs. Moody and Sankey returned from Derry this morning (Oct. 15) to hold their final meeting, ere passing on to Dublin. Mr. Moody presided at the noon-day prayer-meeting. The subject was, "Lessons from the life of Jacob." The meeting was one of great interest. The meeting in the evening was held in St. Enoch's Church. It was exclusively for sinners under anxiety of soul, who professed to be earnestly seeking Jesus. Admission was by tickets, and that, moreover, on personal application.

Readers may judge of the depth of the movement and the measure of awakening power upon the souls of men by the Spirit of God, when I state that upwards of 2,400 persons were so admitted! It was Mr. Moody's last appeal in Belfast to the Christless. I may not attempt to describe the scene! He set before the anxious, sin-stricken multitude, Jesus in all the glory of His sufficiency—in all the attractions of His dying love. He showed Him, as with one foot upon the threshold of the heart He sought admission. Now in faithful and firm words he warned them of the dangers of delay; and now he gently moved them, in tenderness, as one whom his mother comforteth. At length he ceased speaking, that each might hear, in the silence, the voice of Jesus pleading directly. And in the awful stillness of that moment many of that great company of seeking sinners, I trust, were able to say in words expressive of soul-submission, "Speak, Lord, for thy servant heareth."

I think it must have been the most notable meeting in the experience of Mr. Moody. I do not at present remember to have read of any such meeting, as regards the number of the awakened, in modern times. Does it not seem like a return of Pentecostal power, when 3,000 were similarly smitten with soul-concern?

Oct. 17.—Yesterday was the concluding day of the labors of our beloved brethren. The noon-day prayer-meeting was crowded. The great multitude was moved deeply with contending feelings—of joy in God and gratitude for all He had done for so many of them through the labors of His servants; of sorrow because they should see their faces no more till the resurrection morn. The meeting commended them to the grace of God, beseeching the God of all grace to bestow a fresh baptism upon them of power for their work in Dublin.

At the two o'clock meeting in Fisherwick Place Church, Mr. Moody chose as his subject—The gift of the Holy Ghost as a Baptism of power for witnessing and work. As he spoke of power, the Spirit of might seemed to descend upon him.

The meeting in the evening was for the young converts—for all who have reason to believe that they had found Jesus since Messrs. Moody and Sankey came to Belfast. Admission was strictly by ticket. These tickets were only given on personal application. About *two thousand one hundred and fifty tickets were given!* What a rich harvest! How soon gathered! The result of some five weeks' work! I have good reason to believe that even this number fell very far short of the whole number who profess to have received Jesus as the gift of God.

It was a soul-stirring sight to see that vast multitude, including the Christian workers and ministers, numbering more than 3,000. It was like the sound of many waters to hear this multitude sing the new song. As all stood and sung in one burst of praise—

"O happy day that fixed my choice
On Thee, my Saviour and my God,"

the effect was overpowering, filling the soul with a sweet foretaste of the praises of heaven.

Mr. Moody's last word of comfort and encouragement was founded on Rom. xiv. 4, "God is able to make him stand." He closed his address by commending all the new-born souls "to Him who is able to keep you from falling." Hundreds of men not used to a melting mood, with weeping eyes and heaving bosoms, heard him say, as he concluded, "Good-night; we shall meet in the morning when the shadows flee away."

A very touching incident in the service was the singing, by Mr. Sankey, of a hymn composed by a dying youth in Belfast, "Is there room? they say there is room!"

The work of God has begun in a striking manner in many places around Belfast. We are confidently expecting that it will quickly and widely spread. The high mountain, before prayer, will become a plain.

THE WORK IN LONDONDERRY.

One of the most satisfactory features of the visit has been the unanimity and cordiality with which the ministers of all denominations not only joined in the original invitation to Messrs. Moody and Sankey, but also assisted in the furtherance of the work during the present week. Presbyterian, Wesleyan, and Independent seemed to have but one object and one desire — to make the work of revival among the people as general and widespread as possible. The original request to Messrs. Moody and Sankey emanated some months ago from the committee of the weekly mid-day prayer-meeting, in conjunction with the Young Men's Christian Association. All the ministers who were asked to put their churches at the disposal of the committee intimated their willingness to do so. The First Presbyterian Church, however, was selected for holding the meetings, for no other reason than that it was the largest, and therefore best able to accommodate the large numbers likely to be present.

Messrs. Moody and Sankey arrived in Derry from Belfast on Saturday evening, and commenced their labors on Sunday, the 11th, with the same spirit of energy and enthusiasm which carried them through so much in Belfast. Mr. Moody exhibited little trace of hard work or fatigue, though for some weeks past he has gone through an amount of mental and physical toil under which many men would have completely broken down. There were the same freshness and vigor, the same fertility of illustration and pointed application, the same earnestness and simplicity, the same zeal and enthusiasm, and the same intense desire to win souls for his Master. Three services on the Sabbath, and the same number on each of the following three days of the week, with inquiry-meetings each evening, has been his programme here, and he never seemed to fail either in body or mind. He appeared conscious of the shortness of his visit, and seemed to grow more earnest in consequence.

While Mr. Moody faithfully presented the gospel, Mr. Sankey was no less faithful in his lessons in song. He was so admirably assisted by a local choir as to draw a special eulogium from Mr. Moody at one of the noon meetings. He said he had heard a great many choirs assist at these meetings, but he had never yet heard one which sang so sweetly and so well as the one which had been organized to assist in singing the praises of God in Londonderry. On the same occasion he referred to the importance of the Church paying greater attention to the subject of praise. Some were only for singing the psalms, but he thought they should also sing "new songs." A new hymn was just as good as a sermon. They could sing the gospel into many a man's heart. He hoped the Church would feel alive to its duty in this matter of praise, and not be hindered by prejudice, which is the twin sister of unbelief.

The opening meeting was intended for Christian workers, and Mr. Moody dwelt especially on the subject of Christian work, and gave some earnest and practical counsel. On the same day two meetings were held in the First Presbyterian Church; one at four and the other at eight o'clock. The ordinary congregational services were conducted in the church at twelve o'clock, without, of course, any instrumental accompaniment in the praise. At both special services the church was crowded to overflowing, and the gates had to be closed half an hour before the commencement of the service. Indeed, at the evening meeting, the church was filled at seven o'clock, the people crowding in such numbers to the service. Overflowing meetings were held in the Wesleyan Chapel, and were pretty well attended, though better in the evening than in the afternoon.

On Monday, Tuesday, and Wednesday, three services were held each day, including one children's service. Owing to the heavy downpour of rain on Monday, the church was not so well filled as on the other days, when the congregations were very large; but on each evening fully 2,000 found accommodation in the church, filling it from floor to ceiling, while the hundreds unable to gain admittance went to the Wesleyan Chapel, where they were suitably addressed. The concluding meeting on Wednesday evening was especially large, and the services particularly solemn.

With regard to the audiences, they were thoroughly representative. Young and old of all classes, not only of the inhabitants of Derry, but of the surrounding districts, for miles around, attended. Excursion trains on the Irish North-Western Railway and Northern Counties Railway brought many into the town, while hundreds walked and drove many miles, in order to be present at the meetings. The attendances steadily increased to the close, and as the last of the services approached, there seemed to be a general expression of regret on the part of all interested. A noticeable incident in connection with the meetings was the large number of clergymen who were present at them.

The prevailing characteristic of all the meetings was intense earnestness and solemnity, but without any undue excitement. The services seemed to awaken the liveliest interest in the public mind, and to produce a marked impression. The inquiry-meetings after the first night were well attended, large numbers of both sexes remaining for conversation and prayer with Mr. Moody and the Christian workers who were admitted (by ticket) to converse with the anxious. In this respect every precaution was taken that none but duly qualified persons should be admitted. The time occupied at these meetings was brief, but the addresses and conversations earnest and impressive. The upper room was set apart for female inquirers, and the lower schoolroom for males. These meetings are described by those who were present as having been of a most interesting character.

Arrangements have been made for continuing a twelve o'clock prayer-meeting, and a meeting at eight o'clock P.M. daily.

THE WORK IN DUBLIN.

I.

Dublin, Oct. 24th.—At last the prayers of many of God's dear children are being answered by the coming amongst us of these honored servants of our Lord. And to-day we take a slight retrospect of our week's prayer and praise, and work for the gracious Lord.

A general prayer-meeting, preparatory to commencing these special evangelistic services, was held in the Metropolitan Hall, on Saturday, the 17th. It was quite full, though capable of accommodating more that 2,000 people. As one looked over that large assembly, composed of members of all the evangelical denominations of the city, and observed the spirit of unity as well as of earnestness and devotion that prevailed, one could not but feel that we had entered upon a new phase of religious life, and that brighter days are dawning upon Ireland.

For the first time and in connection with this movement have we seen the clergy of all the evangelical churches working cordially together, without the least shade of envy or party spirit, all feeling that they are workers in the same holy cause, children of the same Father, servants of the same gracious Master.

It is right to remark that for many months past we have had in the Metropolitan Hall a weekly union prayer-meeting, in which ministers of the various churches have taken part. Indeed, I believe Mr. Moody requires this, and that he declines to visit any place unless there has been previously much intercessory prayer. Thank God, it was so here, and we are already receiving answers to the prayer, the first droppings of those rich showers of blessings for which we look.

The committee of management procured the Exhibition Palace for holding these services, the largest and most commodious building which has yet, in Europe, been placed at Mr. Moody's disposal. On Sunday last the Christians of Dublin witnessed a sight to gladden their hearts. It has been estimated that, at the first service at four o'clock, from *twelve to fifteen thousand* persons were gathered there. Never before was it put to so blessed a use.

Mr. Moody addressed the vast concourse from the text, "Go ye into all the world and preach the gospel to every creature" (Mark xvi. 16). The audience was greatly impressed by the discourse, as also by Mr. Sankey singing that affecting hymn, "Jesus of Nazareth passeth by."

Every day during the week there was a prayer-meeting at noon in the Metropolitan Hall, which on each occasion was filled to overflowing.

On Monday Mr. Moody presided, and spoke very forcibly on the importance and power of prayer. It was my happy privilege, as a minister of the Irish Church, to preside on Tuesday, and in the name of very many of the Christian people of Dublin, to offer our dear fellow-laborers a hearty welcome. On Wednesday the Rev. F. Dowling (Bethesda Chapel) took the chair, and spoke with much power on "Christian Hope." The Rev. J. Fleming Stevenson (Presbyterian) acted as chairman on Thursday, and brought out with much effect some incidents in Joshua's life. On Friday Mr. Moody again presided, and was listened to with deep attention. To-day Rev. Edward Best (Wesleyan) spoke with much unction and feeling upon "Prayer for Children." On these occasions Mr. Moody and several clergymen and lay brethren led the congregation in prayer. Many went away deeply solemnized, and thanking God for that "sweet hour of prayer."

At the services in the Exhibition building, the weather being beautifully fine, the attendance increased each succeeding evening. On Monday evening, and again on Tuesday evening, Mr. Moody spoke of Jesus coming "to seek and to save that which was lost," interspersing his discourse with many forcible illustrations. The following evening his subject was the powerlessness of the Law to save, and then he set forth Jesus as the only and all-sufficient Saviour. Thursday and Friday evenings were devoted to showing the necessity of Spiritual Regeneration.

II.

WE have never before seen such sights in Dublin as we have seen this last week, night after night, at the Exhibition Palace. It is estimated to hold 10,000 persons. Every night it is filled, and the attention and silence is wonderful. One feels that the Spirit of God is present, and that "a wave of prayer" is continually going up to the throne from the Lord's people.

The second week of this visit has now nearly come to a close; and when the visits of kings and princes have been forgotten, this will be remembered by many, even through all eternity, for the gospel so faithfully preached by Mr. Moody, and so sweetly sung by Mr. Sankey.

Thank God, every day reveals a growing interest on the part of the public at large in their evangelistic labors. Every day their work is extending, widening, and deepening. The inhabitants of Dublin are becoming alive to the fact that we are now in the enjoyment of a great "time of refreshing," and that our gracious God is working powerfully amongst us by the instrumentality of these His honored servants.

Such a sight has never before been witnessed here as may now be seen every day—thousands flocking to the prayer-meeting and the Bible-meeting, and most of all to the evening services in the great Exhibition Palace. It is a sight to fill the heart of the child of God with deepest emotion to stand upon the platform erected in that building, from which Mr. Moody preaches, and to cast one's eye over the vast concourse of people, hanging on the speaker's lips, as in burning words he discourses of life and death, heaven and hell, "Jesus and His love;" and one cannot but ask the question, "What is the magic power which draws together those mighty multitudes, and holds them spell-bound?" Is it the worldly rank, or wealth, or learning, or oratory of the preacher? No; for he is possessed of but little of these (spiritually, indeed, he is richly endowed with them all.) It is the simple lifting up of the cross of Christ—the holding forth the Lord Jesus before the eyes of the people in all the glory of His Godhead, in all the simplicity of His manhood, in all the perfection of His nature, for their admiration, for their adoration, and for their acceptance.

As an Episcopalian minister, I am most thankful to see so many of the dear brethren of my own church, as well as of the other evangelical churches, attending and taking part in these happy services.

One dear brother, an able and godly minister, stated a day or two ago that, by attendance at these services, he seemed to have "returned to the freshness of his spiritual youth."

This is the outline of the past week's work. On Saturday evening, at the service in the Exhibition building, Mr.

Moody entered into a defense of his custom of holding meetings for inquirers after each service, and proved, in answer to objectors, that he had abundant Scriptural warrant for so doing.

On Sunday morning, at eight o'clock, there was a meeting for prayer of the workers connected with this good cause. Mr. Moody addressed them, and spoke many kind words of encouragement. The afternoon service in the Exhibition Palace on the same day was densely crowded, from 8,000 to 10,000 persons being present. Mr. Moody preached a stirring sermon from St. Luke, 4th chapter, 18th verse. The whole audience seemed deeply affected by the sermon, as also by that touching hymn, "The Ninety and Nine."

At the noon prayer-meeting on Monday at the Metropolitan Hall, the subject was, "Work for Christ." Mr. Moody gave some striking instances of awakening which came under his own observation in the inquiry-room during the past week. The Rev. Dr. Marrable (Episcopalian) also supplied some interesting facts, as did also the Rev. Mr. Wilson, president of the Primitive Wesleyan Conference. The Rev. Dr. Craig told of an officer of rank and position, who was one of the first to scale the walls of Delhi, and who, though he was in Scotland during the sojourn there of the American evangelists, yet never attended their services, but who was induced to do so here last week, and the result was that he found a joy and peace in his soul which he had never before known. These are only a few out of very many instances that might be mentioned. To God be all the praise!

All the services have been largely attended; indeed, the numbers seemed to increase from day to day.

I would specially call the attention of your Dublin readers to the Bible-reading at two o'clock each day in the Metropolitan Hall. There indeed is a feast of fat things prepared for them. It is deeply instructive to see the "things new and old" which he draws in rich profusion from the treasury of God's Word. May God bless him, and make him a blessing to thousands!

III.

"Thank God for sending Messrs. Moody and Sankey to Dublin." Such, I am persuaded, is the expression of the feeling in the hearts of thousands in our city to-day; and why should it not be so when many are pressing into the kingdom of God? How many families are rejoicing in prodigals returned, drunkards reclaimed, blasphemers silenced, the careless aroused, the lost restored, sinners converted, and, in a word, Jesus received into the hearts and homes of many! But some may be ready to ask, "Why should this be so now more than ever? Was not the Lord as ready to do this heretofore as now? Is His arm ever shortened that He cannot save?" We answer, "True; but though God is Sovereign, and does all things according to the good pleasure of His will, we can ourselves see reasons for the present success."

In the first place, it is in answer to prayer. God's dear children in England, Scotland, America, and elsewhere, are praying for a blessing on Dublin and Ireland. Now we know that our God delights to hear and to answer prayer. Blessed be His name! that answer is even now being given.

In the second place, God has wonderfully fitted these His servants for the work He has given them to do. It is to be feared that there is a want of directness, if not of earnestness, in our preaching. Now the great characteristics of Mr. Moody's preaching, as Professor Blaikie, of Edinburgh, has well expressed it, "are directness, earnestness, and naturalness," or, as a beloved brother in Dublin described it, "He does not wait for the end of his sermon to make the application, but the Bible in his hands is a quiver, and every passage to which he refers is an arrow, which the Holy Ghost accompanying, he shoots home straight to the hearts of his hearers."

The use of the inquiry-room. I am aware that here I am treading on delicate ground. I know that many ministers and others, either object entirely to the inquiry-room, or are uncertain about it. I had, at first, the same difficulties myself; but from what I have lately seen, I am satisfied it is of great importance to speak, if possible, with each anxious inquirer, while the gospel is still ringing in his ears—while his heart is softened, and his conscience tender. In this way we come to know what are those doubts, and fears, and difficulties which are keeping the poor sinner from Christ.

I regret that I cannot now (Nov. 7)

enter into many details of the work for the next week. I must content myself with saying, in general, that all the services have been carried on as usual, the attendance well kept up. Indeed, the number seems to increase from day to day, and visitors are coming to Dublin from all parts of Ireland to attend these services.

The Bible-reading in the Metropolitan Hall, on Friday, the 30th ult., was conducted by Mr. Moody. It was, as usual, densely crowded. The subject was "Assurance," and he showed from many Scriptures that it is the privilege of the child of God to know that he *is passed* from death unto life, that he hath eternal life abiding in him.

At the Exhibition service on Sunday, Nov. 1st, the crowds were enormous—a most solemn stillness pervaded that vast audience of some 10,000 souls as the preacher gave a connected and most graphic account of the history of our blessed Lord from the hour of His betrayal by Judas to His resurrection and ascension to the Father's right hand. And he dwelt forcibly on the fact that Christians do not worship a dead, but a *living*, Christ, One who ever liveth to make intercession for sinners.

On Thursday and Friday, at the Bible-reading, Mr. Moody spoke upon the person, work, and offices of the Holy Ghost. He strongly urged the necessity of the Spirit's anointing for *service* for Christ as well as for conversion, which should be sought for by continual prayer.

But we are sadly reminded that Messrs. Moody and Sankey cannot stay always with us; like the Master whom they serve, they must visit other cities also—Liverpool, Manchester, etc., and London. If my words could reach the ministers—especially those of the sister Church of England—laboring in those great cities, I would earnestly bespeak for our American brethren a kindly reception at their hands. I would say, Lay aside all prejudice as unjust and unwarranted. Receive them cordially. Trust them. Help them with your prayers and hearty co-operation. They are men of God. The Spirit of God rests upon them. The love of God animates them. They go to help you and not to hinder you in your work; not to make proselytes to any sect or denomination, but to gather in souls to Christ. Their motto with regard to this is, "Let every man abide in the same calling wherein he was called." You will find that you will be greatly refreshed in your own souls. In the effect both upon yourselves and your flocks, you will have abundant reason to bless God for sending them to you.

IV.

BEFORE the children's meeting to-day in the Exhibition building (Mr. Moody was taking some needed rest), I asked Mr. Sankey how he thought the work was getting on in Dublin. "Oh!" he said, "it is getting just like Edinburgh. The blessing is becoming like a great wave. It's easy working now."

For some time, notwithstanding the huge crowds, our brethren felt that they were not reaping heavy sheaves as they had done elsewhere. But the conviction grows upon us that the "set time" to favor us has come. The work is deepening and widening every day. In many families with which I am intimately acquainted, one or more of the members have hopefully turned unto the Lord. I know cases in which I may say the whole family has been brought to seek salvation as the one thing needful. It is very observable, too, how previously existing prejudice has abated, or entirely disappeared, at least in the case of those who manifest any respect for religion. There are, of course, scoffers not a few. But it is truly a matter of astonishment in a city like this, that there is so little of open resistance or even of ridicule.

Our Roman Catholic brethren, as a rule, have acted a noble part. They have been respectful; and, to a certain extent, sympathizing. In this week's number of the *Nation*—an organ at once of National (as it is called) and Ultramontane principles—an article has appeared, entitled, "Fair Play!" which is exceedingly creditable, and which indicates the advent of a new day in Ireland. The editor informs his constituents that "the deadly danger of the age comes upon us from the direction of Huxley and Darwin and Tyndall, rather than from Moody and Sankey. Irish Catholics desire to see Protestants deeply imbued with religious feeling, rather than tinged with rationalism and infidelity; and as long as the religious services of our Protestant neighbors are honestly directed to quickening religious thought in their own body, without offering aggressive or intentional insult to us, it is our duty to pay the homage of our respect to their conscientious convictions; in a word, to *do as we would be done by.*" (The italics are the *Nation's.*) It would surely be a bright and blessed day for our country, if this spirit of mutual respect and toleration were everywhere honestly acted out amongst us. Mr. Moody never makes controversial reference to others. His success in attracting the favorable attention of our brethren of a different faith, has been unexampled in the history of our city.

One very marked feature in the movement is the number of men that are influenced. Many people have remarked the large proportion of them that are inquiring.

A few nights ago an old gentleman, more than seventy years of age, threw himself down on his knees and sobbed like a child. He said, " I was utterly careless about my soul till last night, but I have been so unhappy since, I could not sleep. I seemed to hear ringing in my ears, ' Jesus of Nazareth is passing by,' and if I don't get saved now I never shall be."

Already the influence of this work has begun to tell upon the most remote districts of the country. Parties of thirty, fifty, sixty, etc., are being organized from the most distant parts to Dublin. Many of these carry back with them much blessing. We hear of the young converts witnessing for Christ fearlessly in the trains on their way home from their meetings. " The Lord hath done great things for us, whereof we are glad." But we expect greater things still. I am fully confident, from all the indications I see, that next week will be likely greatly to surpass the previous delightful weeks we have had. The memory of these blessed meetings in the Metropolitan Hall and the Exhibition building, will long, yea, will ever be fragrant in our hearts. I do not think we had ever such an antepast of heaven.

The Public Breakfast given to Messrs. Moody and Sankey yesterday morning, was, in every way, a wonderful meeting. I heard nearly all to whom I spoke on the subject, say that it was the happiest reunion they ever attended. It was a truly catholic gathering. Eminent men among us, under the influence of deep emotion, bore testimony to the spiritual good they had received at the meetings. Ministers testified of the instruction and quickening that had come to them.

No men—ministers, evangelists or others —ever before brought a more interested assembly around them in Dublin than these honored servants of the Lord did yesterday morning in the Shelbourne Hotel. And yet it is not Messrs. Moody and Sankey, but the Christ they preach and sing. It is Christ lifted up that draws all men unto Him. Oh, that we might all learn that we have here the true and only uniting power for Ireland.

V.

THE BREAKFAST AT THE SHELBOURNE HOTEL.

MESSRS. MOODY and Sankey were entertained at a public breakfast on Friday morning by a large number of clergymen, and professional and mercantile gentlemen of all religious denominations, who embraced that opportunity of expressing their confidence in them, and their sympathies with the evangelistic services conducted by them in Dublin. Two of the largest rooms in the Shelbourne Hotel were completely filled by the company, which numbered about two hundred. The object the gathering evidently had in view was the encouragement of Christian unity, which every speaker in the course of the proceedings warmly advocated, in the belief that it is especially needful at the present time, and essential to the further spread of the gospel in this country. Messrs. Moody and Sankey were warmly welcomed by those present, who unmistakably felt that the opportunity which presented itself on the occasion for initiating so great an evidence of superiority to sectarianism was one to be gladly appropriated. The company was thoroughly representative in its character, both clerical and lay.

Sir Edw. Synge Hutchinson, chairman, called on Rev. Edw. Nangle, of Achill, an aged clergyman of the Church of Ireland, who offered prayer, in which he thanked God that we are being drawn out of our sectarianism, which is the great error and vice of our fallen nature; and Rev. Hamilton Magee, who followed, said that many prophets and righteous men have desired to see the things we see, and to hear what we hear, and saw and heard them not.

The Chairman then introduced the speakers in a few very appropriate remarks, in which he said that the efforts of Messrs. Moody and Sankey had his entire confidence. He thanked them for coming over from America, and believed they had done a great work in Dublin. He hoped the effects would be visible after their visit came to a close, in a greater evidence of Christian unity among all denominations of Christians in Ireland. The spirit of unity and concord which had been brought about was delightful, and he thanked God every day for it, adding, " Let us be all out-and-out for Christ."

Rev. W. Fleming Stevenson: If we are here to welcome the two men of God who have visited our city, our welcome is but a shallow thing after all, and it comes rather late, for already their time of departure draws nigh. We are in the midst of a great spiritual movement, which has shown itself not only in Dublin, but throughout the land, in mission weeks, and many other ways; but has gathered most of its force around our two brethren who are with us to-day. There are signs of such a spiritual movement as we have never seen before. Everywhere, in tramcars and omnibuses, on the street, and in the social circle, one inquiry is uppermost, " Have you been hearing of the Lord Jesus Christ?" In one small town, a long way from Dublin, a party of thirty was formed to come up to these meetings, and their report brought a second party of sixty. From Cork, and Limerick, and other places also, visitors come. Never was there a meeting in Dublin so representative as this. Now what are we to do? When these men leave, what is to become of the work to which they have given such an impulse? Having been lifted up by this advancing wave, are we to let another wave put us back into our respective niches? or are we to seek such a baptism of the Holy Spirit as shall knit us together in the work of God? Surely it does not seem impossible that there should be such a unity as shall end in the entire bringing of our people to the Lord. Has not our oneness with Christ another side? Does He not say to us, Have you no sympathy with Me? Will you not gather together with one another round Me, in sympathy with Me? Long after Mr. Moody and Mr. Sankey have returned to America, may they have the joy of hearing that Christians in Ireland are still united in love and work. Prayer is to us now a reality. We pray, and get the answer while we pray. People coming to these meetings unconverted have been saved while prayer was being made for them. One thing more. This work is carried on in entire sympathy with the ministers of the Church of Christ. Can we not continue banded together in one, so that Jesus Christ shall be glorified in our oneness in work and spirit?

Mr. Moody said that was the first meeting of the kind he had ever attended. In a number of places it had been suggested to hold meetings for the promotion of unity, and quite a number had pressed him and Mr. Sankey to have conferences to talk about Christian unity; but the one principle upon which they started was, that they would preach Jesus Christ, believing that He would draw His people together. People had asked him how they had got so many ministers of different denominations into the movement; his answer was, that they had done nothing about it. They had just tried to hold up Christ, and to talk of Him only, knowing that if that did not make friends rally round them, nothing else would. The question had been asked, " What was to be done to keep up Christian unity?" He would tell them. Keep preaching Christ, and don't talk about their church, or creed, or doctrine, and then people would be attracted to them as sure as iron filings to a magnet. By this should all men know that they were Christ's disciples, that they loved one another. He hoped they would preach Christ simply, treating men not as of this denomination or that, but as sinners. He would leave them one word, " Advance." When General Grant, after a career of victory in the West, was put in command of the Potomac Army, which had been before invariably defeated, he was asked to retreat. Retreat had been the constant word, and at his council of war all his commanders were in favor of falling back; but he remained silent, and an hour after, the army were astonished to receive from him the command, "Advance in solid column at daybreak." This was his counsel to them. They might have their differences, but there was the one foe, and they should advance in solid column upon the common enemy.

Mr. Sankey said he blessed God for having been permitted to come with his brother to that land of Ireland. He knew many of their countrymen in his own land, and he loved them dearly. He prayed that the blessed unity which he had witnessed might continue. He believed that many dear men who were still outside the movement would be drawn into it. He knew of some who would be glad to come in; but they had their prejudices, and these should be respected; mistakes had been made in evangelical movements, which to some extent might account for this. Mr. Sankey then sang with deep feeling the hymn, "Here am I, send me, send me."

Dr. Craig (Irish Church) spoke under a sense of the most solemn responsibility. How was this most wonderful work to be carried on? He thanked God and Mr. Moody that he had been revived in his own soul, and he knew that old veterans had again taken their swords to fight the Lord's battles. He knew a man who, fourteen days ago, was an infidel, but had now found the Saviour, and had brought three others, who had also been saved. He had seen a young lady the other day the picture of despair, now with a face shining like an angel. A second Reformation was taking place among their dear Roman Catholic friends. An experienced Christian had said to him that when these men were gone, the work would fall to pieces like a rope of sand; but he believed it would be far otherwise, for God had given the spirit of unity. Dr. Craig went on to make some suggestions as to means for continuing the work.

Lord James Butler said that to say he sympathized with this movement was unnecessary, for all there were met to give expression to that feeling. All should thank God that He had raised up men almost literally to preach the gospel to all nations. They must thank God for raising up other men to organize, and arrange, and take places for these meetings. He also thanked God for the blessing to his own soul through the teaching of these His servants. This work must not cease after their departure. May God himself suggest the means, for all our plans would be vain without Him. It may again be said, in the words of one of England's martyrs, "A flame has been lit in Britain, which, by the help of God, shall never be put out."

Rev. W. Best (Wesleyan) spoke of God's gracious provision of a most suitable building, a valuable committee, a matchless secretary, and a chairman of such Christian spirit. He believed the work was only beginning; he had a profound conviction that the time to bless Ireland had come. Some of the work done by their brethren had already gone before them to America, in souls saved at these meetings.

Rev. F. Dowling (Irish Church), a man of learning, and whose expositions of Scripture are a marked feature of his ministry, said it gave him sincere pleasure to add his testimony to this work. He had read in the papers about the great meetings in Scotland and in Belfast, and had made it a subject of prayer that God would give a like blessing in Dublin. Now he had seen and heard these brethren sing and preach, and he felt far more deeply the blessedness of the work. Never in the past history of their country had there been such a vindication of evangelical truth. The great power was the pure and simple doctrine of the cross. Jesus said, "I, if I be lifted up, will draw all men unto Me," and in the doctrine of the cross He was still lifted up, and men are drawn to Him. If one thing more than another had struck him, it was the honor given to the Personal Word in heaven, and to the written Word on earth, and also to that which is the great mediating power between God and His people, the Holy Ghost. When the Word of God is honored, and the Holy Spirit is honored, there must be great blessing. These men do not make much of themselves. I thank God they came, and pray that, when they are gone, we shall continue to work, and not separate into positions of isolation. I met a gentleman of great intellectual power at these meetings, whom I hardly expected to find there. He said that, as Mr. Sankey was singing, he found the criticism going off at the ends of his fingers. It seemed as though the Bible were to Mr. Moody a great quiver, from which he drew out arrow after arrow, fitted it to the string, and shot it right to the heart; he did not keep the application till the end. He thanked God for the lessons he had learned as well as for the personal influence exercised upon himself. Everything is greater in America than here—higher mountains, broader rivers; it seemed to him we had seen also the American energy and largeness of these men's faith. May God plant deep in our souls to give ourselves in unreserved surrender to carry on this work, while our dear brethren are doing the same in other, and, it may be, distant places.

David Drummond, Esq., gave a most satisfactory account of his treasurership. He had only asked two friends for help, but he had asked God, and the money had come in, in large and small amounts, from all parts, until they had almost all they required.

Rev. Mr. Wilson (Primitive Methodist) said that, in all his spiritual life, he never felt so near heaven as to-day. Souls in his own family had been blessed, and his own soul had been revived. He should be sorry to think this great movement had reached its crisis; the drops had come, the showers would follow. There is a *river*, clear as crystal, flowing from the throne of God and the Lamb. When, in one of our great battles, the Guards were falling by hundreds, and were entreated to surrender, they replied, "The Guards die, but never surrender." This must be our motto, "Hold the fort; for Jesus is coming."

Lord Carrick read Phil. iii. 20, 21, and spoke of the heavenly citizenship of believers, and the change of these bodies of humiliation into likeness to His glorious body, at the coming of the Lord Jesus. He also reiterated the desire expressed by every speaker for the maintenance of unity in heart and labor for Christ.

Brief addresses were then delivered by the Hon. H. Rowley, Revs. Hamilton Magee, Dr. Marrable, Dr. Neligan, C. Nangle, and Mr. Hugh Brown, and the company then dispersed, the proceedings, which commenced at nine o'clock, having been brought to a conclusion about twelve.

VI.

The fifth week of Messrs. Moody and Sankey's happy visit to Dublin has now drawn to a close, their unceasing labors nearly concluded, and we are now in a position to speak with confidence both of the *men* and their *work*.

I.—THE SERVICES IN THE GREAT EXHIBITION PALACE.

These have been most wonderfully successful. On Sunday last the crowds that flowed from all parts of the city and suburbs exceeded anything ever before witnessed in Dublin upon any occasion, or for any purpose. One of the morning papers estimated the numbers present as high as 20,000, but it seems within the mark to say 15,000, and well rewarded they were for coming. Very marked was the stillness which reigned throughout that vast assembly. We cannot but regard this as an answer to prayer; for let it be considered that the doors are thrown open for all to enter, no charge for admittance. How easy would it be for a few evil-disposed persons to disturb the meeting! Yet God has, in His goodness, restrained them, and in answer to prayer: for Mr. Moody's continual prayer in the committee-room, before ascending the platform, has been, "O God, keep the people still; hold the meeting in Thy hand." On each succeeding evening of the week the numbers seem to steadily increase; but on last night they reached their highest point for a week-evening, when the building was quite as full as on former Sundays.

II.—THE NOON-DAY PRAYER-MEETINGS

have continued without any abatement, either in the numbers attending, or in the interest in the proceedings. It is a novel sight in Dublin, but a most gratifying one, to see from 2,000 to 3,000 persons leaving the comfort and retirement of their homes, to enjoy together the hour of prayer. Several hundred requests for prayer from all parts of Ireland, and some from England and Scotland, have been laid before the Lord at each of these meetings. Yesterday the number reached 500. It is also pleasant to relate that many thanksgivings for mercies received in answer to prayer have been presented to the Lord.

III.—THE BIBLE-READINGS

have been deeply interesting and instructive. The first lecture for this week, given on Wednesday in the Metropolitan Hall, was, "God's Faithfulness to His Promises," and Mr. Moody showed from a large number of instances, both from the Old and New Testaments, that God has ever fulfilled His own promise, that "the Scriptures cannot be broken." He strongly recommended Christians to study the Bible with a view to seeing how God has fulfilled His promises in small things as well as in great.

Mr. Moody's subject for Thursday was "Daniel." He handled it, as might be expected, with much ability and graphic power showing how Daniel dared to do what was right, and how God preserved and prospered him. This lecture was so highly appreciated by the audience that he was requested to give it again for the benefit of young men, which he has kindly promised to do at eight o'clock on Sunday morning.

On Friday, the reading was more of a miscellaneous kind, for the purpose of showing the young converts how to read the Bible with most profit, and to draw forth the rich treasures of knowledge and comfort which it contains.

IV.—SOLDIERS' MEETING.

One meeting of special interest, held during the past week, was a tea given by a few Christian friends to about 1,500 soldiers. 100 came by special train from the Curragh camp.

Mr. Moody spoke with great power and point, exhorting them to decide for Christ, and to become "good soldiers of the Lord Jesus." I am informed that many embraced Christ that night as their only Saviour, including at least one officer, who had not before been very favorable to the cause of religion. At the close of Mr. Moody's address, he invited all those who desired that special prayer should then be offered for them, to hold up their hands. For a time no one responded. The request being repeated, one fair, tall, manly young fellow, with an honest face and expansive brow, standing in front of Mr. Moody, took courage, and lifted up his right hand high in the air. After this, one and another in quick succession held up theirs, till quite a number appeared.

MEETING FOR MEN ONLY.

One other meeting I would refer to—to my mind one of the most hopeful and encouraging of all. I refer to the meeting for men only, held every night at nine o'clock.

On Sunday evening last, as on previous occasions, the Metropolitan Hall was filled with men, chiefly young men. There must have been 2,000 present.

VII.

I CAN confidently say that the work here intensifies and spreads every day, I might say every hour. Some of our more timid and cautious friends who had almost never come in contact with a great religious awakening, were fearful, while we were making our preliminary arrangements, that it would be next to impossible to keep up the interest of the people for a month or more; but the fact is, the interest was never nearly so great as it is this moment; and as the time of our brethren's departure draws near, the eagerness to hear their every word and catch their every song is something wonderful to see. As I remarked before, this eagerness does not now proceed from curiosity.

At all the meetings yesterday, the attendance was enormous. It is a very healthful sign of this work, that the Daily Prayer-meeting continues to be so largely attended, although neither Mr. Moody nor Mr. Sankey usually takes a very prominent part in it. The requests for prayer have become so numerous, that it has been found impossible to read even a brief classification of them. The letters have for some days been "spread before the Lord," after the example of good King Hezekiah, the meeting uniting in silent entreaty for the special cases sent in.

Hundreds were obliged, yesterday, to go away disappointed in their efforts to get into the Bible-reading in the Metropolitan Hall. Mr. Moody reserved his best wine to the last. A more suggestive Bible lecture it was never our privilege to hear. We had a compendium of some half a dozen Bible-readings. The great bulk of the people, ministers included, were taking notes. It is given to few preachers to have so many eager reporters. Many a good sermon will be got out of yesterday's addresses. One minister remarked that it was as good as an addition of many a good book to his library. It is calculated that in the evening there were not less than 12,000 persons assembled in the Exhibition building. There is not a Sabbath service in any congregation in Britain in which there is a greater solemnity and decorum than there was in that vast assembly. The sight from the platform of these earnest, and, in many cases, awe-stricken thousands, is one that it will be impossible for us ever to forget. Some one remarked to me, a day or two ago, how significant it was that during the severe weather of last week, even a cough was scarcely heard in that great-crowded glass building. When Mr. Sankey sings, the silence is sometimes even oppressive.

We are now engaged in giving out tickets for the Thanksgiving meeting, to be held on Wednesday evening, the last night Mr. Moody has promised to be with us. The tickets are given only to those who profess to have been brought to Christ during the special services. We are very careful in giving these tickets, though I doubt not there may be many stony-ground hearers.

We have had the help in this work of some of the most experienced ministers of the gospel in our city; and the general impression made on the minds of the brethren who have taken part in it, is of deep and intense gratitude for the many indubitable tokens of the presence and power of the regenerating Spirit of God. About a thousand tickets have been already given out; but many of the converts have not yet applied.

Arrangements have been made for the carrying on of special prayer and evangelistic meetings, after our brethren have left. Leading ministers of all our evangelical churches have thrown themselves heartily into these arrangements. We have felt that it *is* a good thing—good for ourselves, and good for that cause which, with all our imperfections, is dearer to us than life—for brethren to dwell together, and work together, in unity.

Mr. H. Drummond writes that the meeting for young men last Friday in the Metropolitan Hall was larger than any that had preceded it. W.

VIII.

THE happy visit of Messrs. Moody and Sankey to Dublin, which for so long a time has occupied the attention of the Christian public, is now a thing of the past. These men of God are gone from us, but the work remains. That work consists—

1. *In a great general awakening* throughout Dublin and its neighborhood. This is a fact which is patent to all, and cannot be gainsayed or denied. It is a fact that from 12,000 to 20,000 persons have been attracted to the Exhibition Palace every Sunday afternoon since the work began; that the attendance at the services held each evening in the same place, beginning with some 5,000 people, increased each evening till it became as great as on the Sundays; and this notwithstanding an audience of from 2,000 to 2,500 had been in daily attendance at the noon prayer-meeting in the Metropolitan Hall, and on three days in each week at the Bible-readings at two o'clock in the same place. What has been the great attractive power which has drawn together such vast multitudes? Thank God, it was the simple statement of gospel truth—the old, old story of Jesus and His love, plainly and lovingly told.

2. *The bringing in of some* 3,000 *converts to the fold of Christ.* Nearly 2,000 tickets were issued to those who professed to have found the Lord Jesus as their Saviour since these services began. To these must be added the many hundreds who came up from all parts of the country to attend the services, and who found "joy and peace in believing," some of whom are known to myself, besides all those who are still day by day being added to the Lord.

3. *The quickening and refreshing of many hundreds of ministers* in connection with the convention held this week. It was a happy thought to bring so many ministers of the various evangelical denominations together at this time. It afforded them an opportunity of seeing with their own eyes the reality of this great work of God which is going on around us, getting their own hearts warmed up afresh, and thus of becoming, when they return home, more than ever centres of spiritual light and heat in their own parishes and districts.

IX.

THE visit of the American evangelists, Messrs. Moody and Sankey, terminated on Thursday, and with it a series of religious services which have marked the progress of a movement the most remarkable ever witnessed in Ireland. There have been at various times so-called "revivals," which have cast a flood of devotional feeling over the country, but their influence was only transient—they left but little trace of any permanent effect. This new mission has been of a character essentially different, and seemed to possess elements of vitality which were wanting in others. There was nothing sensational, though much that was novel and attractive, in the nature of the services and the mode of conducting them. Mr. Moody, as a preacher, is certainly not superior, if he is not very inferior, in erudition and intellectual gifts, to the average class of educated clergymen. He is eloquent, or he would have no power; but his eloquence is far from being of an elevated style. It is remarkable rather for great volubility and fervor than for the higher qualities of a pulpit orator. It has no pretension to elegance of diction, beauty of illustration, harmonious arrangement, or logical force. His sermons would not stand the test of ordinary criticism. His language is plain and homely, not always very accurate, and sometimes containing colloquial phrase more popular than refined. Thus there will appear to be a considerable balance of disadvantages against him.

How, then, is his marvelous success to be explained? His great earnestness is, perhaps, the secret of it. His heart as well as his head seems to be full of his subject, and he has no difficulty in giving effective expression to his thoughts. The evident absence of any effort at self-display, but rather a sensitive avoidance of it, helps to obtain for him a favorable reception, and he never fails to keep the attention of a vast multitude riveted, and to enlist their feelings by the ready flow of his discourses, in which persuasion and argument were blended with many apt illustrations and personal incidents. He has an inexhaustible fund of anecdote, and in some of his earlier sermons here he appeared to draw upon it rather freely, but he soon came to understand that his audience did not quite relish so abundant a supply, though his stories were generally of some interest, and were told with dramatic effect. He always selected some striking passage of Scripture for his text, and expounded it with great simplicity, but with keen intelligence and a discreet and earnest power, which produced a visible impression.

Mr. Sankey possesses a voice of great volume, and he manages it with much skill, though it has not been properly educated. His utterance is remarkably distinct, and he is able by himself to fill with vocal sound a building in which from 10,000 to 15,000 people are congregated. He accompanies himself with a small harmonium. He takes up some sentiment which Mr. Moody has illustrated, and presents it anew, invested with the attractions and sympathetic influence of music, and so fixes it more deeply in the heart as well as the memory. There is a special collection of hymns, set to airs which catch at once the popular ear. Some of them are original, others are modifications of familiar songs, but all appear to be in the highest favor, though there is no poetry in them, and though even their orthodoxy may be doubted in one or two points. The singing of Mr. Sankey's solos, however, with touching solemnity, had an effect not less marvelous in its way than the united voices of the immense congregation, led by a trained choir, in the delivery of other hymns. There is an individual character stamped upon them which made them appear to express the feelings of each separate person, and not of the whole collective mass.

The services were characterized by a reverence and devotion which were extraordinary, considering that the multitude was composed of literally every creed and class, and that many hundreds who pressed for admission two hours before the doors were opened were attracted only by curiosity, and some by a love of amusement, conceiving that they would find in the proceedings something to excite their ridicule. But the first prayer or the reading of a passage of Scripture, and still more surely the fervid exhortations of Mr. Moody, whose manner, tone, and words brought home to all the conviction that he at least was terribly in earnest, dispelled all ideas of the ludicrous, and made the most light-hearted and careless youths listen with deep attention and apparent interest. There was something very impressive in the breathless stillness which pervaded the vast assemblage covering the whole area of the Exhibition Palace from end to end during the delivery of Mr. Moody's most solemn utterances, or Mr. Sankey's plaintive songs. There were no demonstrations of emotion such as may be seen in other revival meetings—no apparent excitement, but a very marked and universal reverence, and also an enthusiasm which was all the more intense because it was subdued.

Let those who think they can do so, account for the movement, and explain, if they can, what it is which brought together such immense congregations every day for nearly six weeks, and produced such extraordinary effects. The fact itself is memorable and suggestive.

The organization was admirable. There were numerous services of different kinds each day, intended for different classes and conditions of people. Some were in the Metropolitan Hall, but the principal were in the Exhibition Palace, which can accommodate from 10,000 to 15,000 people at least in the Great Transept and the Leinster Hall. There was a platform erected at the angle where the two halls meet, and on this were clergymen of different denominations, who took part in the services; and, as already stated, there was a choir of trained voices. Persons were also appointed to meet "inquirers" after the meetings were over and try to fix in their minds the impressions left by the services. There was no attempt made to win proselytes for any particular church, and not the faintest allusion to any of the distinctive characteristics of sects and creeds. The result was, that Protestants and Roman Catholics, Christians and Jews, Presbyterians, Methodists, Moravians, Arians, and Quakers, were all mingled in the great assembly, and all seemed equally impressed.

The presence of over 750 clergymen of various communions, in answer to the invitation of the Committee who have taken charge of the work, is a significant proof of the success of the movement. At the convention and a private conference held yesterday at the close of the series of meetings, arrangements were made for carrying on the work which Messrs. Moody and Sankey began. The two "evangelists" have gone to England, and intend to make Manchester their next field of operations.

X.

To THE majority of people the fact that between four and five thousand men and women assembled in a public hall at eight o'clock on a frosty morning in December, will be *primâ facie* evidence that they were very much in earnest about the business they had in hand. There were nearly five thousand persons in the Free Trade Hall here this (Sunday) morning, to hear the "American evangelists," Messrs. Moody and Sankey. I arrived at a quarter to eight, under the impression that I was rather early than otherwise. But I hear that at seven o'clock the approaches to the still closed doors of the hall were thronged, and the people waiting patiently under the bleak sky, through which the morning light was struggling. I know that I had to stand during the whole of the service, being one of a crowd wedged in the passages between the closely-packed benches. Every available seat was long ago occupied. The galleries were thronged, and even the balconies at the rear of the hall were full to overflowing. The audience were, I should say, pretty equally divided in the matter of sex, and were apparently of the class of small tradesmen, clerks, and well-to-do mechanics; that was the general class of the morning congregation. But it must not, therefore, be understood that the upper class in Manchester stand aloof from the special services of the American gentlemen. In the afternoon meeting elegantly-attired ladies and gentlemen, wearing spotless kid gloves and coats of irreproachable cut, struggled for a place in the mighty throng that streamed into the hall when the doors were thrown open.

Punctually at eight o'clock the meeting was opened by one of the local clergymen, who prayed for a blessing on the day and the work, declaring, amid subdued but triumphant cries from portions of the congregation, that "The Lord has risen indeed! Now is the stone rolled away from the sepulchre, and the kingdom of God is at hand." Mr. Moody, who sat at a small desk in front of the platform, then advanced and gave out the hymn, " Guide us, O Thou great Jehovah," the singing of which Mr. Sankey, sitting before a small harmonium, led and accompanied, the vast congregation joining with great heartiness. "Mr. Sankey will now sing a hymn by himself," said Mr. Moody; and Mr. Sankey broke in with the first line of the hymn, "What are you going to do, brother?" After this solo, he began to play a tune well known at these meetings, into which the congregation struck with one mighty voice. The hymn would probably excite the unfavorable criticism of Dr. Eadie, if it were proposed to insert it in the Hymnal of the Scotch Kirk, being amenable to some of the objections quoted in a recent newspaper article as having been urged before the Glasgow Presbytery by the reverend Doctor. The words have a martial, inspiring sound, and as the verse rolled forth, filling the great hall with a mighty and musical noise, one could see the eyes of strong men fill with tears.

> Ho, my comrades, see the signal
> Waving in the sky!
> Reinforcements now appearing,
> Victory is nigh!
> " Hold the fort, for I am coming,"
> Jesus signals still;
> Wave the answer back to heaven,
> " By Thy grace we will."

The subject of Mr. Moody's address was "Daniel." One might converse for an hour with Mr. Moody without discovering from his accent that he was from the United States. But it is unmistakeable when he preaches, and especially in the colloquies supposed to have taken place between characters in the Bible and elsewhere. He began his discourse this morning without other preface than a half apology for selecting a subject which, it might be supposed, everybody knew everything about. But, for his part, he liked to take out and look upon the photographs of old friends, when they were far away, and he hoped that his hearers would not think it waste of time to take another look at the picture of Daniel. There was one peculiarity about Daniel, and that was that there was nothing against his character to be found all through the Bible. Nowadays, when men write biographies, they throw what they call the veil of charity over the dark spots in a career. But when God writes a man's life, He puts it all in. So it happened that we find very few, even of the best men in the Bible, without their times of sin. But Daniel came out spotless, and the preacher attributed his exceptionally bright life to the power of saying "No."

After this exordium Mr. Moody proceeded to tell in his own words the story of the life of Daniel. Listening to him, it was not difficult to comprehend the secret of his great power over the masses. Like Bunyan, he has the great gift of being able to realize things unseen, and to describe his vision in familiar language to those whom he addresses. I am afraid his notion of "Babylon, that great city," would barely stand the test of historic research. But that there really was in far-off days a great city called Babylon, in which men bustled about, ate and drank, schemed and plotted, and were finally overruled by the visible hand of God, he made as clear to the listening congregation as if he were talking about Chicago. He filled the lay figures with life, clothed them with garments, and then made them talk to each other in the English language as it is to-day accented in some of the American States.

The story of Daniel is one peculiarly susceptible of Mr. Moody's usual method of treatment, and for three-quarters of an hour he kept the congregation enthralled whilst he told how Daniel's simple faith triumphed over the machinations of the unbeliever. Mr. Moody's style is unlike that of most religious revivalists. He neither shouts nor gesticulates, and mentioned "hell" only once, and that was in connection with the life the drunkard makes for himself. His manner is reflected by the congregation, in respect of abstention from working themselves up into "a state." But this makes all the more impressive the signs of genuine emotion which follow and accompany the preacher's utterance. When he was picturing the scene of Daniel translating the King's dream, rapidly repeating Daniel's account of the dream, and Nebuchadnezzar's quick and delighted ejaculation, "That's so!" "That's it!" as he recognized the incidents, I fancy it was not without difficulty some of the people, bending forward and listening with glistening eye and heightened color, refrained from clapping their hands for glee that the faithful Daniel, the unyielding servant of God, had triumphed over tribulation, and had walked out of prison to take his place on the right hand of the king. There was not much exhortation throughout the discourse, and not the slightest reference to any disputed point of doctrine. The discourse was nothing more than a re-telling of the story of Daniel. But whilst Nebuchadnezzar, Daniel, Shadrach, Meshach, Abednego, Darius, and even the 120 princes, became for the congregation living and moving beings, all the ends of the narrative were, with probably unconscious, certainly unbetrayed, art, gathered together to lead up to the one lesson, that compromise, where truth and religion are concerned, is never worthy of those who profess to believe God's word.

"I am sick of the shams of the present day," said Mr. Moody, bringing his discourse to a sudden close. "I am tired of the way men parley with the world whilst they are holding out their hands to be lifted into heaven. If we are going to be good Christians and God's people, let us be so out-and-out.

Last night I heard him deliver an address in one of the densely-populated districts of Salford. Admission to the chapel in which the service was held was exclusively confined to women, and, notwithstanding that it was Saturday night, there were at least a thousand sober-looking and respectably-dressed women present. The subject of the discussion was Christ's conversation with Nicodemus, whose social position Mr. Moody incidentally made recognizable by the congregation by observing that "if he had lived in these days he would have been a doctor of divinity, Nicodemus, D.D., or perhaps LL.D." His purpose was to make it clear that men were saved, not by any action of their own, but simply by faith. This he illustrated, among other ways, by introducing a domestic scene from the life of the children of Israel in the wilderness at the time the brazen serpent was lifted up. The *dramatis personæ* were a young convert, a skeptic, and the skeptic's mother. The convert, who has been bitten by the serpent, and, having followed Moses' injunction, is cured, "comes along," and finds the skeptic lying down "badly bitten." He entreats him to look upon the brazen serpent which Moses has lifted up, but the skeptic has no faith in the alleged cure, and refuses. "Do you think," he says, "I'm going to be saved by looking at a brass serpent away off on a pole? No, no." "Well, I don't know," says the young convert, "but I was saved that way myself. Don't you think you'd better try it?" The skeptic refuses, and his mother "comes along," and observes, "Hadn't you better look at it, my boy?" "Well, mother, the fact is, that if I could understand the philosophy of it I would look up right off; but I don't see how a brass serpent away off on a pole can cure me." And so he dies in his unbelief.

It seemed odd to hear this conversation from the wilderness recited, word for word, in the American vernacular, and with a local coloring that suggested that both the skeptic and the young convert wore tail coats, and that the mother had to "come along" in a stuff dress. But when the preacher turned aside, and in a very few words spoke of sons who would not hear the counsel of Christian mothers, and refused to "look up and live," the silent tears that coursed down many a face in the congregation showed that his homely picture had been clear to the eyes before which it was held up.

XI.

THE labors of the Evangelists closed with a three days' convention, which was attended by 800 ministers, from all parts of Ireland, besides thousands of the general public. The first day was devoted to discussions on the following topics:— "Praise and Thanksgiving," "How are the masses to be reached," "What can be done to promote the Lord's work throughout Ireland," etc. The second day was signalized by a gathering of over 2,000 converts, to whom Mr. Moody addressed loving counsels, and on the third day there was another gathering of the ministers in Exhibition Palace. And thus terminated one of the most remarkable gatherings ever held in Dublin. Mutual love and courtesy marked all the proceedings. Strangers could not tell to what body of Christians many of the speakers belonged.

The labors of the Evangelists in Ireland were ended, and on Sunday, the 29th of November, at Manchester, they began their new work in England.

THE WORK IN ENGLAND.

THE FIRST WEEK IN MANCHESTER.

I.

November and December.—Our dear brethren have come among us in dark, wintry weather, but there has been no gloom or coldness in any of their meetings, nor have rain or fog diminished the crowds that flocked to hear them. They have evidently come "in the fullness of the blessing of the gospel of Christ," and they have found awaiting them, to all appearance, "a people prepared for the Lord."

Many thousands of Christian people have been praying for Manchester, and hundreds of thousands of prayers have risen to God from Manchester herself for a blessing on the labors of His servants. The preparatory work, indeed, has been going on all the year, especially since the month of April, when united evangelistic services were held in almost all the Nonconformist places of worship throughout the district. These preparatory meetings were brought to a close last Saturday, with a Communion service, in which upwards of 2,000 Christians of various denominations joined.

You have been told something of the meeting for workers on Sunday morning. To those who know the ordinary habits of Manchester, the attendance was astonishing, numbering nearly, if not quite, 2,500 persons. Most of these had walked distances varying from one to three miles, some far more, though the rain fell in torrents through a thick, cold fog.

The work has been going on since, much as it did during the first week or fortnight in Dublin, and in other places. There is no doubt that Messrs. Moody and Sankey have already made a most favorable impression upon a large portion of the Christian public of our city. The charm of Mr. Sankey's affectionate nature has been felt by many, as well as the power of his gift of song. The gifts which fit Mr. Moody to be the leader of a religious movement like the present are recognized by every one. Men accustomed to authority willingly put themselves under his orders. He inspires confidence. All feel at once his practical good sense and singleness of purpose. Among his natural endowments is a power of pathos which must tell everywhere, but will tell especially upon a Lancashire audience. It seems to lay hold of the men even more than of the women. In his energetic, vigorous nature there is a great depth of tenderness, which now and then breaks forth in his addresses with extraordinary power. Above all, he feels and speaks as though he felt that the excellency of the power is of God, and not of us.

The crowds which flock to hear our friends, if they do not increase, continue undiminished. Already not a few have found peace in Jesus through their word. Mr. Moody has more than once said in public, that nowhere, during the first week of his labors, have such meetings been held as in Manchester. Still it would be folly to suppose that the work as yet is more than just beginning. The masses, the general public, are still almost untouched. Manchester is tenacious of the right of independent judgment, and will make up her mind for herself. And more than this, the process with the Lancashire public is somewhat slow. Beneath an apparent mobility, which may easily deceive a stranger, there is a cast of thought and feeling strongly conservative. When one thinks of the enormous population gathered in our city and the circumjacent towns, one cannot forbear the wish that the visit of our friends could be prolonged, at least, a few weeks beyond the too brief month which they have promised us.

II.

The first week of the meetings in Manchester has been full of good omen. The work of God for which we have so long prayed and waited has opened with power. God is bending in blessing over the city. An awakening and reviving breath from heaven has for some time been felt on the face of the churches. For months past strong supplication has gone up to the throne from the noon and other prayer-meetings in various parts of the city; and the churches have been gradually drawing closer together under the influence of the hope of revival. This spirit of union found delightful expression in the Communion services held in two central chapels last Saturday week. Over 2,000 members from many churches gathered around the Lord's table to enjoy a hallowed season of fellowship with each other and with the Head of the Church. The heartfelt greetings between brethren of different denominations told how truly the bond of union in Jesus was felt, and how really the Church of Christ was one, though varied in its outward aspect to the world.

On Sunday week, notwithstanding the drenching rain, the Oxford Hall was filled with Christian workers at eight o'clock in the morning, and hundreds were unable to gain admission. Mr. Moody delivered an inspiring address on "Courage, Perseverance, and Love," as the three requisites of all workers for God. "All the men whom God used in Scripture times were courageous men. God could not use a man destitute of courage. When Elijah fell into despondency under the juniper-tree, God had to find another man; Noah worked for 120 years without seeing results, and yet never got discouraged. We were to be sure God called us to the work. When Moses went out to deliver his brethren before God sent him, 'he looked this way and that way;' but a man whom God has sent never needs to look over his shoulder; straight forward is the word for him." Sharp, graphic, clinging utterances like these sparkled out all over his animating address, which was followed up by Mr. Sankey singing, "Here am I; send me."

The afternoon meetings were still more remarkable. The incessant rain had not abated, yet the overflow, after the filling of the Oxford Hall, crowded the Free Trade Hall, where the service was carried on by various ministers till the American brethren arrived from the other gathering. Mr. Moody's bright and practical exposition of "the gospel" was listened to with lively attention. A mighty interest was gathering, which broke forth with wonderful power on the following evening. That Monday evening meeting in the Free Trade Hall will live long in the memories of those who witnessed it. None could withstand the conviction that the Spirit of God was operating in the solemnized assembly as they beheld, under the influence that swayed the meeting during Mr. Moody's appeals, business men, one after another, rising to be prayed for. The address had been growing in earnestness; the speaker seemed to come into contact with the souls of the people before him. He requested any who wished to be prayed for to rise. He quietly repeated the invitation. One was seen to stand in the left-hand gallery and cover his face with his hands; another in the area. Mr. Moody said solemnly, "There is one risen; thank God for that. Another; and another. Christians, keep on praying. Another. Jesus is passing by. You may never have such an opportunity again. You may never again have so many Christians praying for you." Before many minutes, people were standing in all parts of the hall, amid deep silence, broken only by a hushed response at each new appeal for continued prayer. At the close of the meeting the anxious ones were invited into the inquiry-rooms, where Mr. Moody conversed with them individually. He said afterwards that it was the best meeting he had known on the second day of a series. The crowds unable to obtain access to the Free Trade Hall filled the Oxford Hall, where a solemn service was conducted by the Rev. W. R. Murray. Mr. Sankey came from the larger gathering to speak a few words, and to sing "Jesus of Nazareth passeth by."

The evenings of Tuesday, Wednesday, and Thursday were devoted to meetings for men in the Oxford Hall. Being obliged to attend the overflow meetings, the writer was present only on the latter occasion. The clear exposition of God's way of salvation by faith, and not by works, illustrated and enforced by an admirable and telling use of Scripture and by graphic and pathetic story, wonderfully moved the great throng of men. Many shook with uncontrollable emotion, and much occasion for delightful labor was found in the inquiry-room. A man with whom the writer conversed, rose from his knees, where he had committed "his whole self" to Christ, and said, "I came from Bolton to-day. I did not think I should find Christ." A brother minister brought up another young convert. It was this man's nephew who had just found peace. The two greeted each other with joyful surprise.

Afternoon meetings for women have been held in the Rev. A. McLaren's chapel, Oxford road. It is strange to observe them thronging the road on their way to the chapel, and still more strange to see them occupying all the available standing-room in the spacious building. Not less than 2,000 women were present on Tuesday afternoon. These meetings, like all the rest, increase in power as they proceed, and on Thursday, when Mr. Moody entered the lecture hall, he found it filled with weeping, kneeling inquirers. Many left with the joy of pardon on their spirits.

The noon prayer-meeting has, with one exception, been held in the Free Trade Hall, with an attendance of from 2,000 to 3,000. In these meetings may be found the soul of the movement. It is the daily united cry to God which brings upon the city the power of the Holy Spirit for conviction and conversion. On Thursday, dealing with the objection that this work is not of God, Mr. Moody said, "What do these noon prayer-meetings mean? What do men come here by hundreds, I might say by thousands, to pray for? A genuine work of God. And will He give us a counterfeit? If we ask bread will He give us a stone? The Shunammite fell at the feet of Elisha and said, 'As the Lord thy God liveth I will not leave thee.' She wasn't going to trust in that old staff, nor in the servant. She would trust only in the master; and well it was for her, or she would never have got back her child. And the prayer-meeting clings to the feet of God. We will not have the staff; we will not trust in the servants, but only in the Master himself; He can and will raise the dead." In this conviction we unfalteringly concur. The voice of the Son of God is being heard, and they that hear it live. In His majesty Christ is saying, "I am the Resurrection and the Life; he that believeth in Me, though he were dead, yet shall he live;" and the spiritual resurrection we are persuaded will go on till there is marshaled for God an exceeding great army of the living.

III.

December 4.—We are drawing to the close of this first week of the special meetings in this city, and I send a little account of what I have seen of them.

First of all, we have to praise our faithful God for the abundant answers to prayer already experienced in the gathering together and quickening of so many of His people day after day at the noon prayer-meeting in the Free Trade Hall. All to whom I have spoken testify to the spiritual refreshment and power they have received at these hallowed seasons of intercourse with God.

One feature of the noon meeting here is particularly striking, contrasted with what I have observed elsewhere, and that is, the very large proportion of *men*, who, in this busy city (one of the busiest, I suppose, in the world), leave their business to come and spend an hour in the middle of the day at the prayer-meeting.

Another marked feature has been the spirit of prayer poured out on those who took part in the meetings. Is it not a token for good when God is putting such deep, earnest longings for spiritual blessing into the hearts of His children, when the burden of every heart seems the same, and one yearning desire is heard in every petition for the revival of God's work in the hearts of His own people, and among the unsaved multitudes of this great city? I believe God is about to do a mighty work of grace in Manchester. Although but a few meetings have been held, we have already had abundant proofs of the Lord's presence and power in the salvation of souls.

The first evening meeting was held on Monday, at the Free Trade Hall, which was crowded in every part; and there was an overflow meeting, conducted by some of the ministers, in the Oxford Music Hall, at the same time. The meeting in the Free Trade Hall was a most solemn one; and when, at the close of his address, Mr. Moody requested those who wished to become Christians that night to signify their desire by standing up, quite a number did so, and afterwards came to the inquiry-room, where Mr. Moody spoke with them alone.

As the building is not large enough to accommodate all who wish to come to the meetings, Mr. Moody decided to have a meeting for women, every afternoon, at three o'clock, and to preach to men in the evening. This plan has been carried out since Tuesday, with very blessed results. The women's meetings are held in a spacious chapel, (Union Chapel, Oxford Street,) which accommodates 1,500 to 2,000 persons; and at the close of yesterday's meeting a large number came into the inquiry-room as seekers for salvation. This afternoon, again, there was a crowded women's meeting in the same place, and the number of inquirers was remarkable. Thank God, many left professing to have found peace in believing.

The evening meetings for men only have been most interesting and encouraging. Every night there are numbers seeking and finding salvation in the after-meetings. One case which came under my own notice was so interesting, that I must give you particulars of it.

A young man came into the noon prayer-meeting on Wednesday, and afterwards to the evening meeting in the Oxford Hall, under deep concern about his soul's salvation. He is the child of Christian parents, who, after praying for the conversion of their children for years, at last passed away to their rest without seeing the desire of their hearts granted. One of the sons settled in Dublin and another in Manchester. During Mr. Moody's visit to Dublin, the brother living there was induced to attend the meetings, and was led to trust in the Saviour; and on Tuesday last he wrote to the brother in Manchester telling of his own conversion, and urging him to attend Mr. Moody's meetings here. By the same post, the Dublin brother's Christian wife, who was in the north of Ireland, wrote to the same effect, so that the Manchester brother got these two letters on Wednesday, and was so troubled about his soul in consequence, that he at once got a newspaper to see where and when the meetings were held. He attended the two meetings as I have said; remained for the after meeting, and that very Wednesday night professed to find Christ, and went home rejoicing in the Saviour.

Mr. Moody remarked at one of the noon meetings that he had not seen anywhere more real and deep conviction of sin than in some of those who had come in to the inquiry-room at the close of the men's meetings the last few nights. All this is most cheering, and leads us to expect still greater things for Manchester. One who knows the city well tells me that everywhere men are being stirred up to inquiry about the great question of the soul's eternal interests.

Dec. 5.—The meeting in the Free Trade Hall last (Friday) night was the best I have seen here. The hall was crowded to excess, and the presence and power of God were most manifestly visible. It was one of the most solemn meetings I ever attended, and at the close of the first meeting, when Mr. Moody announced that an inquiry-meeting would be held in the Oxford Hall, a large number went to that building, and the Christians present had the joy of pointing many anxious, seeking souls to the Lord Jesus.

The following circular has been issued by Mr. Moody:

TO THE CLERGY OF MANCHESTER AND SALFORD.

Having come to Manchester with my friend, Mr. Sankey, for the month of December, with the one object of preaching Christ, it has been a matter of disappointment that not more clergymen of the Church of England have attended our meetings.

As God has granted large blessings where unity has prevailed, we earnestly trust that you will join in seeking a blessing for Manchester.

Manchester, Dec. 4, 1874. D. L. MOODY.

You will be glad to hear that we have had most cheering news from Dublin about the progress of the work there since Mr. Moody's departure. The Lord is still working mightily, especially amongst the young men, whose meeting is carried on every night, in the Metropolitan Hall, with ever-deepening interest and blessing. One young man stood up recently to tell how the Lord had saved him. He commenced his testimony with the words, "This day week I did not believe there was a God." Mr. Moody had spoken to him while he waited for a friend at one of the meetings, but he went away angry. He came to another meeting wishing to speak to Mr. Moody, who asked him to call on his wife. He subsequently called on Mrs. Moody, and had a long conversation with her, but went away apparently unchanged. Now the news comes that after Mr. Moody left Dublin he was brought to the feet of Jesus at one of the men's meetings, and afterwards stood up, as I have said, to tell to others of the blessed change which God had wrought in him.

To-day Mr. Moody went to Liverpool to meet the committee who are making arrangements for his visit to that town. It was decided to erect a large wooden structure, capable of holding 8,000 people. A piece of ground has been obtained for the purpose in a most central situation

(Victoria-street), and as the building cannot be finished this month, Mr. Moody will not go to Liverpool until February, spending January in Birmingham and Sheffield.

IV.

MANCHESTER, I rejoice to say, is now on fire. The most difficult of all English cities, perhaps, to be set on fire by anything but politics, is now fairly ablaze, and the flames are breaking out in all directions.

Yesterday (Dec. 6) the Free Trade Hall, within whose walls scenes of no common interest and excitement have often been witnessed, presented a spectacle such as those who beheld it will not easily forget. The Rev. Dr. McKerrow, my venerable predecessor in the ministry, assured me that he had seen no such sight, even in the most excited political times, during the forty-seven years of his life in Manchester, as that which he saw there on Sunday afternoon.

The building was densely crowded. Not an inch of standing-room was unoccupied. Long before the appointed hour, hundreds found it impossible to gain admission. And Mr. Moody—in what terms shall I describe his address? Theological critics might have said there was nothing in it; but only eternity will reveal how much there came *out of it*. I should not be surprised if hundreds of conversions should result from that single mighty appeal. Taking for his text the first question addressed to them, "Where art thou?" he brought it home to the bosom of every hearer with a power and pathos that were simply irresistible. Having referred to the case of a young man who had cried out in the inquiry-room on Friday night, "Oh, mother, I am coming!" the young man himself sprang to his feet, and exclaimed in tones of impassioned earnestness, "THAT WAS ME!" The effect was electrical. Not an eye but was suffused with tears. The whole vast assembly was impressed with a profound sense of the presence and power of the Holy Ghost.

The meeting for young men in the evening was equally wonderful, no fewer than seventy-one having remained behind as anxious inquirers, not a few of whom went home rejoicing in the peace of God that passeth understanding.

There is only one sentiment, I feel convinced, in the hearts of all God's children in this vast community in regard to this great work, and that is, a sentiment of devout thankfulness to our heavenly Father that He has sent among us two such men, full of faith and power, and yet eminent for humility and lowliness of mind. "The Lord hath done great things for us, whereof we are glad."

Dec. 11.—The meetings of that memorable Lord's day gave a tone of solemnity and a character of power to all the meetings of the week. The tide rose steadily, day by day, until it became full, overflowing the bank in all directions—a very spring-tide of blessing; and only eternity will reveal how many immortals are now launching out upon its waters in the bark of a simple trust in the Son of God. Oh, that in the end an "entrance may be ministered unto them abundantly into the everlasting kingdom of our Lord and Saviour, Jesus Christ!" May every soul whose hope of salvation is now being fixed upon Jesus, when the storms of temptation and sin are all past, be found "safe within the veil!"

The evenings of Monday and Tuesday in the Free Trade Hall, will long be remembered by the thousands who were present. Mr. Moody delivered his famous discourses on Heaven. Much as we have read and heard of the fervor and unction that characterize them, we were not prepared to find these apostolic qualities in so superlative a degree as that which marked them on this occasion. The second was especially interesting and delightful, treating as it did of the society and the treasure of heaven; and the contrast drawn by the preacher between these and the treasures and society of this world, seemed to strike the minds of the vast audience with all the force of a revelation; constraining many a heart, doubtless, to resolve to seek henceforward "the things that are above." The appeal with which it closed, for power and pathos, exceeded, in our judgment, anything that he himself has uttered.

And then the discourse on Hell, on the evening of Wednesday, coming as it did immediately after the addresses on Heaven, was certainly one of the most solemn and impressive utterances that have been heard within those walls. Every eye was riveted on the speaker. The projected shadow of the great white throne seemed to fall and rest upon every countenance.

Even the fervent exclamations in which some of our friends indulge at religious meetings, and which had been just a little too fervent the night previous, were hushed, and scarcely a sound broke the awful stillness with which, for nearly an hour, the people listened to the oft-repeated charge, like so many claps of thunder, "Son, remember!" In bygone revivals such heart-smiting, conscience-stirring, soul-firing words as those which poured from the preacher's lips, would have caused hundreds to start to their feet, and cry out with frenzy, "God be merciful to me a sinner!" But in harmony with the prevailing character of this awakening, the conviction of sin produced on that occasion seemed to be too deep and too sacred to find expression in mere excited exclamations or physical prostrations, and were known only to Him who seeth in secret! God was in the midst of us of a truth. The Holy Spirit came, as of old, with the force as of a rushing mighty wind, and filled all the place where we were sitting. The powers of the world to come were brought nigh to every conscience in a manner never to be forgotten. We seemed to be looking across the gulf that divides time from eternity, and beholding the torments of the self-destroyed victims of a broken law and a rejected gospel. No wonder that the inquiry-room was full that night of inquirers of the most anxious description, and that the after-meeting, over which we presided, was larger and more earnest than any that has yet taken place. Doubtless the heavens blossomed into song overhead, and the angels of God rejoiced over many souls turning from sin and Satan unto the living God!

On Thursday Mr. Moody was, for the first time, absent, having gone to London to visit his friends. The noon prayer-meeting was uncommonly well attended, considering the murky atmosphere which wrapped our city, but we missed the ringing voice and hearty appeals of our friend. In the evening a very large audience assembled in the Oxford Hall, to hear addresses from the Rev. W. H. Aitken, M.A., and the Rev. Alex. McAuley, of Liverpool, both of whom spoke in such a manner as to hold their hearers spell-bound for upwards of an hour. On Friday, Mr. Moody returned, and in the afternoon gave the second of his deeply-interesting and most instructive Bible-readings, which have been so highly appreciated wherever he has been.

The subject was, "Confessing Christ." Passage after passage of Scripture was quoted and illustrated, all bearing directly upon this primordial duty, until one felt that by no possibility could a single undecided hearer present justify, on Scriptural grounds at least, his remaining in an undecided state for another hour longer. The preacher's running commentary on the gospel narrative of the healing of the man blind from his birth was peculiarly interesting, as well as singularly felicitous, and proved a fine illustration of the duty he was inculcating. We trust many in that hall were prompted to imitate the example of that subject of the gracious power of Jesus, by boldly testifying, as he did, to the reality of the change which has taken place in their hearts, and saying, "One thing I know, that whereas I was blind, now I see."

On the evening of the same day, despite the inclemency of the weather, the Free Trade Hall was again crowded with an audience composed of persons on whose faces one could easily read their preparedness to hear the word of the Lord. Taking for his subject the parable of the marriage feast, the preacher dealt with the excuses commonly urged by those who, in reality, "will not come to Christ that they may have life." One excuse after another was considered, and shown to be a refuge of lies, to be swept away hereafter, if not here, by the storm of God's righteous judgments. The word of the Lord in the hand of the evangelist was as a two-edged sword, piercing to the dividing asunder of soul and spirit, and of the joints and marrow, and proving a discerner of the thoughts and intents of the heart. A more searching analysis of the state of a human soul in vain seeking to excuse itself from accepting the invitation of the King of heaven, and coming to the gospel feast, it has never been our fortune to hear. The thought and the prayer were uppermost in our mind, "Every refuge of man's invention has been exposed and demolished. Oh, that sinners may now flee for refuge to the hope set before them!" God be thanked, many did flee to that hope on Friday night. In the inquiry-room we conversed with several who owned that the mask of hypocrisy had been torn from their faces, that they saw themselves in the light of God's holy law, and that their only hope was in Jesus Christ. To God alone be all the praise!

V.

On Saturday evening the Oxford Hall presented a spectacle which those who witnessed it will not soon forget. In response to Mr. Moody's invitation, some 3,000 persons, professedly Christians, and chiefly young men, assembled to hear him counsel them regarding Christian work. The heartiness with which they ever and anon broke forth into song before he made his appearance, and the manliness with which they sang, especially "Dare to be a Daniel," indicated that they were ready to receive with gladness the word of command from the lips of the great Organizer. He spoke briefly but effectively. He told of the work done by the young converts elsewhere, especially in Glasgow, in connection with the evangelization of the masses. He made particular reference to the noble army of volunteers that rose to their feet in that city when the appeal was made to them, "Who will work for Jesus?" And then, when he made the same appeal to themselves, calling upon all who were ready to work for the Master to stand up, almost the entire body of young men—a grand and inspiring sight—sprang to their feet. One could not help exclaiming, "God be thanked! there's hope for our city! Manchester, with such a host, may yet be won for Christ!" By a special arrangement, as it seemed, of Providence, Mr. Reginald Radcliffe was present, and immediately put before them a definite plan for making a great gospel attack, so to speak, upon the city. He suggested that an ordinance map of Manchester should be cut into small squares, each representing a district, and that two or three young persons should undertake to carry the gospel, in the shape of a tract or otherwise, to every house, great and small, within that district, so that no single dwelling should be omitted. The plan appeared to approve itself to the judgment of the meeting, all the more so that he told us how successfully he had carried out a similar one in Edinburgh and Liverpool in years gone by. The Lord grant it abundant success!

The workers' meeting, yesterday (13th), was the largest since Messrs. Moody and Sankey came to Manchester. The address was most powerful. A forcible appeal was made to Sabbath-school teachers in this city; but one conviction seemed to exist in the minds of the vast audience of 5,000, "Let us arise and work."

Had Mr. Moody come to deliver only this address, his mission had not been in vain. In the afternoon, from 15,000 to 17,000 struggled for admission. Various meetings had to be held in the Free Trade Hall, Oxford Hall, and Cavendish Chapel; all crowded as they never have been before. As many more halls of the same size could have been filled. From twenty to thirty meetings were held in the streets of the neighborhood, where addresses were delivered by ministers and laymen. At every meeting the Lord was present to heal. Anxious inquirers were very numerous. Great numbers professed to find the Saviour.

The meeting for young men in Oxford Hall, at eight, was also crowded to excess, hundreds being unable to obtain admission. Mr. Moody spoke as if tongues of fire hovered over his head.

VI.

The spiritual movement in this city is now a fact—a solemn but joyful fact, which must be observed even by those who take their stand outside as mere spectators, with marvel; and, indeed, skeptics marvel.

"It is a most strange phenomenon," said one to me, who is a clever journalist, "to see such multitudes brought together by mere curiosity, and this curiosity increasing day by day, when there is nothing to be seen or to be heard that is fitted to excite curiosity." So it is. A striking feature of these meetings is the absence of all excitement. The thousands who usually flock to our hall, when once seated, are impressively still; it is a grand, encouraging sight to watch this sea of human faces eagerly waiting for the word of life. Mr. Moody puts no effort forward to attract; he stands before his audience quiet; he never introduces himself; you see at once he wants you to listen to his message. His words are most simple and earnest; there is nothing elaborate, or strange, or new, not even his illustrations. But as his words fall from his lips, hearts are moved. If you watch the audience you can see faces changing expression; you can read there shame, contrition, confession, hope faith, peace—as the case may be. The truth comes home! There is power! No man could do it! It is God's power! It is the Lord's doing.

Christians have been drawn together as

we have not known here before; and though there remains yet much that is to be desired, still we are encouraged and hope for greater things; we know we cannot make unity by arrangements and efforts; the Lord's laborers have learned to realize more than ever that the work is God's, not ours; that He works mightily with His power, if we do not hinder and are willing, as Mr. Moody puts it, to be simple channels, just as those dusty, rusty, crooked-looking ' gas-pipes. And many who have been hitherto too ignorant or indifferent, or too cowardly to work, have now come forward and said, " Here am I, send me."

A dear friend, from Liverpool, who is almost daily with us, has used the opportunity and organized a scheme by which every house in this city shall be visited. I will only add that hundreds of our visitors are already busy visiting and speaking and singing in the sick-chambers of isolated sufferers, in the desolate homes of the godless, of Him who came to seek and to save that which was lost. The reports of the visitors are most cheering.

For all this let us praise the Lord!

VII.

THE time is drawing unpleasantly near for the departure of our brethren, whom the better we know, the more abundantly we esteem and love. The unassuming character of the men, the simplicity of their aim, their unwearied earnestness and devotion to Christ, and the revival of spiritual life they have been the means of bringing us, have endeared them to many Christian hearts. They have stayed with us to our profit and joy, and they will not leave us without the accompaniment of our fervent prayers.

The noon prayer-meeting in the Free Trade Hall has steadily kept up its numbers. The large proportion of men who find time in this commercial centre to consecrate an hour to prayer at mid-day, is a striking feature of the meeting. The first twenty minutes are generally spent in reading the requests for prayer, and presenting them in silent and audible supplication to God; a large proportion of these requests bear upon intemperance. This noon gathering affords an opportunity for Christian workers from all parts to give tidings of the progress of the work of God. The other day, Mr. Moody read a telegram from the venerable Mr. Somerville, who has gone on an evangelistic mission to Calcutta, reporting the conversion of thirty-one persons at a special service held by him in the theatre there on the previous evening. Last Monday, the Rev. G. Stuart, of Glasgow, told how solidly the work is continuing in that town, and how it is in contemplation to purchase Ewing Place Church for £20,000, for evangelistic purposes, growing out of Messrs. Moody and Sankey's labors. He also related several remarkable instances of answers to the prayers offered at the Glasgow noon prayer-meeting. On Tuesday, the Rev. A. McLaren followed up Mr. Moody's address by a brief and telling speech, in the course of which he strongly urged prayer for the consolidation of the growing union now observable among the churches of Manchester.

The meetings for Christian workers in the Free Trade Hall on Sunday mornings at eight o'clock have imparted a great stimulus to Christian labor. Never shall we forget Mr. Moody's address on "Daniel" last Sunday morning. The hall was crowded to excess; between 5,000 and 6,000 persons brought together at that early hour, in the depth of winter, testifies to the power with which the awakening has laid hold of the city. The character of Daniel was exhibited with graphic skill; the varied scenes of the first six chapters of the book were vividly portrayed; every actor in the story became instinct with life and humor, and the lessons were rapidly and sharply drawn in a way not likely to be forgotten. The scene of Belshazzar's feast was powerfully sketched; and while Daniel read out the mysterious writing on the wall—read it easily, for it was "his Father's handwriting"—the breathless silence which fell upon the vast throng in the hall told with what reality the scene was presented before them. The whole story involved a running satire upon the yielding temper of the present day; and the address constituted a powerful appeal to young men, which we have never known surpassed. At the close Mr. Sankey sang "Standing by a purpose true," and the audience joined with unmistakable enthusiasm in the chorus, "Dare to be a Daniel."

The meetings for parents and children, held every Saturday at noon, in the Free Trade Hall, are gatherings of great attractiveness. An interesting episode occurred last Saturday when, at Mr. Moody's suggestion, a collection was made—to be repeated next Saturday—for the purpose of giving a New Year's present to every orphan child in Manchester and Salford.

The gospel meetings on Sunday afternoons and week evenings are still as thronged as ever. The numbers at the inquiry-meetings increase; many have been led to the Saviour. So permeated with Bible truth is the teaching given in Mr. Moody's addresses, that inquirers perceive the way of salvation with unusual quickness; Christ is presented to them, and they simply and immediately close with Him. Last Sunday afternoon, Mr. Moody addressed the great assembly in the Free Trade Hall, from the seven following "Beholds": "Behold, I was shapen in iniquity." "Behold, I bring you good tidings of great joy." "Behold the Lamb of God." "Behold, now is the accepted time." "Behold, now is the day of salvation." "Behold, I stand at the door and knock." "Behold, he prayeth." It was an address of thrilling solemnity. The crowded meeting which at the time filled the Oxford Hall, was addressed by the Rev. J. Rawlinson and W. Hubbard. It may interest readers to learn that a band of workers has been organized to visit every house in Manchester and Salford, with a card, bearing on one side the hymn, "Jesus of Nazareth passeth by," and on the other, the following

ADDRESS BY MR. MOODY.

"Behold, I stand at the door and knock: if any man hear my voice and open the door, I will come in to him, and sup with him, and he with Me" (Rev. iii. 20). A woman in Glasgow got into difficulties. Her rent was due, but she had no money for the landlord, and she knew very well that he would turn her out if she did not satisfy his claim. In despair, she knew not what to do. A Christian man heard of her distress, and came to her door with money to help her. He knocked, but although he thought he could hear some one inside, yet the door was not opened. He knocked again, but still there was no response. The third time he knocked, but that door still remained locked and barred against him!

"Some time after he met this woman in the streets, and told her how he had gone to her house to pay her rent, but could not get in. 'Oh, sir!' she exclaimed, 'was that you? Why, I thought it was the landlord, and I was afraid to open the door!'

"Dear friends! Christ is knocking at the door of your heart. He has knocked many times already, and now He knocks again by this message. He is your best Friend, although, like that woman, perhaps, you think He comes with the stern voice of justice to demand from you the payment of your great sin-debt. If so, you are sadly mistaken. He comes not to *demand*, but to *give!* 'The *gift* of God is eternal life.' He knows you can never pay the great debt you owe to God. He knows that, if that debt is not paid for you, you are forever lost! He loves you, though He hates your sins; and, in order that you might be saved, He laid down His life a sacrifice for the guilty. And, now, He comes! bringing the gift of salvation to the door of your hearts. *Will you receive the gift?*

"D. L. MOODY."

Encouraging signs of union amongst the different religious bodies are coming to the surface. Various attempts have been made during the year by the committee which conducted the preparations for the visit of our brethren to secure the co-operation of ministers of the Church of England. The invitation, which emanated from the meeting of ministers of the Established and Free Churches on Friday afternoon, Dec. 18, calling a meeting of all the ministers of all bodies in Manchester and Salford for conference on Tuesday, Dec. 22, at ten A.M., was well responded to; about 150 clergymen and ministers assembled in the Town Hall, and after free interchange of opinion and frank statement of difficulties, resolved on continuing the noon prayer-meeting after the departure of Messrs. Moody and Sankey. The previously existing committee was requested to enlarge itself, so as to embrace all the ministers and clergymen of Manchester and Salford. These, at Mr. Moody's suggestion, will meet monthly; and a sub-committee representing each denomination, will carry out the determinations of the larger monthly assembly. It has also been resolved to purchase the museum in Peter Street for the Young Men's Christian Association, for £30,000, which sum will, it is hoped, be raised without great difficulty. The building will then become the home of the noon prayer-meeting, and the centre of the united

Christian effort, which now appears to be fairly inaugurated in Manchester.

December 27.—The collections above alluded to for presenting a New Year's gift to the orphan children of Manchester and Salford have amounted to £146. This morning, notwithstanding the frost and fog, the Free Trade Hall was again crowded at eight o'clock. Mr. Moody spoke from Dan. xii. 3, urgently enforcing personal effort as the great means of "turning many to righteousness." This afternoon I had to leave the crowded meeting in the Free Trade Hall to attend the overflow meeting in the Oxford Hall.

VIII.

A FEW yards from the Free Trade Hall, on the same side of the street, stands a dingy-looking old public building. It was formerly used as a natural history museum, but since the erection of the magnificent Owen's College, and the consequent transference of its contents, the old museum has been unused. The Young Men's Christian Association have long been looking for some suitable building as a centre for their operations in this important city, with its 70,000 young men; and now the necessity is felt for a place to carry on the daily prayer-meeting, and other united evangelistic efforts, after Messrs. Moody and Sankey have left; so it has been decided to purchase the old museum building, and use it for these purposes. It was secured accordingly on Monday last; and, in a couple of days, part of the building, giving accommodation to about 500 persons, was seated, lighted with gas, and heated; so that, on Wednesday night, Mr. Moody used it as an inquiry-room, after the meeting in the Free Trade Hall, and we had the joy of seeing it full of anxious souls. This was a blessed consecration of the building for a higher and nobler object than ever it had been used for before.

This (Saturday) evening there was a thanksgiving meeting in the Oxford Hall, at which Mr. Moody presided, and, in his opening address, expressed his thankfulness to God for the happy spirit of unity and love which now prevails among the different sections of the Church of God in this city. The walls of separation have been over-stepped, party spirit laid aside, and all are uniting together to exalt Christ, and bring sinners to Him.

After his address, Mr. Moody invited any persons in the hall who had cause to thank God, to stand up and express their gratitude. It was most touching to hear one after another stand up to declare what great things God had done, either for themselves or for loved relatives and friends, during the last three weeks. Ministers of the gospel tell of new life and blessing in their work. Then a father tells, in tones tremulous with emotion, of dear children brought to Christ and of the "happiest Christmas" ever spent in his house. Then a prodigal son tells of his new-found joy in the Saviour's love. When he sits down an aged man rises with streaming eyes—it is the father of the last speaker. What cause of thankfulness he has as he tells of the letter which brought him the news of his son's conversion to God! And so the time swiftly passed away, and at the close many were found anxiously inquiring the way of life, and desiring to share in the joy they heard others speak of.

Nothwithstanding all the excitement and bustle of "Christmas week," the meetings have exceeded the expectations of most of us, both in the numbers attending and in the blessing vouchsafed.

On last Sunday the crowds at the Free Trade and Oxford Halls, as well as those unable to gain admission to either, were quite as great as on the Sunday previous.

On Monday, Tuesday, and Wednesday there have been three meetings daily (at 12 noon, 3 afternoon, and 7:30 evening) in the Free Trade Hall, as well as the men's meetings, conducted by Mr. Drummond, every night in the Oxford Hall. Every night scores of anxious inquirers have remained to be spoken with personally, and very many have gone home from each meeting professing to have found peace and rest of soul by believing in the Lord Jesus Christ. The ministers and other Christian workers who have been engaged at these after-meetings in pointing seeking sinners to the Saviour, all testify that they have never seen such a wonderful work of grace in this city.

This work is not only seen in the bringing large numbers of the unconverted into the fold of Christ, but also in the revival and refreshing of the children of God, and the uniting together different sections of the Church of Christ in the common object of seeking the salvation of perishing souls. Large numbers of the clergy of the Church of England, who did not see their way to join this movement at the outset, are now entering most heartily into the work along with their brethren of other communions. Several meetings of the clergy and ministers of all denominations have been held with the object of promoting this Christian union and carrying on the blessed work after Mr. Moody and Mr. Sankey have left.

The scheme for the visitation of every house in Manchester, is working well, and with the happiest results. The following is the plan adopted : A Christian architect, who has entered most heartily into this service, has cut up the large scale Ordinance Map of Manchester into about fifty districts, each of which is under the charge of a superintendent, who is supplied with a sufficient number of visitors to reach every house within the limits of his district. A leaflet, containing the hymn, "Jesus of Nazareth passeth by," and a short address by Mr. Moody is left at each house; but it is understood by the visitors that this paper is only to be used as an *introduction*, for the purpose of gaining admission to the houses, so as to have personal conversation about eternal things with each individual, as far as possible. Some of the visitors have already given in most cheering reports of the marvelous way in which the hearts of the people seemed open to receive their visits, showing that the Lord is in this movement, and is preparing many hearts for the reception of His own blessed message of salvation.

The committee at Sheffield have made arrangements to do the same work there before Mr. Moody and Mr. Sankey come, and have got 30,000 of the same leaflet for that purpose. It will probably be taken up by Birmingham and Liverpool also, and why should it not be done in London itself? It seems a bold thought, but if a thousand Christian men and women can be found to visit the half-million inhabitants of Manchester, surely London could furnish eight thousand earnest laborers for the same glorious service.

Mr. Moody left for London on Wednesday night, and returns (D. V.) on Saturday. He remains here until Wednesday next, which will be his last day in Manchester for the present, but he will probably return again for a day or two next month. The Sheffield committee have secured the Albert Hall for the meetings there, and it is proposed to commence with a "watch night" meeting, beginning about half-past ten o'clock on the last night of the year.

Dec. 26.—There was an immense concourse at the children's meeting in the Free Trade Hall to-day. Mr. Sankey presided, and Mr. G. Beith, Rev. Mr. Kernock, etc., took part. A collection was made for the purpose of presenting a New Year's gift to every orphan in Manchester. The total amount collected was £146.

The subject of conference, held in the Town Hall the same afternoon was, "The inquiry-meeting, and how to deal with the anxious." Mr. Moody opened the discussion, and was followed by various speakers, all of whom concurred in recommending that instant faith in Christ should be urged in the case of anxious souls.

In the evening Mr. Moody gave a powerful address on "Faith," and there were many seeking Christ.

About 3,000 parents and children were present in Free Trade Hall on Saturday, the 19th, at noon. A letter was read by Mr. Sankey from the children's meeting at Edinburgh. It was printed, and a copy given to all present last Saturday.

IX.

THE mission of Messrs. Moody and Sankey to this country dates from July in last year.

It was the late Mr. Pennefather, of London, and a Mr. Bainbridge, of Newcastle-on-Tyne, who induced the two American preachers to extend their mission to this country. After long hesitation they accepted the invitation; but before they reached Liverpool both Mr. Pennefather and Mr. Bainbridge were dead.

"We arrived in York on Saturday night in July, 1873," says Mr. Moody, "and did not know a soul in the place." They soon made friends, however, though during their stay in the city their fame had not gone beyond the circles of the chapel congregations whose ministers had lent their pulpits to the strangers.

From York they went to Newcastle-on-Tyne, and here they began to attract public attention, great crowds gathering around them, not alone in Newcastle, but in all the towns on Tyneside which they visited in succession. From here they went to Edinburgh, and were received with an enthusiasm that was surprising to lookers-on. The local clergy came forward, and not only offered their pulpits, but supported the strangers with their presence whenever they appeared in public. Mr. Moody preached and Mr. Sankey sang twice and thrice a day; but every day thousands were, for lack of room, turned away from the doors of the halls and churches where the services were being held.

The growing wave of enthusiasm carried them through Dundee, and other towns of the North, and appeared to culminate in Belfast and Dublin, whither they next directed their steps. The impression created in the latter town will be best understood by mention of the fact that the meetings were held in the Exhibition Palace, which is capable of holding 14,000 persons, and which was always crammed to the doors when Moody and Sankey were announced to appear. From Dublin they came hither, and commenced work under circumstances that were all the more disheartening with the memory of the eager throng in Dublin still fresh in their minds. At first the Oxford Hall was found more than large enough for all who cared to assemble; and when the Free Trade Hall was adventured upon, there were a good many empty benches. But day by day the excitement rose, and if there were any hall in the city that would hold 15,000 people, it would certainly be filled at any one of the meetings.

As to the practical issue of the work, it may be mentioned that in Dundee a body of young men have united for the purpose of carrying on the work in that town. Their scheme is to prepare a breakfast, to which they invite all homeless people who can be found in the streets, and after comforting them with coffee and filling them with bread and butter, they talk to them in a friendly way about their present mode of living, and try to lead them into the way of doing better. Again, it is said, that in Belfast a number of clerks in warehouses and offices—as many as seventy are from a single establishment—have formed themselves into an association, and have devoted themselves to the work of bringing all their fellow-clerks to a "knowledge of Jesus." In Liverpool £5,000 have been raised for the erection of a temporary edifice, in which Messrs. Moody and Sankey may conduct their services when they visit that town. This money, like the rest of the large sums required to meet the expenses of the tour through Great Britain and Ireland, comes from unknown hands, at least to the extent that it is privately and quietly subscribed, without appeals from the pulpit or publication of lists of donations by the press. There has, I am assured, been only one collection made from congregations gathered to hear Messrs. Moody and Sankey, and that was in Dublin on Hospital Sunday.

X.

SUMMARY OF THE WORK IN MANCHESTER.

MESSRS. Moody and Sankey left us, for the present at least, on the afternoon of Thursday, the last day of 1874. For four weeks, in the darkest, coldest, and dreariest season of the year, have these men of God toiled among us with an amount of diligence and zeal such as I never saw equaled, far less surpassed; and what has been the result? That is the question that shaped itself in my mind. A complete answer to it would cover page after page of this journal. Only eternity will disclose the amount of good that has been done through their instrumentality. To speak figuratively, we have had summer in the depth of winter. The Sun of Righteousness has shone forth most brightly and genially, even while the material sun has been hid from view amid fog and darkness. From the lips of hundreds the song might have been heard, "Lo, the winter is past, the rain is over and gone; the flowers appear on the earth; the time of the singing of birds is come, and the voice of the turtle is heard in our land."

In speaking of definite results, so far as these can be ascertained, I may be forgiven if I begin with the ministers of Manchester. If one class has been blessed more than another during these four past weeks, it has been the regular Christian ministry. I am sure I speak the sentiments of all my brethren, who have thrown themselves heart and soul into the movement, when I say that we have received nothing less than a fresh baptism of the Holy Ghost. Our own souls have been quickened. Our faith in the adaptation of the glorious gospel of the blessed God to the wants and longings of the human spirit has been deepened. Our sense of the magnitude and responsibility of our offices as heaven's ambassadors, charged with a message of reconciliation and love for the guiltiest of the guilty, and the vilest of the vile, has been greatly increased. We have had demonstrated to us in a way that at once startled and delighted some of us, that after all, the grand levers for raising souls out of the fearful pit and the miry clay, are just the doctrines which our so-called advanced thinkers are trying to persuade the Christian world to discard as antiquated and impotent. These are —the doctrine of the atoning death of Jesus Christ; the doctrine of a living, loving, personal Saviour; and the doctrine of the new birth, by the Spirit and the Word of Almighty God. One of our ablest ministers, at the noon prayer-meeting, on the last day of the year, solemnly declared that, whereas the first of these cardinal verities had not been fully realized by him before these services commenced, he now felt it to be a spring of joy and satisfaction to his soul such as language could hardly express. And then how shall I speak of the gladness that has filled our hearts when we heard, as we did almost from day to day, of conversions in our congregations, of parents rejoicing over sons and daughters brought to Jesus, of young men consecrating their manhood and strength to God, and of converts offering themselves for any department of Christian service.

If our dear friend, Mr. Moody, had accomplished nothing more than the quickening of the ministers of this great centre of population, and stirring us up to greater devotion to our glorious vocation as "laborers together with God," his visit would not have been in vain. Give us a revived ministry, and we shall soon see a revived church.

Next to the Christian ministry, I believe the great army of Christian workers have shared most largely in the blessing. Perhaps the most remarkable, in every respect, of all the services held by the evangelists during their stay here were those on Sunday mornings in the Free Trade Hall. With the exception of one of these mornings, the weather was as severe as any we have had in this exceptionally severe winter, and yet the vast building was densely packed, at the early hour of eight, with audiences presumably composed of Sunday-school teachers, tract distributors, district visitors, missionaries, evangelists, etc., drawn not only from the city and borough, but from the whole surrounding district. The fruits of these wonderful meetings are already apparent. I question if there be a single Christian agency in all Manchester that has not been the better for them. From that one meeting, as from a great fountainhead, streams of blessing have flowed, are flowing still, and, I believe, will continue to flow, that will spread life and beauty over the whole field of Christian work, such as we have not witnessed here before. Teachers went straight from the hall, in many instances, to their classes, with their souls fired with love for their scholars. Missionaries received fresh impetus and courage for their peculiarly difficult work of going from door to door, knocking for admittance in the name of Jesus. Visitors of tract districts felt stimulated to greater diligence in the discharge of their important duty, as the bearers of those silent monitors from house to house that have so often brought "light into the dwelling." Above all, drones felt rebuked, and ceased to be drones. Recruits in large numbers were enlisted in the name of our Lord and King. Many who had been languidly sighing out, "My leanness, my leanness!" were constrained to cry out, "My laziness, my laziness!" and to add in all seriousness, "Lord, what wilt Thou have me to do?" In short, could our American brethren repeat these addresses in that great hall once every year, they would do for our various Christian organizations what requires to be done periodically for the machinery of our mills and factories—overhaul them completely, renew and improve much of their belting gearing, and render their operation at once more vigorous and more productive.

The noon prayer-meeting has also been largely blessed. Like some old Eastern well, it has been daily visited by hundreds, who have refreshed their souls with the water of life, and returned to their businesses and their homes feeling that the "sweet hour of prayer" was the sweetest of all the hours of the day. And the requests for prayer that have been presented, —who shall number them?—who shall even classify them? Above all, who shall say what revelations they afforded of the yearning solicitude, the agonizing supplications, the impassioned cries, that exercise the souls of immortal beings, in every relation and condition of life, in this world of distance and darkness? Whatever some may think of this novel feature in the mode of conducting a prayer-meeting, I feel sure, from observation and experience, that it has imparted new life and interest to a much-neglected institution. These requests have given reality and intensity to the prayers that were offered. They drew out our sympathies towards our fellow-Christians, in connection with trials and wants such as never entered our minds to conceive. They made us feel that "one touch of Nature makes the whole world kin," that "as in water, face answereth to face, so doth the heart of man," that we are all members of the one family called by the one name of Jesus Christ. They did more than that—they gave us glimpses of the fullness that is in our Redeemer, out of which so many thousands may draw, "and grace for grace"—"enough for all, enough for each, enough forevermore." And, in hundreds of cases, they have not been in vain, if we may judge by the fact, so frequently brought out at these meetings, that thanksgivings have been publicly made for abundant answers to them, sometimes vouchsafed in very wonderful ways. Parents have stood up and given thanks for the conversion of their children, and children for the conversion of their parents—brothers for the conversion of sisters, and sisters for the conversion of brothers—teachers for the conversion of their scholars, and ministers for the conversion of some even of their church-members. "And now, O Lord, we thank and praise Thy glorious name!" "Not unto us, O Lord, not unto us, but unto Thy name, be praise, for Thy mercy, and for Thy truth's sake!" "Praise the Lord, O Jerusalem; praise thy God, O Zion, for He hath strengthened the bars of thy gates; He hath blessed thy children within thee!"

The afternoon Bible-readings have been greatly relished by thousands. At these Mr. Moody surprised and delighted many of us ministers by his wonderful acquaintance with the Word of God. Whatever the subject in hand, whether the Blood, confessing Christ, the Holy Spirit, grace, faith, or assurance, he proved himself to be a very giant in Bible knowledge; and though the immense audiences, comprising some of the best of our citizens, did not come provided with the Book so generally as they might have done, I have reason to believe that in hundreds of cases they went home to it with souls hungering after righteousness, and determined to become better acquainted with the word of life.

THE EVANGELISTIC MEETINGS.

What shall I say of these in closing? They have been blessed to vast numbers. In the inquiry-room, I have met with many who stated that they had never had the way of salvation so plainly put before them as by Mr. Moody. In not a few instances, too, Mr. Sankey's beautiful and touching solos, especially "Jesus of Nazareth passeth by," "Almost persuaded," and "Prodigal child," have proved to be arrows of conviction, entering the heart in the most unexpected manner, and leading to conversion. And what shall I more say? for the time would fail me to tell of all the blessed fruits, already apparent, of the extraordinary efforts of these dear men of God. Suffice it to say, in a sentence, that all classes of the community—old and young, rich and poor, learned and ignorant, ministers and laymen, masters and servants, teachers and scholars—have received a large blessing from the religious services conducted by the American brethren, and are deeply sensible, I trust, of the mighty debt of gratitude under which they have been laid. The Lord bless them, and make them blessings, wherever they go!

XI.

The closing week has been the most joyful of all. The tide of blessing, which has been steadily rising, has this week reached its flood; the earnestness of the preacher and the eagerness of the people have seemed alike to intensify, and the unconverted have been called to take refuge in Christ with a vehemence of entreaty which has exerted a mighty influence on the assemblies. During these five weeks God has answered the prayers of many years, and we cannot but feel that what has been going on in the city has made Manchester peculiarly interesting to the dwellers in heaven.

At nine on Wednesday evening, about 2,000 men reassembled in the hall, to hear what Mr. Moody had to say on the subject of the Young Men's Christian Association. Mr. Herbert Spencer occupied the chair, and gave a brief address, intimating that it was in contemplation to buy the Museum for the Young Men's Christian Association, for £30,000. Mr. Moody delivered an inspiring address, in which he enlarged on the spiritual advantages of the Association, and urged the straining of every effort to reach the young men of Manchester, and to secure the building in question for the Association. A collection towards the object, made at the close, realized £1,800, £1,000 of which was given, I believe, by the chairman. This amount, with what has been received before, including £500 given last week by Mr. J. Stuart, makes a total, at present received or promised, of £8,000.

On Thursday morning, Mr. Moody addressed a crowded meeting in the Higher Broughton Presbyterian Church, and then came on to the noon prayer-meeting in the Oxford Hall, where he read and commented on the earlier part of the 103d Psalm. He said he had to bless the Lord for what He had done for him. It had been the best year of his life. He had been more used by God than in all the seventeen preceding years. He did not know of one sermon he had delivered that had not been blessed to the conviction or conversion of some souls. It was a delightful meeting. Every word uttered was set to the tune of "Bless the Lord, O my soul!" When one minister rose to say, "I have to praise God for the conversion of the brother of dear friends of mine, who have prayed for him twenty-five years; for the conversion of the sister and of the servant of another friend; for the salvation of three persons in my own congregation, for the dispelling of the doubts of a young man who traveled 150 miles to these meetings—all which blessings have been given in the course of the present week;" when another minister rose to say he had never met with so much of scriptural teaching concerning the way of salvation, and the clear direction of inquirers to Jesus, as in Mr. Moody's addresses; and another to say that the last ten days had been the happiest of his life—that he had derived an inspiration, had discovered how to preach Christ, had enjoyed sweeter communion with Jesus, and felt like a man whose chains were broken—they only uttered what many could have endorsed, as a description of the blessings they themselves had received.

Our beloved brethren left in the afternoon for Sheffield, whither our prayers follow them. They are to return, however, for Friday and Saturday, January 8 and 9, and then we hope not only to have a repetition of the blessings we have so abundantly received, but to hear glad tidings of similar grace bestowed on the neighboring town.

THE WORK IN SHEFFIELD.

I.

On Thursday afternoon, our beloved brethren, whose visit has been looked forward to with much earnest desire by the Lord's people here, arrived from Manchester, and held their first meetings, the same evening, in the Temperance and Albert Halls.

Considerable excitement was manifested, a few days ago, when it was reported that the clergy of the Church of England had withdrawn from the executive committee, and that, in consequence, Messrs. Moody and Sankey had refused to pay their expected visit. It is, however, a matter of deep thankfulness to God that this difficulty has been overcome. The clergy have rejoined the committee, and everything is working smoothly. I am glad to be able to state that the difficulty was only one of a mere technical kind, arising from a proposed scheme of house-to-house visitation which interfered with parochial boundaries; and in their letter of withdrawal, the clergy stated that their only motive in doing so was to remove the hindrances to the visitation, and that their feelings and sympathies were unchanged as regards Messrs. Moody and Sankey's mission. When, however, it became known that, in consequence of their action, Mr. Moody had declined to come to Sheffield, the visitation scheme, which formed a part of Messrs. Moody and Sankey's personal work, was abandoned for the present, and the clergy, as I have said, rejoined the committee, and are now working most heartily with the ministers of other denominations, for the furtherance of the one blessed object of leading perishing souls to Christ.

II.

THE work has opened here most auspiciously; the two meetings held on New Year's eve were crowded, and the impressions produced were most solemn.

The first meeting was held in the Temperance Hall at nine o'clock. Mr. Sankey sang a new hymn written by Dr. H. Bonar expressly for him, "Rejoice, and be glad! the Redeemer has come." The air, which has been set to these words, is peculiarly appropriate, a bright, joyous melody.

The impression produced by his singing was very striking; those who had been merely curious or altogether indifferent, seemed attracted, and earnest attention and even, in some cases, silent weeping, took the place of carelessness. Mr. Moody spoke on the subject of "Work," dwelling chiefly on Isaiah vi. 8: "Here am I; send me." His address was well fitted to stir the Christians of this town to be up and doing. In concluding, he appealed to all to come forward heart and soul, "and let us have a fortnight of faithful, prayerful work for God." The watch night service was particularly solemn. The Albert Hall, where it was held, was crowded, many having stood before the doors an hour before they were opened, in order to make sure of admittance.

Messrs. Moody and Sankey were accompanied onto the platform by a large number of ministers of all denominations, amongst whom were the following: the Vicar (the Rev. Rowley Hill), Rev. R. Stainton, Rev. J. Smith, Rev. R. Poole, Rev. R. Green, Rev. J. Flather, Rev. P. Whyte, Rev. J. Calvert, Rev. H. H. Wright, Rev. M. Washington, Rev. G. J. Watts, Rev. W. Cobby, Rev. B. Trotter, and many others. The Vicar offered up a fervent prayer for the Divine blessing on the work in Sheffield.

One most interesting feature in this service was Mr. Sankey's singing of "Jesus of Nazareth passeth by." It might be the novelty of his style, or the associations naturally arising at the near approach of the new year, but I certainly have never seen such an effect produced. I have heard him in all the towns they have visited in Scotland, and also in Manchester; but I never heard him sing so pathetically more especially in the last stanzas:

"Too late! too late! will be the cry,
Jesus of Nazareth has passed by."

Mr. Moody spoke from Luke xix. 10 "For the Son of man is come to seek and to save that which was lost." As illustrating this verse, he graphically narrated the two stories immediately preceding his text, that of the opening of the eyes of blind Bartimeus, and the conversion of Zaccheus. It was only a re-telling of the stories, but given in that way peculiarly Mr. Moody's own, making his listeners part and parcel of the story, as if the whole thing were enacted just in the Targate, and Jesus were just passing the hall-doors. He connected the two stories by throwing out the thought that as Bartimeus was on his way home to tell his wife, Zaccheus met him. "Why, isn't that the poor blind beggar—it's like him; but it can't be he, for his eyes are open."

"Yes, it is I."

"What has made your eyes open?"

"Jesus of Nazareth did it."

"Where is He? I must see Him."

"He's just on the road to Jericho."

Away Zaccheus runs; and because he is a little man, he gets up a tree, to see well. Jesus stops, looks up, calls him, "Zaccheus, come down." This was one instance of sudden conversion. Some don't believe in sudden conversion; but here Zaccheus was not converted when he went up the tree, yet he came down a converted man. We are told he received Jesus gladly. From these incidents, he proved how willing, how eager Christ is to save all. What have we to do? Nothing! blessed be God. If we had, we would never do it. Only accept. What had Zaccheus to do? Only come down, only obey.

He concluded by drawing the attention of the audience to the fact that the old year was fast dying—only a few minutes—and what if the new year should come and find us where we were—lost! Oh, let each of us take it, the offer is here; will you have it? Salvation—ay, even before this year is closed you may be saved. As there are only a few minutes of this year remaining, let us finish the old and begin the new on our knees.

The whole audience then sank on their knees, and the new year found them bent in silent prayer. Mr. Moody asked that those who were unsaved might stand up, that they might be prayed for. For a time none were willing to do so, but on Mr. Moody's asking a second time "if there were none in the hall wishing salvation," a few stood up, and the Christians were asked to pray for them.

Just then the bells began to ring in the new year, and the Rev. R. Green engaged in prayer for an outpouring of the Holy Ghost on the town of Sheffield, and most particularly on the special meetings to be held. Mr. Moody also engaged in prayer. This was one of the most solemn scenes I have ever been privileged to witness. While the audience were bent in prayer, the most intense stillness prevailed, broken only by an occasional sob. After singing the Doxology, the meeting separated.

The streets were made lively after the meeting with vigorous singing of hymns, as bands of Christians wended their way home.

New Year's Day.—The noon-day meeting was held in the lower Cutlers' Hall. It is a great pity this hall is so dark and confined. The platform is nothing but a mere box. It is to be hoped the ensuing meetings will be held in some hall more suited for speaking. Mr. Moody chose as his subject, "Faith," as illustrated by the healing of the leper and that of the man sick of the palsy, in the fifth chapter of Luke. Mr. Sankey engaged in prayer, and also sang his beautiful solo, "Whatsoever a man soweth that shall he also reap."

III.

FRIDAY evening meeting was also held in the Cutlers' Hall. After the opening hymn, "Free from the law, O happy condition!" had been sung, the Rev. F. Kellet engaged in prayer. Mr. Sankey sang "The Ninety and nine;" before singing it, he asked that whilst he was singing, Christian friends might pray for the salvation of those who were wanderers.

Mr. Moody's address was on the "Gospel" (Mark xvi. 15): "Go ye into all the world and preach the gospel to every creature."

At the conclusion of the meeting, Mr. Moody said there would be a prayer-meeting for those who loved the Lord, and asked any who were in anxiety about their souls to go into the side room. He explained that an inquiry-meeting was simply to give an opportunity for a little private talk about salvation to any who wished it. He had found in all his experience that he could do more good in five minutes' private talk with a man, than in five hours' talk from the platform.

The noon-day meeting on Saturday was intended particularly for parents and children. It was held in the Temperance Hall. Mr. Moody spoke the truths of the glorious gospel in such a simple style that the youngest child could not fail to catch his meaning. He kept their attention fixed by judicious questions, such as, "What did Jesus come to this world for?" "To save us." "Who did He come to save?" "Sinners." "Are there any sinners here to-day?" "Yes." "Are there many?" "Yes, we are all sinners." "What will take away our sins?" "The blood of Jesus."

In this way he interested the little ones, and at the same time his address must have been most impressive to all present, more especially to those who were parents, as he dwelt very strongly upon the duty of seeing that the children were led to Jesus in their youth.

IV.

Sunday, January 3.—Truly this has been a day of blessing for Sheffield. The meetings have been attended with most blessed results.

The morning meeting for Christian workers was not, perhaps, so well attended in point of numbers as might have been expected, but the Christians who had come out at this early hour were right-down hearty workers. As Mr. Moody said, "He would rather have a moderately small meeting of such earnest Christians than have it packed with thousands of careless people." His address was on "Work" (Mark xiii. 14): "To every man his work." Faithfully and earnestly did he lay it before his audience, that there was for each one some appointed work, and if we neglect it we must answer for it. He also showed the joy of working for the Lord, and its reflex effect on our souls in building up and comforting our hearts.

At the afternoon meeting, the Albert Hall was densely packed half an hour before the time; the lower Albert Hall was thrown open for the overflow, but even then many had to go away disappointed.

Mr. Moody addressed this large gathering from Rom. ii. 23: "For there is no difference." Many must have been startled by the plain way in which he put this truth before them, that all are alike in God's sight. He does not divide by classes or ranks, rich and poor. The only division before God is saved and unsaved.

Many were evidently struck to the heart; some whom we heard scoffing at the commencement, were in tears at the conclusion of his address.

When Mr. Sankey followed by singing, "Free from the law, O happy condition!" it seemed to produce a deep impression.

The Sunday evening meeting was glorious. The hall was again densely packed. Mr. Sankey sang his solo, "There were ninety and nine." Mr. Moody then gave his address on "Regeneration," from the words, "Ye must be born again" (John iii. 7). His thrilling words must have gone to the heart of many grieved ones who had been awakened by the former address to a sense of their lost condition. The result of the whole proceedings in Sheffield since the coming of our dear brethren, must be considered highly satisfactory, and as affording great cause for thankfulness.

Sheffield has been cold and indifferent to religious matters, but we hope that now when it has been shaken the blessed result of this work will be a new zeal for the service of the Lord, and a desire to win souls for Christ.

V.

THE past week has been a time of pentecostal blessing in Sheffield. The town has been stirred to its depths; more, perhaps, than any of the towns in England the evangelists have yet visited, considering the short time they have been here; and the Spirit of the Lord has been manifestly working in many hearts that have hitherto been strangers to His power. Messrs. Moody and Sankey have been the subject of common conversation; and while, no doubt, much of it has been nothing more than curious talk, yet there remains a large and solid substratum of good, and, I trust, lasting, result. And yet there does not appear to be any disposition to unduly magnify the human agents in the work. The mouths of those who do not sympathize with the movement have been wonderfully silent; and the overshadowing presence and power of God have revealed themselves, even in this respect.

The *Sheffield Methodist* devotes a large proportion of its space to a record of the meetings, and says, in a sketch of Mr. Moody:

"Some say that the secret of his success is to be found in Mr. Sankey's music and singing. Some say it is to be found in that tact which draws large numbers of the most Christian ministers and laymen around him. Our impression and belief is this: he is full of the love of Christ and true gospel simplicity; is filled with the Holy Ghost and with faith; fears nobody, loves everybody, has full confidence in his plan of working, and in a deep and well-developed Christian experience; carries in his own bosom that divinely-written commentary on the truth of God which causes him to pour forth streams of spiritual light over the minds of his crowding auditors. The consequence is, the hand of the Lord is with him, and multitudes believe and turn to the Lord, both men and women—yea, and children also.

"We have been told that some of Mr. Sankey's hymns are rather childish. Our answer was, that the older, wiser, and better we grow, the more childlike we become. Praise God for sweet hymns, in which both very old and very young Christians can happily join."

The *Sheffield Independent*, of Saturday, reports several of Mr. Moody's addresses at considerable length. Of course, the burning earnestness and homely bluntness of his delivery, or the wonderfully moving charm of Mr. Sankey's singing, cannot be reproduced by any verbal description; and these will always attract the multitude, both careless and Christian, although the words used may be familiar.

The meetings on Monday and Tuesday of last week do not call for any special notice. They were crowded—sometimes to excess—and in every way indicative of most hopeful results.

On Wednesday, Mr. Moody and Mr. Sankey were both at Manchester, and on Thursday, as the Albert Hall was engaged for a ball in the evening, the evangelists did not hold any public service, but from three to ten o'clock P. M. a special meeting for inquirers, in the Temperance Hall, was attended by a constant stream of anxious ones, who were pointed Christward, there, as I trust, to lose their burdens.

Mr. Moody spoke no less than *four* times on Friday, on each occasion with much power, and with signs following. It need scarcely be added that Mr. Sankey's solos, including such favorites as "Only an armor-bearer," "Dare to be a Daniel," "Whiter than snow," etc., deepened the influences produced by Mr. Moody's impassioned discourses. Indeed, it is made more and more manifest that the special gifts of each evangelist have been most happily wedded together for the common purpose they have in view.

At the closing service in the evening there was no diminution either in the attendance or the interest. It was chiefly intended for the young converts, who were admitted by ticket, and crowded a large part of the area of the Albert Hall. Both the galleries were also crammed long before the hour of commencing. It was a glorious and inspiring sight to look on such a vast sea of human faces, all lit up with eager expectation, and all assembled to hear the simple story of the Saviour's grace and power. The scene was more impressive still when, at the appointed hour, Mr. Moody and Mr. Sankey having quietly crossed the front of the platform, and taken their seats, the whole assembly rose and joined in singing the hymn,

"Ring the bells of heaven, there is joy to-day,
For a soul returning from the wild."

And afterwards, in that jubilant old hymn that used to be sung at revival meetings fifteen years ago, and is ever fresh and new:

"O happy day! that fixed my choice
On Thee, my Saviour and my God;
Well may this glowing heart rejoice,
And tell its raptures all abroad."

After Mr. Sankey had sung "Whiter than snow," Mr. Moody spoke with his accustomed pungency, simplicity, and power, chiefly addressing the young converts. Surely they will never be able to forget his words of affectionate encouragement and caution, as he pointed out the dangers that would inevitably come to them in their Christian life, and the unfailing source of strength amidst them all. Then came his parting words, evidently painful alike to speaker and hearers. "I have learned to love you," said Mr. Moody, and the earnest gaze and tearful eyes before him testified, more loudly than words, how his love was reciprocated, and his labors and counsels prized. I was forcibly reminded of the scene of Paul's farewell meeting with the elders at Miletum. I verily believe that many hundreds of young converts would, one and all, have fallen on Mr. Moody's neck, and kissed him, sorrowing most of all for the words which he spake, that they should see his face no more. One little fellow, at the close of the meeting, came to me in great distress when he found that Mr. Moody had left without having given him a shake of his hand.

Before the meeting was dismissed, Mr. Sankey sang a parting hymn to the tenderly pathetic tune of "Home, sweet home," and the vast crowd lingered long in the hall where Christ had won so many sons and daughters within the past two weeks.

The work among the young men has been taking root during the week, and now that the counter-attraction of Messrs. Moody and Sankey's services is wanting, it is believed that the meetings specially for young men will be largely attended and much blessed. Mr. Drummond remains in Sheffield during this week, to assist in consolidating the work among this important class of the community. It has been a "Happy New Year" for Sheffield, and the faith of the Lord's people prompts them to hope that "still there's more to follow."

Upwards of eighty—clergymen of all the evangelical denominations in the town, and the other members of the committee—met Messrs. Moody and Sankey at breakfast in the Imperial Hotel, on Saturday morning, to bid them farewell. Reporters were excluded, but I understand that the unanimous expression of the company was one of gratitude to the evangelists for their untiring and successful labors in Sheffield, and for the spirit of cordial co-operation among the various divisions of the Church that their visit had so blessedly generated. Practical as he always is Mr. Moody used the occasion to urge upon the committee the necessity of rearing a central and suitable building in the town, where all those interested in the continued success of the work could meet on neutral ground, and carry on the meetings. An influential committee of laymen was appointed to take immediate steps for carrying out the suggestion, so that the good work may go on.

Messrs. Moody and Sankey left for Birmingham on Saturday afternoon, attended by the prayers and good wishes of hundreds in Sheffield whose hearts have been made glad, and whose lives have been illumined through their instrumentality. May our loving Father in heaven have all the praise!

THE WORK IN BIRMINGHAM.

I.

THERE is every outward indication that the wave of spiritual awakening and quickening, now passing over our land, is about to make itself felt in this very large and important centre of the world's industry also. Messrs. Moody and Sankey began their labors here on Sunday last, January 17, and they are to remain in the town for two weeks. Birmingham has been called "the toy-shop of the world," and its immense population—which amounts to nearly 400,000—is largely made up of the artisan class. Experience has shown that wherever the American evangelists have gone—though their services have been attended in some towns by numbers of those in the higher ranks of society, socially considered, and also by a sprinkling of the very poor and degraded—the bulk of those coming under their influence have belonged to what we are accustomed to call the middle classes. There is, therefore, a wide field in Birmingham for the efforts of our American brethren, and the meetings on Sunday were such as to encourage the hope of much success.

Their first meeting was held at the somewhat early hour of eight, but long before, just as the grey dawn was breaking up, streams of people were moving with hurried feet from all directions to the place of meeting, the Town Hall. By the time that Messrs. Moody and Sankey appeared on the platform, the fine hall was crowded, passages and all, with some 3,000 people. The meeting was advertised for "Christian workers," but there did not appear to be any restrictions as to admission. I would fain cherish the hope that those present on Sunday morning who have hitherto done any work for Christ, went away with a deeper determination to devote themselves to it, inspired by the stirring words of Mr. Moody, and constrained by the heart-melting tones of Mr. Sankey's sacred songs.

The whole audience joined heartily at the commencement, in singing "Hold the fort," an evident proof that the hymns used at these services have now become almost household possessions. Then Mr. Sankey sang, amid the utmost silence, the rousing hymn, "Here am I, send me."

Mr. Moody's address was directed specially to workers, and was well fitted to awaken the slumbering energies of the Church. Mr. Moody is very careful in his addresses to lose his personality in his theme, but the characteristics of Christian workers, on which he insisted, are all remarkably apparent in his own character. They were "courage," "love," and "enthusiasm," and one could not fail to be impressed with the notion that he was speaking the things that in his inmost soul he knew and acted out. His wonderful magnetic power was shown when he related some of his oft-told illustrations, which seemed to lose none of their wonted effect by repetition. Numbers of the local clergy and ministers were on the platform.

Half-past two was the specified hour for the afternoon service in the Town Hall, but I believe the building was surrounded by crowds waiting admission about mid-day; and when I reached the hall, some time before the hour, ingress was almost impossible. At the church which stands opposite, the ordinary service was going on, and it too was speedily filled with disappointed crowds, while hundreds went away. I succeeded in getting into the hall with much difficulty, just as Mr. Sankey was about to sing for a closing hymn, "The Ninety and nine." His few touching words before he commenced to sing, and the pleading tones of his rich, full voice, as he sang of the lost one brought back at such a terrible cost, evidently moved and thrilled many hearts, and after the benediction was pronounced, everybody seemed unwilling to depart.

Mr. Moody's theme, I learned, was "the old, old story" of the cross—the "good news;" and its effect may be judged from a remark made to me at the close of the meeting—by a Methodist local preacher and class leader, who, he said, had been converted thirty-five years—that he had never seen such a service in Birmingham before.

After the audience had slowly filtered out, a large number of people who had been unable to gain admittance, rushed in, but as there was nobody apparently appointed to speak in such an emergency, and Messrs. Moody and Sankey had gone, they were obliged to retire. One roughly-clad man, to all appearance a common laborer, who had come in after the meeting was over, seemed much disappointed. He had walked, in the rain, nearly six miles, in order to hear Messrs. Moody and Sankey, and arrived too late to gain an entrance. He said he had to walk back again and preach the same evening. He was somewhat relieved when he succeeded in obtaining a ticket for the workers' meeting next Sunday morning; but I suspect he will have to start from home before Birmingham is awake, if he is to make sure of getting inside the Town Hall.

Such a gathering has seldom, if ever, been seen in this town, as was witnessed in the Bingley Hall on Sunday evening. Birmingham has the reputation of being a hot-bed of political agitation; and on one occasion, I am informed, this stupendous building was filled to overflowing to hear John Bright; but it is a new thing for it to be crowded with 10,000 souls to hear the grospel preached and sung. There must have been at least that number inside the doors, and how many were excluded I cannot say; but the service was somewhat disturbed ever and anon by the clamoring multitude outside knocking at the doors for admission. It was a sight truly gladdening to behold, and never to be forgotten.

For an hour or so before Messrs. Moody and Sankey arrived, the time was occupied in singing hymns, and as soon as they reached the platform, Mr. Moody asked all to join in singing the doxology, "Praise God from whom all blessings flow." It was repeated at his request, with a more overpowering volume of sound than before.

Mr. Sankey sang "Jesus of Nazareth passeth by," and subsequently, "The Ninety and nine," and his voice rang through the immense building with wonderful effect.

Mr. Moody delivered a powerful and affectionate address on "The Gospel," in continuation of his afternoon's address on the same subject. He seemed as if he could never tire of dilating on the absolute freeness and fullness of the offer of salvation, and his illustrations, as usual, were very telling and appropriate.

Altogether it has been a memorable day in Birmingham. At none of the meetings, however, was any provision apparently made for inquirers; but this will no doubt be seen to as the necessity arises, which I trust and believe it will.

Mr. Moody announced that there would be services in Bingley Hall every evening except Saturday, at half-past seven o'clock, and a noon prayer-meeting every day in the Town Hall.

Thus the movement has taken root in Birmingham, and great and glorious results may be confidently expected.

II.

NEVER before in Birmingham have any preachers drawn such vast numbers of people as these brethren are doing at this time. Thousands are flocking daily to hear them from the districts around. The whole community seems stirred up. That which seems to be uppermost in men's minds, is the present marvelous gatherings that are daily taking place. There is no lack of opportunity for the Christian to put in a word for the Master, for wherever you go, whether in the counting-house, shop, refreshment-room, train, omnibus, and even as you walk along the street, the one topic is the doings of these wonderful men of God. If you want to get a seat at their meetings, you must be there fully one hour before the time, and a stranger entering the town must be struck with the determination of those who daily seek these gatherings.

Every day this week hundreds have been turned away from the noon-day meetings held in the Town Hall. Meetings are now being held in Carr's Lane Chapel every afternoon at three o'clock, and here again it is necessary to be there some time before the service commences. In fact, yesterday I was there at two o'clock, and the body of the chapel was then filled. It is estimated that three thousand people are in this building every afternoon.

To convey to the mind of the reader the sight which presents itself on entering Bingley Hall (the place of evening meeting) is impossible. Sloping down from the galleries which run round the building, other galleries have been erected, and the whole building, from the speaker's platform, looks like one vast amphitheatre. The crimson cloth which drapes the galleries adds to the general effect, and makes the hall (said to be one of the dreariest-looking buildings in the Midland counties) look very comfortable. The immense sea of faces is singularly impressive, especially when from 12,000 to 15,000 people are listening eagerly to catch the words that fall from the speaker's lips.

The question may be asked, What effect is this movement having upon the people in general? I reply, good every way. The stirring addresses given by Mr. Moody to Christians from the very first morning, are bearing fruit. They are beginning to look about, and realize that thousands around them are living without Christ. Many Christians have spoken to me of the fresh energy with which they have been stimulated, through attending the meetings. As for those who nightly throng Bingley Hall, the best test of the work I can give is, that whereas at first the after-meetings were held in a neighboring church, the anxious ones have now become so numerous, that they are obliged to remain in the hall, while earnest Christian workers, with Bible in hand, pass from one to another, and open to inquirers the way of life.

All this proves to us the great power of God, and what He can do by two men who give themselves wholly up to Him. The work "is marvelous in our eyes," but it is not less marvelous that their physical strength does not give way under their unceasing labors. While Mr. Sankey is greatly gifted with power to use his voice in singing the gospel, Mr. Moody has a way of marvelously picturing, in the most vivid manner, Bible truths. From the humorous he can come down to the pathetic, and so move his hearers to tears, and withal there is a "holy boldness" which is seldom to be met with in the preachers of the present day. May the Lord bless abundantly the efforts of these men, who have produced such an unusual and powerful effect upon Birmingham.

III.

THE *Morning News* says: "Never before in the history of Birmingham, I believe, have two men drawn such large numbers of people together as Messrs. Moody and Sankey have done, time after time, during the whole of last week and yesterday. The Town Hall, Carr's Lane Chapel, and Bingley Hall have been entirely filled at most of their meetings, uncomfortably crowded at some, and all but full at one or two others. Since commencing their labors here, they have held twenty-two services, namely, four in Carr's Lane Chapel, six in the Town Hall, and twelve in Bingley Hall. No doubt in many cases the same persons presented themselves at the meetings again and again; but it is probable that the audiences were, for the most part, different on each occasion. At the four meetings in Carr's Lane Chapel some 12,000, at the six in the Town Hall about 24,000, and at the twelve in Bingley Hall, at least 120,000 persons must have been present, making a total of 156,000 men, women, and children, to whom, during the last eight days, they have preached and sung the gospel. Nor does the interest in the men and their work as yet know any abatement, it being likely that the services to be held this week will be as numerously attended as those of last week."

IV.

AMIDST all the cavil of unbelief, and other opponents, thousands can testify, day by day, to the *reality* and *power, widely* spreading and *deepening* blessing upon their souls. Sinners have been converted to God, and believers edified. Whole congregations, both in churches and chapels, have felt its animating power. The clergy and ministers of various denominations have rejoiced together in this blessed work of the Lord, and felt its quickening influence. Many of the Lord's servants have met together for the first time, and felt their hearts drawn out in brotherly love and sympathy, enabling them to overlook various minor differences of creed and church government.

The noon-day prayer-meeting was first held in the Town Hall, which large building was filled long before the appointed hour. A very solemn and prayerful spirit seemed to pervade the masses—the stillness was quite impressive, and the great bulk of the people seemed to enter most deeply into the importance and solemnity of the occasion. The numbers at the noon-day prayer-meeting were probably quite 3,000. Afterwards it was changed to Bingley Hall, where thousands more might be accommodated.

The afternoon Bible-reading is also well attended, and greatly enjoyed by many. The evening meetings have gone on, steadily increasing, until at length I suppose some 15,000 must have been congregated together. The attention of these great masses (assembled an hour before the time) was well sustained by singing—and, as a brother clergyman said to me, on the platform, "we never heard such singing of the good Old Hundreth Psalm before, and probably may never hear the like again"—as it burst forth from the hearts and lips of this vast assemblage. Oh! it was a touching sight, and a telling sound—such as Birmingham itself had never witnessed before—15,000 met together, night after night, to listen to the loving, sympathizing, fervent preaching of JESUS CHRIST, the Saviour of sinners! And the audience felt it! The Holy Spirit of God seemed working in our midst—alike on preacher and hearers—and many were the hearts moved.

At 7:30 Messrs. Moody and Sankey entered the building. The service began by singing, then prayer was offered, another hymn or two were sung, a portion of Holy Scripture read, another hymn, and then followed the address. Numerous anecdotes were related, as if not only to illustrate certain points, but also to rivet the attention, and then, as the preacher's heart and tongue seemed set on fire, all these little adjuncts were submerged in the one glowing, burning theme—salvation for lost sinners—yea, a present and immediate salvation for every one that believeth in Jesus! As I sat near the preacher, I could read the meaning of the big drops upon his brow, and how his whole frame was moved, not with selfish passions, seeking personal admiration, but steeped in the love and the spirit of his Master. One great object was kept steadily in view—the glory of God in the salvation of sinners through Jesus Christ, and the intense longing that thousands might share with him the blessings and the joys of THIS GREAT SALVATION! Almost breathless stillness chained the audience.

Numbers stayed for the after-meetings; the females in the side galleries, the males in the Scotch Church adjoining. On the first Monday evening Mr. Moody himself undertook the men, but finding the numbers so large, he sent up to the platform for assistance. Undoubtedly personal interviews are the best.

We have reason to believe that many found pardon and peace in Jesus, and are spreading their happy and holy influences around. The singing appeared to be improving night after night, as the vast masses gradually learned the tunes and hymns. Mr. Sankey's solos were powerfully and sweetly sung, and his clear utterance and distinct enunciation of syllable after syllable gave a great effect and pathos to the whole.

Many of my own people are deeply interested, and though our parish is a suburb four miles from Birmingham, numbers continually attend; rich and poor seem thoroughly to appreciate and enjoy it. I have also noticed clergymen coming in from all parts.

And on Tuesday, Jan. 26, the day of the convention, it was supposed that from one to two thousand ministers of various denominations attended the gathering, which began that day at ten o'clock and continued till four P. M. Truly it was a great evidence of the divine blessing, as the delegates from Edinburgh, and Dublin, and other cities, told how the work was still progressing in their respective cities, after Messrs. Moody and Sankey had left, and in some places ripening in a most marvelous manner. Indeed, a letter reached me only yesterday, telling me of a brother clergyman in Dublin, who had a list of *sixty* persons in his congregation, who had apparently been brought to Christ through attending the meetings of Messrs. Moody and Sankey.

Verily the Lord is blessing the evangelistic labors of our dear brothers in Christ—Moody and Sankey. I do not pretend to endorse every utterance, or to see with them exactly, eye to eye, on every point. But I do see, and I do greatly rejoice in their being raised up by God to proclaim, so touchingly, and so successfully, the utter ruin of sinful, fallen man, and his recovery solely through FAITH in JESUS CHRIST!

V.

THE ALL-DAY CONVENTION

on Tuesday was in every way a successful meeting. It was attended by immense crowds throughout the day, and many well-known ministers and others were present from London and various towns in the provinces, as well as Scotland and Ireland. Mr. Moody presided throughout the day with his usual tact and energy.

The first hour was fitly devoted to praise, and Mr. Sankey's opening address was followed by powerful testimony to the value of the services by our brethren in Scotland and Ireland. All the speakers concurred in saying that a new song had been put into their mouths.

Mr. Moody occupied the next hour with an address on "Work;" and his trenchant words, uttered in the presence of so many Christian workers, were potent with blessing, in stimulating them to do more than ever for the Master in their widely-separated vineyards.

"How to conduct Prayer-meetings" was the next topic, and a most important one it is. We cannot better describe many of the prayer-meetings we have been accustomed to attend in past years than by comparing them to "wet blankets." They have been characterized by so much frigidity and routine, that we do not wonder the attendance has mostly been small. Mr. Moody will have done us British Christians a great and lasting service if he has been enabled to show how our prayer-meetings may be made broad and deep channels of blessing and happiness both to Christians and the careless world round about us. We look for this result.

More important, perhaps, was the subject of the next hour, "How to reach the masses." Whoever will solve that problem will earn the unspeakable gratitude of all who sigh for the conversion of the nations to Christ. The rousing addresses of Mr. Chown, of Bradford; Mr. Newman Hall, of London; Mr. R. W. Dale, of Birmingham; Mr. Fletcher, of Dublin, and others, all men of large experience, will, we trust, have contributed somewhat to this desired end.

Mr. Moody was as practical as ever in his answers to the questions sent in; and, if those who sent them will only apply those answers, we are inclined to think the hour devoted to the "Question Drawer" will be the most fruitful of any.

In the evening a public service was held in the same place; hundreds were unable to gain admission. The Rev. Newman Hall, of London, delivered an address, earnestly entreating all present to forsake sin and come to Christ. Mr. Moody, in his discourse, urged on his hearers immediate decision for Christ.

Mr. Sankey's singing of sacred songs seems to make a deep impression upon the great congregation.

At the meeting in Bingley Hall on Friday evening, Mr. Moody said: I was very dejected last night. Our meetings have been so much blessed that an effort was put forth to get Bingley Hall for another week. When we got home last evening, we found a dispatch from a gentleman, saying we could not have the hall. I was greatly depressed all day. Now, however, I have just been told we may yet obtain the hall for another week. But the committee are wavering a little, as they have some fears the people will not come out to the meetings next week. We have good committees wherever we have been; but we have never had a better committee than the Birmingham one, and I know they will come to a wise decision. But if you are anxious about your souls, you'll attend the meetings. We'll get several gentlemen to speak, and we hope you'll rally round them and the committee. We have had great blessings in other towns; but I think we never met with anything that came up to this—to our meetings in Birmingham. I must say I've never enjoyed preaching the gospel more than I have done since we came to Birmingham. We've reached so many people. I only wish we could have such a hall wherever we go. I think if we could only take up Bingley Hall, we would carry it round the world with us, as a place in which to preach the gospel to all men. But I would like you Birmingham people to go with us. Well, then, if we do our best to get speakers for another week, will you do your best to get hearers for the speakers?—(Many cries of "Yes," "yes.") Well, keep your promise. Why, almost any man could speak in this hall to such a meeting as this. The very sight of you is enough to make a dumb dog bark. I'll telegraph off to Liverpool and London to send us all the help they can. There will be a service on Sunday afternoon, when one of your own ministers will preach. On Monday night you'll have a thanksgiving service. Come to it to thank God for having answered our prayers to bless these meetings. Has God not answered your prayers?—(Cries of "Yes," yes.") Then on Tuesday we'll get some one else to speak. On Wednesday there will be the usual services in the churches and chapels. On Thursday night there will be another speaker. On Friday I will come back, on my way to Liverpool, and we'll have a meeting for all the converts. Now, let all rise who will support the committee and attend the different meetings. [Almost the entire audience stood up in response to this appeal.] Yes; the committee are quite satisfied. We'll go on then. Pray there may be hundreds and thousands converted next week. If things do not always please you, don't complain; just pray. Pray for a great blessing next week.

VI.

THE FAREWELL MEETING.

SERVICES were held in Bingley Hall last week as follows: On Sunday afternoon Mr. Sankey conducted, and Rev. F. Gallaway, and Messrs. J. B. Gould (American Consul), and W. H. Greening (a member of Birmingham School Board), delivered addresses. On Monday night Mr. R. W. Dale, M.A., and the Rev. Newman Hall (London), were the speakers; on Tuesday, the Rev. Newman Hall; on Wednesday, the Rev. Donald Fraser (the well-known Presbyterian minister of London); and last (Thursday) evening, Mr. W. P. Lockhart, Liverpool. All these meetings have been well attended, from 5,000 to 7,000 persons having been present at each.

The Birmingham *Morning News* gives a long report of two services held there by Mr. William Nobbs, the converted policeman, and says: "Plain and unaffected, disdaining alike oddities of gesture, mannerisms of speech, or adventitious, and it might be added meretricious, helps of all kinds, William Nobbs, with the same quiet force with which he might have narrated his evidence in a police-court, conducts his sermon. He has something important to say, and he says it in a clear, straightforward manner. To him the Judge is always present. The little Bible he holds in his hand is grasped much after the same fashion as if he had just sworn to speak 'the whole truth, the sole truth, and nothing but the truth,' before a jury."

Messrs. Moody and Sankey's farewell service at Birmingham was held on Friday evening, when the Bingley Hall was once more crowded to its utmost, nearly 1,600 converts' tickets being applied for. It would be manifestly premature to assert that this number of people have been converted during the previous three weeks' services. As Mr. Moody said at the Conference in London, on the same day, they did not desire to reckon up the number of converts, because they could not judge of the reality of the cases. At the same time we think it very probable that many have been brought savingly to believe in Christ who did not apply for converts' tickets. In any case, the progress of the movement in Birmingham has been such as greatly to encourage and cheer our American brethren and those who helped them in their labors; and we respond to Mr. Moody's hope that it may "continue for a year."

Mr. Moody's address to the converts was, as usual, most fitting. Pointing them to the true and only source of abiding strength, he warned them of the dangers ahead; and we hope his words will dwell in their hearts for many a year to come. His parting sentences were the expression of affectionate regard, and it was plain, from the demeanor of the audience, that the parting on their side was a most reluctant one.

Mr. Sankey sang the farewell hymn with great pathos and feeling; and on leaving the hall both he and Mr. Moody were beseiged with friends anxious to receive a parting shake of the hand. They proceeded to Liverpool on Saturday.

A CORRESPONDENT writes concerning this meeting: "We shall never forget that address." Such was the almost involuntary exclamation of a well-dressed mechanic who was standing by us in the aisle of Bingley Hall. And truly the work of the Lord in this town is such as has never before been seen here. We were praying and expecting great things, but the blessing has exceeded our expectations; never before have the people of every class been so moved and such glorious results followed. A week having elapsed since Mr. Moody left us, we are enabled to speak in a measure of results. First, the life of the ministers who have taken part has been largely increased, so that the testimony of many of the hearers last Sunday was, "Our minister preaches like a new man;" then the renewed life of the churches is already manifesting itself in the desire to work either in Sunday-schools or tract districts; and besides this the people outside are more disposed to hear the gospel, many coming into our churches last Sunday, and in more cases than one when notice was given out after the service that inquirers would be spoken to, numbers varying from twenty to sixty passed into the vestry, and many rejoiced in a new-found Saviour. In our own chapel we have had services nightly, conducted by our brother, Mr. Nobbs, of Gloucester, and such a gracious time we never knew before, on each occasion the chapel being crowded, and many coming after into the vestry and going home happy in Jesus. Our hearts are indeed full of praise; should we be silent the stones might well cry out, "But we will bless the Lord from this time forth, and forevermore."

VII.

I venture to follow my letter of a fortnight since with a second upon the work of God through Mr. Moody's faithful uplifting of Christ, and Mr. Sankey's singing of the songs of peace. I know of no one of the many blessed hymns which has more struck the heart and arrested attention than that sweet one whose chorus begins, "Oh, 'twas love, 'twas wondrous love, the love of God to me." This love and its manifestation is the theme of every sermon, and, of course, God owns it. Ministers wonder at failure, and try to discover the cause; a week of services such as Birmingham has had for the last fortnight, I think must answer the question, "What is the cause of failure?" for we have seen in the crowded meetings, in the overwhelming number of anxious ones, in the utter breaking down of strong men, the secret of success. The wondrous love of God has been the weapon which has been used; failure in using this weapon has been the cause of failure in result. Never has Birmingham been so mightily moved; in the workshops Sankey's songs are sung, and men who cared for none of these things are anxiously inquiring after the good news. Oh, may our God carry on the work begun with mighty power.

VIII.

I must write you a few lines, according to promise, of what the impressions made on my mind are by the few days spent in Birmingham in connection with the work of our friends, Messrs. Moody and Sankey.

The first thing I will notice is the evidence the work gives of the power of the simple preaching of the cross of Christ. The singing, no doubt, attracts many; but the power is not in the singing, but in what is sung. It is the old story of a Saviour's love to a lost world that thrills the hearts of those who hear it sung, and that helps to swell the vast audience that daily, and often twice a day, and on Sundays four times a day, draws together vaster crowds than, perhaps, were ever drawn together consecutively for any political or scientific object.

Christ meets every need, heals every disease, and wipes away every tear; and it is the simple declaration of this that gathers the crowds now, as it did nineteen centuries ago on the mountains and plains of Galilee. The infidel is prone to say that Christianity is dead, and does not meet the need of the age. These meetings show that it is alive, and still meets the need of all who come to the healing stream. As in the days of our Lord's personal ministry thousands followed who went away, and were none the better, so it is now. But when hundreds stand up at the close of an earnest, simple gospel address, as we have witnessed at these meetings, desiring to be prayed for, and wishing to become Christians, there is a power from God which it must rejoice every true-hearted child of His to witness or to hear of.

The next thing I would remark on is the directness of the preaching of Christ, the lifting up of the Lord himself, that characterizes the ministry. Dogmatic teaching will not serve; we need a personal Christ, brought near to the personal need of the sinner. And this is a marked feature of the work, and one which all engaged in the preaching of the gospel will do well to ponder, and see whether the secret of the general want of power felt be not that theories of truth have taken the place of the living truth itself. Mr. Moody lifts up a living Christ, and makes his hearers almost think they hear His footfall, and listen to His voice. Christ lives in his ministry.

Firstly. Christians are led to realize the fact that they are, firstly and chiefly fellow-disciples of the same Lord, having in the gospel by which they are saved, and which they preach, a common ground of evangelistic effort, which, while it does not call for a compromise of the smallest truth, holds paramount the living truth of a common Saviour and a common salvation. This is no small gain in this day of narrow sectarianism and Pharisaic exclusiveness. May all be willing to say, as our brother Moody said on one occasion, "If I have one drop of sectarian blood in my veins, I would gladly pour it out!"

Secondly. An uncompromising testimony against a worldly Christianity that would seek to take as much of this world as would not absolutely rob the soul of the world to come. Unflinchingly our brother exposes the shame and sorrow of a world-bordering life. He would give his Lord all, under the influence of the grace of the all that has been received; and this in a manner that will be very simple and very new to many of his hearers, who may have been Christians like Lot in Sodom, almost, perhaps, ever since they were converted. We shall not readily forget the gold chain that had become so tarnished in a man's pocket, from contact with some lead there, hat he was giving it back, thinking it a fraud. And Mr. Moody added, "The gold will be tarnished by the lead, but the lead will not take up the gold.' The world draws down one that seeks its fellowship, but is not drawn up by him.

Thirdly. We would notice the great prominence given to working for Christ. He remembers the work, "To *every* man HIS work," and, in the deep earnestness of his own conviction of the joy of working for such a Master, would he press it on all, and lead many to find in it a remedy for coldness of heart and lifelessness of walk, and many a life-long sorrow.

Mr. Moody's addresses to Christian workers are full of plain, common-sense directions, wherein is no mystery, and which commend themselves to the judgment of all faithful men. There may be a little danger of a forgetfulness of that communion with God which can alone give work a tone and character which will make it what the work of Him was who spent His days in labor, and His nights in communion with God. This element is, perhaps, a little wanting, but to most the call is sorely needed, "Rise, and be doing;" "Work, for the night is coming." May that little song, "Nothing but leaves," linger long in the hearts of all who heard it sung, and lead to many *sheaves* and fewer *leaves*.

Fourthly. I think we see that there *is* a power in the gospel to reach the masses. Hard and careless many are, but there are soft moments in the hardest heart, and anxious forebodings not unfrequently fill the most careless soul. There are hours and days, weeks and months, of divine visitation, when many will gladly listen, if a fitting person and a fitting place be provided. They know not what they want; they know not where to find it; but we, who know the sinner's want, and know where it is to be met, are responsible for so presenting it, and for so providing for its presentation, that the subject and the place shall fall within the reach of needy and often thirsting souls.

After much had been said about reaching the masses, Mr. Moody wound the question up in these few words, "Go and fetch them."

But may God keep His people from seeking to reach those outside by leveling up and leveling down, as if to meet God and the world half-way. Let the heights and the depths remain, as they ever will, in fact, remain, and seek the power of the invisible, almighty arm that lifts out of the one, and places the sinner's feet firm on the other, in one mighty act, that lifts from the dust to the throne. There are no steps here.

The ADDRESSES *and* LECTURES *in this Edition are reprinted from Verbatim Reports.*

OF

D. L. MOODY,

WITH A

NARRATIVE OF THE AWAKENING

IN

LIVERPOOL AND LONDON.

(*SUPPLEMENTARY ISSUE.*)

NEW YORK:
ANSON D. F. RANDOLPH & COMPANY,
770 BROADWAY.

CONTENTS.

PART II.

THE WORK IN ENGLAND.

LIVERPOOL:

    I. LIVERPOOL MOVED, ... 1
    II. OPPOSITION AND SCORN, ... 2
    III. THE PRAYER OF THE EVANGELISTS, . . 3
    IV. GLORIOUS WORK, ... 4
    V. PHYSICAL AND MENTAL EFFORT, . . . 7
    VI. GATHERINGS IN THE CIRCUS, . . . 8
    VII. CLOSING ADDRESSES, ... 10

LONDON:

    I. THE BUILDINGS, ... 11
    II. OPENING SERVICES.—SCENES OUTSIDE, . . 12
    III. THE FIRST MONTH IN LONDON, . . . 22
    IV. LETTER OF REV. R. W. DALE, . . . 24

PROGRESS AND CONCLUSION OF THE WORK IN ENGLAND:

    I. INTRODUCTORY REMARKS, ... 36
    II. NOON MEETINGS, VICTORIA THEATRE, . . 37
    III. CAMBERWELL-GREEN HALL, ... 40
    IV. CLOSING MEETING AT BOW-ROAD HALL, . 43
    V. FAREWELL AND THANKSGIVING MEETING, . 44
    VI. THE FAREWELL MEETING AT LONDON, . . 45
    VII. IN NORTH WALES, ... 48
    VIII. IN LIVERPOOL—FAREWELL TO GREAT BRITAIN, 51
    IX. THE DEPARTURE, ... 57
    X. "TWO AND TWO," 58
    XI. IMPRESSIONS OF THE RECENT REVIVALS IN EDINBURGH, 60
    XII. MEETINGS AND ATTENDANCE, . . . 61

SERMONS AND ADDRESSES:

    I. GOD'S HUMAN INSTRUMENTS, . . . 62
    II. CHRIST SEEKING THE LOST, . . . 67
    III. SAVED OR LOST, ... 72
    IV. MAN SEEKING FOR GOD, ... 77
    V. THE CALL TO SELF-EXAMINATION, . . 82
    VI. THE NEW BIRTH, ... 88
    VII. A SERMON ON ONE WORD, . . . 93
    VIII. THE MASTER'S PARTING COMMISSION, . . 100
    IX. POPULAR PRESENT-DAY EXCUSES, . . 105
    X. ABOUT HEAVEN, ... 112
    XI. THE BLOOD, ... 117

NOTE.

I.

THE present Publishers issued in March last, a NARRATIVE OF Messrs. MOODY & SANKEY'S Labors in Scotland and Ireland, and also in Manchester, Sheffield, and Birmingham, England.

This supplementary issue, besides containing in full the article of the Rev. Mr. DALE, originally printed in the *Congregational Quarterly*, and which is altogether the best analysis and review of the character and labors of the Evangelists which has yet appeared, also sketches the work in Liverpool, and its progress in London, until the close of the first month. As the work in London has been conducted substantially in the same manner as elsewhere, it was thought that in the place of a continued narration, the American reader would prefer to possess in a permanent form a full report of some of the Addresses and Lectures of Mr. MOODY. Those included in this volume were delivered in London during the first two weeks of March, and are the verbatim reports of the (London) *Christian World*.

MAY, 1875.

NOTE.

II.

THE additions now made to the issue of May include an account of the last days in London, and the final service at Liverpool.

OCTOBER, 1875.

THE WORK IN ENGLAND.

LIVERPOOL.

The labors of Messrs. Moody and Sankey at Liverpool began on Sunday, Feb. 7th, and closed on the 7th of March. For their accommodation, Victoria Hall, a wooden structure of enormous strength, had been erected. The internal dimensions of the building are 174 feet long by 124 feet wide, with a seating capacity for 8,000 persons. The wide passages rendered the capacity of the building ample for 11,000 hearers. Inquirers rooms adjoined the platform. The building was constructed within forty days, and cost £3,500.

The Friday preceding the arrival was observed as a day of preparation on the part of many of the churches, and the first meeting of the Evangelists was on Sunday morning, at eight o'clock, for Christian workers. This was followed by the afternoon and evening meetings, when the work was begun, and subsequently carried on in the same manner as elsewhere. It is not necessary, therefore, to detail it here. The following letters will sufficiently indicate its progress and the state of public feeling in regard to it.

I.

As of the apostles of our Lord it was said, "These men, who have turned the world upside down, are come hither also," so of our brethren, Moody and Sankey. It is joyful to see that the success which attended their self-denying labors elsewhere, promises to be equaled, if not surpassed, in Liverpool.

All Liverpool is moved by them; but as yet, not all with the most desirable feelings. Some seem actuated by a spirit of embittered hostility, and do not hesitate to write and speak of these servants of Christ, what has not the shadow of truth. This very opposition is, however, doing good. God makes "the wrath of men to praise Him." I have known of some who entered Victoria Hall bitter enemies, and left it attached friends to the movement. Many flock to the meetings, apparently from idle curiosity, and thousands under spiritual anxiety, whilst God's people rally round the evangelists with an enthusiasm and hearty good-will which is cheering to observe. Mr. Moody seems to have lost none of his popularity in address. His style is plain, peculiar, and strikingly forcible. No sameness of figure or phraseology, but new thoughts, new subjects, and new illustrations, producing deeper impressions, and drawing greater crowds than the one preceding it. There is no other satisfactory reason which can be assigned for his unprecedented popularity, but that the mighty power of God is with him.

The good work in his and his brother Sankey's hand seems like the noble river pouring its waters down the mountain, reaching the valley, deepened, widened, and expanding itself by the numerous tributaries which join it, it flows on with irresistible majesty, bearing before it every barrier of man, and yet not a ripple on its placid surface.

Perhaps, of the marvelous work of God's grace going on in our midst, the "after-meetings," or "inquiry-room," is the most interesting and remarkable. Here are found representatives of all ages, from the very young, of only ten years, to the aged. All classes of character are discovered there, from the virtuous and moral, to the regardless and abandoned.

"Can such a wretch as I be saved?" was the question asked me by a seaman whom I knew twelve years since, the son of a humble, but truly pious man. Oft had he wept over his thoughtless son, but faith never failed him. He still said James will be all right yet. What was my joy may be conceived, but cannot be expressed, when the above question, "Can such a wretch as I be saved?" was asked, and a strong seaman caught my hand.

"Yes, James, you can, and God is this night answering the prayers of your dear good father."

"But, sir, I am such a sinner."

"No matter, if every sin you have committed was as aggravated as all the sins of your life combined, and that you committed ten thousand sins, for every one of which you are guilty, the Blood of Jesus Christ, God's Son, can cleanse from all sin."

(1)

"Oh, but I am so ashamed, I cannot look up." Such was the feeling of the poor prodigal; but the loving Father was not ashamed to take His sinful son to His heart and home.

"Will you give yourself to Christ?"

"I will," was his reply; "and I go to sea to-morrow, sailing for America as I never did before."

THE SCOFFING INFIDEL BROUGHT TO GOD BY A HYMN.

Such I have known to be the case. At last Monday evening's meeting, an intelligent young man informed me he came into that hall to scoff at all he heard. "I believed only in God and the devil; the latter I served well, and as sitting laughing at the fools (as I then thought) about me, that beautiful hymn, 'Safe in the arms of Jesus,' was sung. A sudden thrill passed through my whole frame, and then like a dart ran through my very heart. My feelings were awful, but I listened to the next verse, and felt there is a Saviour. Who is He? Where is He? Instantly I realized the truth, Jesus is the Saviour. I threw myself into His loving arms, and here I am now, rejoicing in Him."

"Blessed be God," I said, "for such news. Now, brother, go home and tell your friends what great things God hath done for your soul."

"Will you pray?" he said.

We went together to the throne, and then he said, "God bless you. I will now live and work for Jesus."

The devil lays his plans, and no doubt thinks they are well arranged, but whilst he proposes certain events, God disposes of them in a very different way than Satan expected. He works by his servants, as God works by *sanctified souls*.

THE DECEIVER DISAPPOINTED.

Of this I have had an instance.

"I am under a dreadful temptation," said a young man to me.

"What is it?" I asked.

"I was given drink by a man professing to be a Christian, and whom I have heard preaching the truth to me and others, but who is opposed to Moody and Sankey, and I was sent here by him to give annoyance. Now I am brought to Christ, in place of dishonoring Christ in this meeting, what am I to do to this man?"

"Pray for him," I said, "and God will give him to you as a star for your crown. Tell him plainly his state, and bring him here with you next night."

"I knew a lady who went to a religious meeting an avowed infidel, sent there by two sisters-in-law for a similar purpose to that which brought you this night here. She was brought to Christ, and sent back to them full of Jesus, and was the means of their saving conversion; and now all three are rejoicing in the great salvation effected by Jesus, the Son of God, for every penitent, believing child of Adam."

Truly the Lord is doing great things for us, "whereof we are glad." But faith induces us to expect still greater blessings, feeling that we have only yet got the first drops of the showers. We look for the latter rain, praying that Liverpool may be made in every home the habitation of righteousness, that the Word of God may be in every hand, and its precious truths written on every heart, and holiness to God may characterize all its inhabitants."

II.

It may emphatically be said of them, "They came, they spoke, they conquered." For twenty years I have been more or less mixed up with the evangelistic work of the town, but never have I met with more opposition and scorn to any movement than the present.

The erection of the vast hall to hold 10,000 persons, was looked on as monstrous folly. As it was being built, the talk was, To what purpose is this waste? But now what was called Moody's folly, is seen to be God's wisdom.

Men who wrote, spoke against, and laughed at it, now speak with bated breath, come and hear, and go with changed thoughts. "Nothing succeeds like success," is an old world's adage, and in this is proved to be true:—6,000 at a midday prayer-meeting; 6,000 at the afternoon Bible-lecture; 10,000 at the evening meeting, with the inquiry-rooms full, are something that even the Exchange has to admit. But beyond this, there is the mighty working power of God's Spirit working and acting, which no tables can register or numbers record. "'Tis not by might, nor by power, but by My Spirit," was the key-note of the preparatory meetings, which has been steadily kept before all the workers.

Looked at in a plain matter-of-fact light, we ask, What brings the people together?

Preaching, teaching, singing, can be heard, more artistic and eloquent, almost anywhere, we are forced to admit. It is the power of the Spirit in making plain words burn and simple singing touch.

The part allotted to me in the great work has enabled me to see and test much that is going on. And this I can say—there is wheat; there is chaff. The wheat is sound, and will be a glorious, bountiful harvest. The chaff will be blown away. Wheat and chaff always grow together. Never have we been privileged to see so much real, genuine work—anxious faces, tearful eyes, aching hearts. The general feeling is that souls are being born again, even though they have not courage to stand up or walk into the inquiry-rooms.

Last Thursday night, Mr. Moody, after a telling address, went into the inquiry-room, and his place was occupied by a layman, who wielded the sword of the Spirit with amazing power right and left. His words, powerful and well-chosen, fell with force, and told on the vast audience that seemed spellbound. Many seemed to be convicted of sin, and hurried into the inquiry-room.

Liverpool needs the prayers of the Christians of Great Britain at this crisis. Every arrangement that man can make has been made for the well carrying on of the work. But what we fear are unwise helpers and the Sanballats who would come in and mar the work.

III.

"LET Thy work appear unto Thy servants, and Thy glory unto their children. Establish Thou the work of our hands upon us; yea, the work of our hands establish Thou it."

Such has been the prayer of our beloved friends, Messrs. Moody and Sankey, and those who have supported them in their work here for the last ten days, and the Lord has heard and answered them of a truth. In some places which our brethren have visited, the fruits have begun to appear almost as soon as the seed was sown. In others, and notably Dublin, the apparent results were some time delayed, though the blessing of many conversions to God was afterwards abundantly given. In Liverpool, already, the issue of the meetings has been most encouraging.

In the opening services, Mr. Moody remarked that many people thought the Victoria Hall was a bad investment, but that, if souls were born there, perhaps some of them would like to have a little stock in it.

On the evening of Monday week, the first opportunity was given for anxious inquirers to be spoken with. Mr. Moody had just delivered a most heart-searching address, in which he had shown man's unutterably lost condition without Christ, and many refuges of lies had been laid bare. But he did not leave the sinner there. With all simplicity and affectionate earnestness, he held up a crucified Saviour, and once again it was shown that the good, old-fashioned gospel—stripped of all rhetorical dead-weights and conventionalisms that often prove hindrances instead of helps—had not lost its power. Mr. Moody's earnest invitation to those who were anxious about their salvation, to stand up, and afterwards to meet him in the inquiry-room, was responded to by hundreds, who were not deterred from showing their anxiety by the curious gaze of many thousand spectators.

Many striking instances of conversion have occurred, and other cases have come under my own observation in which backsliders have been led to return to their first love. One day at the noon prayer-meeting Mr. Moody told of an interesting case of conversion he had met the night before. A young, stalwart man, who was to sail for America next day, had come into the meeting. He had been pricked to the heart by Mr. Moody's pointed appeals, and found his way to the inquiry-room, and here, as he believed, to lay his hitherto unforgiven sins on Jesus. Later in the evening he called on Mr. Moody at his hotel, and received a letter of introduction to any of the Christian friends in America he might meet. He was accompanied to the hotel by his brother, who had come from the country with him to see him sail, and who seemed overjoyed to think that one so nearly related to him was taking Christ with him ere he left his native shores.

The experience of Monday evening has been repeated every evening since, more or less, and I have not the least doubt but great numbers have been led to see and feel their lost condition, and to cast their all on Christ, who have not openly avowed it. We may reasonably suppose that curiosity has mostly passed away, yet it is no easy task sometimes to induce the vast audiences who throng into every corner of Victoria Hall, to leave, after the benediction has been twice pronounced.

At the evening meetings the hall is always crowded with something like 10,000 people, and if it were not that the committee kept a great part of the passages clear to allow of access to the inquiry-room, every inch of standing ground would be occupied. The attendance at the noon prayer-meetings averages 4,000 to 5,000, the audience, of course, not being so mixed as those in the evening. One gratifying circumstance, however, in connection with the noon meeting should be noted, and that is, the presence of so many of the Liverpool merchants and business men. I have heard it stated that between twelve and one, when the noon prayer-meeting is held, 'Change is half deserted, and it has been remarked that no other source of attraction has ever drawn so many of these busy men away from their money-making for an hour in the middle of the day. May they carry away some truth that will cling to them when they are tempted to forget God in their haste to get rich! The requests for special prayer have been very numerous and so varied that it would be impossible to characterize them. The notices of the secular press, while not expressing any hearty sympathy with the movement, have been very fair and honest, as a rule, considered as simple reports of the proceedings.

A very happy feature of the work here, as elsewhere, is the sympathetic co-operation of many clergymen and ministers of various denominations. They appear on the platform and take part in the services, as well as in the personal dealing with the anxious. This is matter for thanksgiving, though some correspondents of the Liverpool papers assert that the very presence of the evangelists here, and the admitted need there is for their labor, is a slur upon their own zeal and fitness for the work of evangelizing the destitute and depraved masses of the town.

I am not seeking to defend any apathy that may have been shown by ministers or Christians generally with respect to the moral condition of this town, which by universal consent is most deplorable. But all must agree that, after the ministers and their flocks have done their utmost, there is a crying need here for special effort. And it is gratifying to know that the worst are being, to some extent, reached by the present movement. At one of the noon meetings, important testimony on this point was given by Dr. Owles, of the Liverpool Medical Mission. He stated that he had met with some few among the very poor of this town who had already been present at the meetings, and he had heard of many others who were desirous to come. In the inquiry-room on Monday night there was one little fact which was most encouraging to some of those who were laboring in the lower part of Liverpool. Among the first batch of those who were present, numbering somewhere about twenty-five, there were three well-known faces from the district of Scotland road, and each of them had brought another friend with him. During the past two evenings they had conversed with several souls. In some the impression had evidently been very slight, but in others he might say, with equal certainty, it had been very deep; and in one or two instances within his own personal knowledge souls had returned on the second evening to state that they were truly saved. He might say, too, that, on speaking further with them, the impression of one or more intelligent Christians was, that their cases were real, that they had truly distinguished between the touch of the crowd and the touch of faith, and that virtue had gone out from the Great Physician which had healed their souls. The inquiry-room, however, was no necessary test of what the real work was. Many who were naturally somewhat forward were ready to go there under very slight impression; whilst others who were more reserved, and in whose souls the work had gone far deeper, would go home, and would not dare to speak even to their nearest friends. He trusted, however, that as the work went on, very many would come with the cry, not only in their hearts, but within their lives, "What must I do to be saved?"

IV.

"THIS is glorious work; this is *reality*." Such was the remark that reached my ears one evening last week as I was passing through the inquiry-room adjoining Victoria Hall. There, I thought, is the whole movement in a nut-shell. The more I see it, and the more I ponder over it, I am impressed with the feeling of *reality* that pervades this work as it is now going on in Liverpool. Endless are the surmises, and very ludicrous some of the guesses, as to the secret of its wonderful success. The

Liverpool critics (and their name is legion) are fairly puzzled. They cannot dispute facts, though they are not always careful to ascertain what the facts really are, and seem to have a wonderful aptitude at twisting them. But there is much that they cannot help seeing and knowing, and they are at a loss to understand how two simple, common laymen have been able to do what hundreds upon hundreds of highly cultivated and refined theologians have not got within sight of. I sum it all up in the one word *reality*.

Mr. Moody has often been described, and criticised, and dissected, both by friends and foes, but I think sufficient stress has been laid on his predominating characteristic of *reality*. His gospel is the same as one hears in most places; yet it is different, because it is so real. Never mind if his weapons are not of the most orthodox kind; they accomplish the desired object all the more, perhaps, just as the youthful David's sling and stone went straight to the mark. To follow up the parallel, Mr. Moody is not content with sending his message straight to the hearts of his hearers, but he follows it up, as David did, when he completed his victory over the Philistine. As he said the other day, he pulls up his net anon to see what he has caught. This is the highest test of his reality, and the one that has evoked the greatest criticism. But it is the one that has all along contributed most to the success of the movement.

During the past week the slain of the Lord have been many. Every evening has seen fresh groups scattered over the inquiry-room, with tearful eyes and troubled hearts, drinking in the affectionate words of invitation, or the plain words of appeal, addressed to them by Mr. Moody and his co-workers. People who know least about it may affect to shrug the shoulder at the inquiry-room, but one or two visits there would do them good, and probably convince them how indispensable it is to success in this work. I hope one result of this awakening in our land will be that every minister of the gospel and every one who seeks to speak to his fellow-men about salvation, will not only cast out the net, but will draw it up every time.

The leading attraction of the meetings last week was Mr. Moody's Bible lectures. On Tuesday and Wednesday he gave two lectures on "The Blood," and on Thursday and Friday two lectures on "Heaven."

These were delivered each day at three o'clock in the afternoon, and again in the evening, so as to enable a large number of persons to attend them. On each occasion the hall was crowded; so that on a moderate computation, the seed of the word of God relating to these two most important subjects was sown in the hearts of some 60,000 or 70,000 persons, many of them from a long distance. In the words of the hymn we may ask, "What shall the harvest be?" The day shall reveal it.

The lectures are a treat of no ordinary kind. As expository discourses they are most valuable, and reveal, to some extent, how Mr. Moody has got, to use a common phrase, "the Bible at his finger ends." Probably few of his thousands of hearers ever before had such a correct estimate of the value of the doctrine of "the blood," or, as Mr. Moody calls it, "the scarlet thread" that runs through the Scriptures, like the thread that holds together a string of precious pearls. Mr. Moody traced the doctrine, from the slaying of the beasts in Eden, with whose skins God clothed our first parents, recorded in Genesis, down to the Revelation where the redeemed sing the song of Moses, and the Lamb that was slain. The lectures on "Heaven" must have left the impressions and ideas of that "prepared place for a prepared people," clearer to the minds and dearer to the hearts of the listeners than ever.

But these lectures have a wonderful hortatory as well as expository value. As Mr. Moody held up the sacrifice offered on the cross, "once for all," and dwelt on the exceeding preciousness of the blood of Christ as a sufficient atonement for sin, many a head was bowed, and many a heart melted, that had hitherto been steeled against the story of a Saviour's love. Again, when Mr. Moody, speaking on "Heaven," showed the utter worthlessness of earthly treasure when compared to the "prize" for which Paul looked and longed, the arrow of conviction went home to many a heart. His remarks on the necessity for many Christians throwing out a good deal of "ballast" before they could rise to a higher spiritual life, were, I think, very timely, and capable of application in these money-getting and money-worshipping days.

It is a gratifying fact that the attendance at the evening meetings chiefly continues to increase. During the first week of the services the Victoria Hall was almost sufficient to hold the crowds of eager listeners; at any rate, the overflow was not considered so great as to necessitate the opening of other places. Last week, however, overflow meetings were held, sometimes in two and sometimes in three different places.

One evening, I went to St. John's Church, where I found W. H. M. Aitken and the Vicar of the church conducting the service after the model of the services in Victoria Hall. The body of the church was filled partly with the overflow from the hall, and partly with those who had been induced to enter by personal solicitation, and by hearing a group of young men singing hymns in the churchyard. It was a motley company, and a great majority consisted of those who, from their dress and appearance, do not often find their way to God's house. There were numbers of men such as one sees lounging at street corners and about public-houses, many young girls in working attire and without bonnets, and a number of rough, neglected-looking street Arabs. Their behavior, with one or two exceptions, was most orderly and attentive. A good sprinkling remained at the close to be conversed with, and many of them were enabled to lay their sins on Jesus, or, as the speaker said, to accept the fact that God had laid them there nearly nineteen hundred years ago.

It is interesting and refreshing to notice how all grades of society and all ages are represented among the anxious who throng the inquiry-room at the close of Mr. Moody's addresses. From the richly-dressed lady to the poor waif of the street, with scarce enough of clothing to cover his nakedness; from the boy and girl of eight or ten years, to the horny-handed, grey-headed working-man, with all the intervening stages of life, there you find all, burdened with the same sense of sin, and afterwards rejoicing in the same Saviour. Truly, we are all one in Christ.

The noon prayer-meetings continue to be well attended, and are chiefly remarkable for the accumulated testimony that is given to the good effect of this movement in outlying towns and country districts. The meetings have been attended during the past week by large numbers of Welsh ministers and others, and with their proverbial fire and energy, these warm-hearted laborers in the Lord's vineyard, among their native hills, will become retailers of the quickening and refreshing influence they have received in Liverpool. At one of the noon meetings some most interesting accounts were given of good work among the sailors here, who had attended the Victoria Hall services.

The special work among the young men, which has been carried on in other towns where the evangelists have been, is being organized here also. On Saturday evening there was a meeting for young men, chiefly to make arrangements, at which Mr. Moody was present. In the meantime the meetings will be held in Newsome's Circus, and shortly it is expected that the Concert-room of St. George's Hall will be available.

Sunday last was another day of much sowing of the precious seed of the word, and reaping too. The early meeting for "workers" was some 8,000 strong. Mr. Moody's address was a continuation of those he had delivered on the two previous Sunday mornings—"To every man *his* work." His remarks were chiefly directed to work in the Sunday-school, in which he said the whole Church of God could be engaged. He spoke of the good that even little children could do. He would a good deal rather have a little miss some thirteen or fourteen years old to tell the other children of the love of Jesus than an old man with no fire in his heart. He enforced his appeals by some striking and appropriate incidents, of which he seems to have an inexhaustible store. He prayed that all those present "might have a passion for souls."

This service was not quite so largely attended as on the preceding Sunday, but by the time Mr. Moody's address commenced, the hall was quite full. It was a somewhat saddening thought that so many thousands of people in this town, who most of them have not the slender excuse of want of respectable clothing, should admittedly and regularly absent themselves from the public worship of God. Yet it was pleasant to think that they were so far convinced of the importance of spiritual things as to come to Victoria Hall to hear more about them. Mr. Moody simply, and in that wonderful realistic way in which he describes things, told the story of Christ's agony, betrayal, shameful maltreatment, trial, and crucifixion. The heart must have been hard indeed that could remain unmoved, and the whole congregation seemed deeply to feel the surpassing interest of the story recited by Mr. Moody. Numbers rose at his invitation, indicating their desire to become Christians, and the inquiry-room was filled at the close with those whose hearts had been touched, and who desired a sense of God's pardoning love, through the infinite merits of the Crucified One.

The afternoon meeting for women was a wonderful sight. The hall was packed to excess, and many hundreds failing to gain entrance, an overflow meeting was held in Newsome's Circus. Mr. Sankey sang the solo, "Mary Magdalene," amidst the most profound silence, and the pathetic and beautiful words of the hymn brought tears to many an eye. Mr. Moody spoke on "What Christ is to us," a most pregnant and powerful address on a theme that he said it would take all eternity to exhaust. As at other times, Mr. Moody asked those who wished to be prayed for to rise up, and hundreds upon hundreds responded in all parts of the house. A more touching or cheering sight I never witnessed. Mr. Sankey sang, "Almost persuaded," and Mr. Moody said that there were so many anxious, it would be impossible to speak with them; so he asked them to go home, and at five o'clock to take God's Word, and kneel down pleading his promise, and commit themselves to Him. All the Christians in the hall would be praying for them at that hour. He prayed that they might be altogether persuaded.

Mr. Moody repeated his afternoon address to an immense audience of men in the evening, and in the course of it made strong reference to the great curse of Liverpool, the drink traffic, amid the approval of the vast congregation. He asked them to show their detestation of it by becoming abstainers. There were hundreds of inquirers at the close. A deeply interesting meeting of about 7,000 young men was held in the Circus from nine to ten o'clock, conducted by Mr. Henry Drummond. These meetings are to be continued every night.

V.

Work in connection with these special services, if we avail ourselves of ours privileges, means much toil. Mr. Moody has scored the word "duty" out of our vocabulary, and inserted "glorious privilege." Those who take up this work, and carry it out faithfully, find that each meeting, especially in the evenings, involves some four hours' physical and mental effort, making due arrangements for the comfort of ten thousand visitors, looking up and after the numerous cases of special inquirers. I can liken it to nothing so much as work in the trenches before a besieged city, in which every nerve and energy, spiritual and physical, has promptly and wisely to be put forth—parties sallying out, either singly or in company, to trace out and capture the anxious and inquiring.

Our great hinderers in this are the Christian lookers-on and curiously inclined; they feel an interest in the fight with the powers of darkness, but, from various motives, do not help. Such will persist in filling up the benches, to the exclusion of hundreds who ought to be brought in. There is a large amount of selfishness in the Church, very apparent in our meetings. We do not know how to deal with it, taking up, as it does, the best seats, and monopolizing much room. Mr. Moody and others have spoken from the platform about it, and tried to stir up the conscience, but in vain. They are almost worse than Meroz; for they not only do not help, but they hinder.

The house-to-house visitors report that the very poor, those to whom every hour is daily bread, say that it is no use going to the hall; they cannot get in; and they cannot afford to leave work at five o'clock, and wait two or three hours for the meeting, which those who have no employment do, to get the seats with backs. Christians had much better be holding prayer-meetings elsewhere, for the Spirit's power on the word, than keeping out those who know not the truth, but would come to hear it.

Those who know Liverpool best, all say that those who can face a Liverpool audience, and pass the crucial test of its critical investigation, must have something more than ordinary in them. This is now being done, with the usual accompaniment of respectable and rough rowdyism doing all it can to blacken and wrest plain-spoken truth.

The old slave-driving element is largely developed here, and is not confined to the back slums and dark corners of the town. Many a tale of shame might be related of how Liverpool has treated honorable grey heads, that have come on missions of philanthrophy and love; but we forbear. The Master went that path, so all His followers must. The disciple is not above his Master here, as in days of old.

VI.

The nightly gatherings in the Circus, from nine to ten, have been well sustained during the past week, and have been fraught with interest. Mr. Henry Drummond invariably presides, and conducts the proceedings with much tact and discretion. He throws aside all formalism, and endeavors to give the meeting as much of a family and social aspect as possible, in order to remove the natural diffidence that most young men feel in making any public statement about their conversion, which may be very recent, or their spiritual experience, which may not have been very deep or well defined. While the meetings are free to all who may feel disposed to speak, any attempt to raise controversy on disputed points of doctrine is vigorously repressed. Such a thing, however, seldom occurs, and would obviously be out of harmony with the object of these meetings, which is to encourage the young converts to make public confession of their faith in Christ, in the hope that the simple story of their conversion may lead others to the Saviour. Sometimes a few broken sentences from a young convert, telling how he lost his burden at the cross, have more effect on the hearers than could be gained by an hour's ordinary preaching. An ounce of testimony, modestly and truthfully given, is often worth more than a ton of theological disquisition.

Hitherto these meetings have been such as to warrant the belief that a solid and lasting work of grace has begun among the young men of Liverpool. In few, if any, of our large towns are the temptations to evil more numerous and more seductive.

The larger and more public meetings in Victoria Hall have been continued during the past week without any diminution in the attendance (except at one or two of the noon meetings, when the weather has been very severe), or the apparent results.

Many interesting statements have been made at the noon prayer-meetings with regard to the progress of the work in places which Messrs. Moody and Sankey have already visited, and in remote country districts which this wave of revived spiritual life has reached.

At the same meeting a letter was read giving some cheering intelligence of a movement among the engine-drivers and stokers on the North-Western Railway.

At the Wednesday noon meeting, a minister stated that at the conclusion of the previous Sunday's service, a barman came to him and told him that he feared he could not go on with his occupation and serve God. Moreover, he said that he had his father dependent upon him. He told the man to trust in God, and recited cases in which God had not forsaken those who had so trusted in Him. After a few minutes' struggle the young man was able to throw himself entirely on the Lord, and he thought that he had left the church a believer in Christ.

At the meeting on Thursday at noon, another minister mentioned a circumstance that came to his knowledge, showing that whole families had been recently led to the Lord. About ten or twelve days ago a young lady in the hall decided for Christ, and since then her only brother had given himself to the Lord, two sisters had become Christians, and five brothers-in-law, as well as others in the same circle, making fifteen persons who were now rejoicing in Him.

Another speaker said that they were aware that a house-to-house visitation was going on in connection with the services. They had heard some complaints of visitors only putting a tract into the letter-boxes, and not making any efforts to speak with the people in the houses. He hoped such persons would remember that that was not the primary object of their work, but it was desired that they should give some practical and verbal testimony of the Lord Jesus Christ. He knew of an instance in which a visitor called at the residence of a wealthy lady in town. Contrary to her expectation, the visitor was admitted, and the lady said that she thought the visits were to be limited to the poor. Before the visitor left, however, the lady was in tears about the state of her soul. Another case occurred in a poor district, the visitor being received by a woman who asked her to go in to see her husband, who would not go to hear the word of God. But, notwithstanding the invitation, the visitor went away without saying a word to the poor man. To those who are willing to undertake visitation, districts would be readily assigned.

Mr. Sankey said that when they were in Glasgow they heard a great deal about their not reaching the lapsed masses. He did not hear much of similar complaints in Liverpool. That, however, was not their chief object. It seemed to him to be the duty of the Church to go after the masses.

He hoped those Christian friends who had got themselves fired up at the meetings, would make it their life-work to reach those people who were perishing in the lower places of the town. It would be better for such people to go into it than that Mr. Moody should do so, who, however, often did work of that kind in his own city.

On Friday some valuable testimony was given as to the tangible effects of the work in Liverpool. It was stated that one class reached had been those who, though religiously trained, had, during these special meetings, seen a new meaning and power in the truths with which they were familiar. Many sailors, and ship captains, too, had come to the meetings and been guided into the true haven of rest and peace. Then there were many workingmen who had plunged into the depths of intemperance, and whose insulted and injured wives, after being driven from their homes, had been compelled to support themselves and their children for years together. These wives, in this day of grace, had sent letters to their husbands, extending their forgiveness and imploring them to come to Victoria Hall and seek forgiveness of the Saviour. Some of them had come and found that forgiveness, and gone back to lighten their homes again with a new lustre and joy.

Allusion was made by one of the speakers to another class, one much too large and full of strange and painful interest, consisting of those who have in past years made a profession of love to Christ, but have wandered

"Away on the mountains, wild and bare,"

and have been glad to take of the husks that the swine did eat. It had often been asked whether the converts connected with this revival would stand the test of time, and endure the temptations of the world. When the question is put, as it often is, by a Christian brother, I ask another: "Brother, have all *your* converts stood fast?" I can only confess that, during the forty years but one that I have preached in this town, I have missed a great many from the fold; but I have found some of them in that inquiry-room. The first night the inquiry-room was needed, I lingered on the platform, not intending to go into the room, when a message came to me, "You are wanted immediately; an inquirer wishes to see you." I went, and I had not seen that face—I will not tell you whether it was man or woman —for twenty years; and I found that soul had wandered away, and had kept out of my sight with perfect success. The first conviction was to go and tell him by whose hands they had been received into the Christian Church. Many a wanderer has come, and Christ alone knows how many more He will welcome back to His all-forgiving arms, and fill our hearts with a gladness they have never experienced before."

And so the great work flows on steadily, unhindered in the least, as I believe, by the newspaper opposition of the "carping critics."

Mr. Moody's Bible-lectures last week, though (with one exception) perhaps not so full of interest as those of the week before, have been very largely attended and evidently enjoyed.

One of the most interesting meetings of the week was the "children's service" on Wednesday afternoon, at which Mr. Moody and Mr. Sankey were both present. So many little ones it has never before been my lot to see gathered under one roof. Some of the daily papers put down the numbers in Victoria Hall at 12,000, with an overflow meeting of about 2,000 in the Circus. Think of such a number of young, impressible natures brought at one time under the sweet sound of redeeming love! Mr. Moody's address, founded on the book with three leaves, black, red, and white, was a sort of running interchange of simple yet searching questions, and answers very promptly given. The singing by Mr. Sankey of some of his solos was greatly enjoyed by the youthful audience, and when they all joined in the chorus, or sang other of the hymns right through with great heartiness, and as with one voice, we had yet one more proof of how universally and, I trust, inalienably, these sweet gospel songs have become household possessions throughout the kingdom.

The evening meetings during the week have, as usual, been crowded for some time before the regular hour for commencing the service. The overflow meetings have been held in the Circus. Mr. Sankey has generally been present in the course of the evening at both places. Mr. Moody's gospel addresses at the evening meetings have been characterized by much simplicity and power, and the result has been seen in the crowds of both sexes who pass nightly into the inquiry-room. In the words of one who has been closely associated with the work during the past week, they have been "flocking into the Kingdom by scores." I understand from the same source that several of our much-to-be-pitied fallen sisters have been reclaimed through the agency of the meetings. Would to God that every poor drunkard and profligate in Liverpool might "come to himself" and return to his Father like the prodigal of old. Let us thank God for what He has done, and ask Him in faith to bless this special agency yet more abundantly.

As the days and weeks roll past, and the same scenes are so often repeated, it is difficult to find fresh terms in which to describe "these wondrous gatherings day by day." The four meetings on Sunday last may briefly be stated as a repetition of those on the Sunday before. All crowded to the utmost capacity of the great hall, and, in some cases, especially at the afternoon and evening meetings, multitudes turned away for lack of room.

The service for "non-church-goers" at eleven o'clock was a fresh illustration of the power of Christ's wondrous love, or "compassion," to melt the hearts of the most supine, and to move the consciences of the most sin-stricken. The arrows of conviction went home right and left, and there was a large ingathering of souls at the close. Mr. Moody used, by way of illustration, a very touching chapter of personal family history that brought tears to many eyes.

At the three o'clock service for women the hall was filled to overflowing an hour before the time. The women are quite as determined in their efforts to get in as the stronger sex, and some say not quite so well behaved under the trying conditions of a crowd. I suppose, however, there must be some allowance made at this special season, and if one could be certain that they are all as anxious (as Mr. Moody said he hoped) to press into the kingdom, a little roughness of demeanor may well be overlooked. To my mind, these Sunday afternoon meetings for women have been the most wonderful of all, and certainly not the least important, when we consider the power for good or evil that must be exerted by so many thousands of our mothers and sisters. I must say these meetings have proved that the women are not only quicker in their apprehension of the truth, but more honest and courageous in avowing their apprehension of it. At the close of Mr. Moody's searching address on "Excuses," a very considerable proportion of the audience promptly stood up to show that they wished to excuse themselves no longer from accepting the gracious invitation to the marriage supper of the Lamb. Mr. Moody spoke to the inquirers that filled the inquiry-room, in language and by illustration so beautifully simple and apt, that it is almost impossible to conceive any difficulty could have remained in their minds. At the same time Mr. Sankey addressed, in a very artless, homely, and touching way, a large body of anxious inquirers who remained in the hall.

Mr. Sankey's singing at this service was peculiarly appropriate and effective. At the opening, he sang that solemn and tender invitation to the feast, "Yet there is room," and when Mr. Moody had ceased speaking, and the whole assembly was hushed in silent prayer, he broke the death-like stillness, by singing, in subdued and pleading tone, "Almost persuaded." His rendering of this hymn, which in some parts could only be compared to a wail of sorrow at lost opportunities, sent a deep thrill through the hearts of those thousands of listeners.

The inquiry-meetings for men, at which Mr. Moody re-delivered the address on "Excuses," was another season of pentecostal power, and the Holy Ghost was present to wound and to heal, to kill and to make alive.

VII.

MR. MOODY, before leaving Liverpool, addressed an immense meeting in behalf of the Young Men's Christian Association, and laid the corner-stone of the new building, and in the evening held a service at Victoria Hall for young men under 35 years of age. On the two following days, there was a Convention, at which Mr. Moody presided, and which was very largely attended.

The closing services were held on Sunday, the 7th of March, at eight A. M., for Christian workers; at eleven A. M., for young converts and inquirers; at three P. M., for women only; and at eight P. M., for men only. Each of these services was very largely attended. For two hours before the proceedings commenced hundreds of people besieged the building, eager to secure admission. Mr. Sankey was not present, but Mr. Moody delivered appropriate addresses, exhorting his audiences to perseverance, and commending the efforts of the ministers who promised to take his place in a series of services, to be held this month in the capacious building in which they were then assembled.

THE WORK IN LONDON.

The prescribed limits of this narrative will not permit anything like a complete or consecutive account of all the daily scenes and incidents in connection with the labors of the Evangelists in London. The following extracts from the religious journals will enable the reader, however, to obtain some glimpses of the work of ten or twelve days, and the impression made on the minds and hearts of men in all ranks and conditions of life.

I.

The four months' evangelistic work in London with which Messrs. Moody and Sankey propose to bring their sojourn in Britain to a close, was begun on the evening of Tuesday, March 9th. During the first month the meetings are to be held at the Agricultural Hall, Upper-street, Islington. The following month, it is proposed, will be spent at the West-end. The third, in all likelihood, will be devoted to the East-end, where the meetings are to be held in a building at present in course of construction for the express purpose. This edifice is situated near the junction of the Mile-end and Burdett roads, and is expected to be ready by the end of April. The southern quarter of the metropolis will thus fall to be visited last; though we believe it has not yet been finally decided whether "the leafy month of June" will be allotted to the East-end or to the Southside. The use of the Agricultural Hall has been secured by the committee for ten weeks, extending from the 28th of February to the 9th of May, at the rent of £50 a week. The arrangements made to adapt this immense edifice to the purpose of the meeting have necessarily been on a most extensive scale, and have involved a large amount of expense, and the exercise of no inconsiderable skill; for, like Bingley Hall at Birmingham, it is simply a great shed designed for the exhibition of prize cattle, and especially for the famous Smithfield Show of fat stock, which annually, in the depth of the winter season, attracts so many of our country cousins up to town. In the body of the hall, 12,000 new cane-bottomed chairs have been placed, to supplement the 2,000 already belonging to the establishment; and there are besides forms capable of accommodating 2,000 persons. The lighting of the place has been effected by means of large gas chandeliers hanging from the vaulted roof, with lines of gas jets along the sides of the building. The thousands of burners that bead the walls, and which, with the chandeliers, yield an abundance of light, run in straight lines save at the centres where they rise in three semi-circular arches. The acoustic properties of the hall are greatly aided by an immense sounding-board over the speaker's platform. At the centre of this platform there is a small dais, covered with red cloth, and having a slight rail round it, and a little book-board at one corner. This is for the president of the meeting. On his right are the seats for the choir, and Mr. Sankey's American chamber organ, the latter placed by itself in a projecting square. The seats on the left are for the committee and others taking part in the service. A broad strip of red cloth runs round beneath the lines and arches of light, and this, besides serving as a pleasant bit of color, bears appropriate passages in white lettering. The first of these, on the right of the platform, is—"Repent ye, and believe the Gospel;" the first on the left—"The gift of God is eternal life." The Central Noon Prayer-meeting Committee which is the body charged with the management of Messrs. Moody and Sankey's services, has for the present its headquarters fixed at the Agricultural Hall. The meetings commence each evening, except Saturday, at 7:30; doors open at 6:30. On Saturday, when the Evangelists rest, there will be no service. A noon meeting is to be held each day at Exeter Hall from twelve to one; doors open at eleven. This meeting will be regularly conducted by Mr. Moody, and Mr. Sankey will also take part in the service. On the Saturday, however, as we have just indicated, they will ot appear at Exeter Hall.

II.

OPENING SERVICES. — SCENES OUTSIDE.

TUESDAY evening, the 9th of March, will long be memorable in the north of London, as the occasion of the first of the services in the Agricultural Hall. Long before the hour appointed, large and small groups were wending their way to the north from all quarters of London—each group well provided with the blue-grey covered hymn-book, or the more pronounced yellow, red, or blue music books. Of all ages, from the white-headed grandsire to the babe in arms — of all stations, from the dignitaries of the empire to the low-class workmen and laborers—of all grades, from the highest Christian working ladies of the land to the lowest women of evil lives—the twilight met them seeking the Hall in thousands; and after the Hall was filled, the doors closed, and the adjoining hall filled also, thousands upon thousands came, saw the closed doors, and turned away to give place to other disappointed thousands following them. The infidels were present also in foolish force outside the Hall, distributing handbills full of the most false and malignant misstatements, pretending to describe Sunday discourses of Mr. Moody *which have not yet been given.*

"Moody and Sankey's Hymns, one penny! with all the music, one shilling!!" "Moody and Sankey's photographs, sixpence each!" "Life of Moody and Sankey, with likeness, one penny!" "Wooden image of a tumbler, christened Moody for the occasion, one penny!" "Italian organman playing 'The gate's ajar,' christened Sankey, and requested to accompany himself." All these together formed such a crowded and ever-shifting illustration of "Vanity Fair" as John Bunyan never dreamed of. Many policemen to keep the way; multitudes of young men full of fun and joking, multitudes also of evil women and girls gaily dressed joining in the ribaldry; the two together forming a mass of well-dressed but disreputable blackguardism; proving to demonstration that the American Evangelists had come exactly where they were sorely needed at last.

Omnibus-men, cab-men, tram-car-men, board-men, and loafers of every description, took part in the universal carnival. Oaths, jests, slang, ribaldry, and mockery were all let loose together; but not one serious face, not one thoughtful countenance,—not an apparent thought of God's judgment, or of eternity, in all the vast changing multitude shut outside.

After the service within had ended, and partly during its continuance, detachments of choirs belonging to the neighboring missions had stationed themselves adjoining the Hall, and occupied themselves in singing the "song and solos," and delivering addresses of the briefest character. Some of these groups were too close together; and the effect was exceedingly bad, as the songs were inextricably mingled, and thus caused to suggest anything but serious thought. This, however, was speedily remedied by the incorporation of the choirs, when better work was done. But all seemed in vain; the very spirit of mockery seemed to possess the great majority. There was nothing like spiteful opposition, much less of interference; rather the singers and speakers were regarded as amiable enthusiasts, who had rashly delivered themselves to the merciless mockery of a London mob.

Was there any good done by these open-air services? Certainly: if only for the unflinching but temperate stand made by the Lord's servants in showing themselves on the Lord's side; and, doubtless, when this persistent bitter wind will allow of earnest speaking taking its full share in the work, much better things will fall to be reported.

THE SCENE WITHIN.

Long before the hour appointed to commence the service, the enormous Hall was filled to its utmost capacity, and the doors were shut. The building consists of a vast central space open to the roof, which is arched—constructed of iron and glass, and sides with a gallery running round the building, which is oblong in form. One side to the gallery is boarded in for inquiry-rooms; in the centre of this side, a platform, holding 500 persons, is erected, and from the front of the platform the addresses are given. Seats are provided for about 12,000 persons, beyond the seats already in the building. Large temporary galleries are also erected at each end; so that the whole vast audience is in full sight of the speakers on the platform; and the view of this vast audience is a sight that is majestic from its very magnitude.

THE SERVICES.

When, upon the word given, the vast multitude arose to sing together, the effect was wonderful; not so much for the magnitude as the full, deep, rolling volume of sound issuing from voices uttering music known most probably to every one in the building.

To pass the time pleasantly, various hymns were sung until the time appointed; when Mr. Moody took the president's place, and Mr. Sankey sat down to his instrument. There was some applause as they entered, as also at the conclusion of Mr. Sankey's solo; but both were immediately hushed by those who remembered they were met for the worship of God.

The commencing word was Praise — "Praise God, from whom all blessings flow." "Praise God," said the chairman, "for what He is going to do for London." Then earnest prayer followed, and the 100th Psalm was sung. Silent prayer followed. Then Mr. Moody led an audible prayer, and Mr. Sankey sang "Jesus of Nazareth passeth by." The whole audience joined in singing "Rock of Ages," and then followed Mr. Moody's address of the evening.

He read 1 Cor. i. 17 to end; but after speaking for a short time, was interrupted by a man who was not sober. "When a man is in liquor," said the president, "and makes a noise, he ought to be removed; we will therefore sing the 61st hymn while this is done." Seized upon energetically by five strong hall stewards, the drunken one was speedily removed, the stewards receiving an injunction to "be careful" as they passed the platform. Then Mr. Moody resumed and concluded his address. The meeting concluded by all the people singing "Hold the fort," with such vigor and effect, that the sound was heard and recognized in the neighboring streets. Then prayer and benediction closed the evening, as there was no after-meeting.

THE DAILY PRAYER-MEETING.

The next day, Wednesday, March 10, at noon, the first daily prayer-meeting was held in Exeter Hall, Strand. Long before the hour of commencing, the Hall was filled to overflowing in every part but the platform; and it seemed lamentable to keep hundreds of seats empty for those who did not come, and shut outside those who were anxious to occupy them — the noise of disappointed applicants being distinctly audible during the meeting. Precisely at noon Messrs. Moody and Sankey and the Committee appeared, and the meeting commenced by singing "Sweet hour of prayer," followed by requests, silent and audible prayer, and the singing of "The Great Physician." Mr. Moody's address followed from Jer. xxxii. 27, "Is there anything too hard for me?" in which he said, "This is God's challenge to Christians to call upon Him — to cast out all 'ifs,' all doubt and unbelief, and rely joyfully upon the Lord God who made heaven and earth." He also read a most touching extract from the first letter received in London concerning a child, who had proposed to wait for their coming, to be a Christian; but had found the pearl of great price, and been "called home" before they reached the city. Mr. Sankey then sang "The ninety and nine," and the meeting was thrown open. Two or three brethren followed in prayer, and the meeting closed.

THE SECOND EVENING.

On Wednesday evening, *March* 10, the second meeting was held in the Agricultural Hall. The attendance was not nearly so large as on the first evening, resulting from the fact that Mr. Moody requested the doors to be closed at half-past seven, thus preventing many thousands who were unable to attend so early from gaining admission. The services commenced with prayer and singing. Mr. Moody then read part of Ezek. xxxiv. and Luke xv., commenting as he went on, then announced the coming meetings on the Lord's day, at 8 A.M. for workers, at 3 P.M. for women, and at 8 P.M. for men. Tickets would be issued for all these meetings. Then silent prayer, and singing "Lord, I hear of showers of blessing."

Mr. Moody then spoke from Luke xix. 10, "The Son of man is come to seek and to save that which was lost." It was speedily apparent that great blessing from on high was present in that meeting. The address was full of power; anecdote, illustration, Scripture entreaty, persuasion, succeeded each other again and again, with lightning speed and force, while the vast audience listened intently. As the interest heightened, and story after story was told, many could be seen wiping the tears openly, apparently unconscious of what they were doing. The graphic picture of the meeting of Bartimeus and Zaccheus, after the former had been healed, was thoroughly enjoyed; and the quiet hit at those "who don't believe in sudden conversions," in the statement that Zaccheus "was converted between the branches and the ground," was greatly enjoyed. The story that followed, of "the young man converted on his mother's grave," gave occasion for an impassioned appeal to turn to Jesus then and there. Silent prayer followed the conclusion of the address; and, amid a hush that was almost awful, the sound of music floated on the air, and Mr. Sankey sang softly "Come home—come home." Every head bowed, thousands earnestly praying, while the soft music seemed to enter into the very souls of that mass of humanity, bowing and swaying even the hardest to thoughts of repentance and prayer. Then Lord Radstock concluded with prayer, and the hymn, "I hear Thy welcome voice," was sung as Mr. Moody went from the Hall to the first inquiry meeting in London. Many hundreds followed him, but whether workers or inquirers did not at the time appear, and it is far too early yet to speak of results.

Thursday, March 11.—Mr. Moody presided for the second time at the noon prayer-meeting. There was also a falling off in the attendance here compared with the day before; but the great Hall was nearly filled, and would doubtlessly have been filled to overflowing during the service had the doors remained open. The first hymn was, "Lord, I hear of showers of blessing," followed by one new to London, entitled "Wondrous love," but which will assuredly become a special favorite. Silent and audible prayer followed the *classified* requests for prayer; and after, Mr. Moody spoke in explanation and defence of "inquiry meetings," instancing many such meetings from the Scriptures, and asserting that the inquiry meetings ought to be credited with four-fifths of the work done. He was just a very little bitter in saying, "I don't know what some men would do at a Pentecost," or his earnestness seemed intensified to bitterness; but this disappeared when he spoke of a boy of fourteen with a Bible under his arm, whom he had met in the inquiry-room the previous night, and asked as to his presence there? The boy replied that he was a Christian, hoped to meet some little boy like himself to tell about Jesus. Afterward, the boy was seen kneeling with another in a corner. Mr. Sankey also spoke earnestly in defence of the inquiry-room—asking objectors to visit and see for themselves instead of finding fault beforehand; adding, warmly, "It don't take half a man to find fault." The meeting ended as usual; but after its close, there appeared to be an impromptu *reunion* of nearly all the evangelical workers in London; the resemblance being almost perfect to one of the evening conferences at Mildmay Park.

THE THIRD EVENING SERVICE.

This was much more largely attended than the second, every seat in the Hall being occupied, and the galleries well filled. The choir sang several hymns before the service commenced with the well-known "Maggie Lindsay" hymn (as it is called here), "The gate ajar," followed by prayer. "The Great Physician" was next sung, and the reading followed from Luke x., being the parable of the good Samaritan, in which the priest and Levite were used as types of churchmen and dissenters to the credit of neither party. "Rock of Ages" was next sung, and Mr. Moody resumed his discourse of the preceding evening from Luke xix. 10. Much better order was observed than at the commencement of the previous evening, the meeting being admirably controlled. The address was most solemn and searching in character, concluding with an exhortation to immediate and final decision. Mr. Moody ended his discourse by prayer. Then "Safe in the arms of Jesus" was sung, then silent prayer; next, "Guide me, O Thou great Jehovah!" then the benediction and the inquiry meeting.

Friday, March 12.—At eleven o'clock in London, the weather was of the most severe and trying description—hail, rain, snow, slush, and a bitter east wind over and through all. Nevertheless, the Great Hall for the noon prayer-meeting had a glorious gathering, and there were quite as many present as on the previous day; but notably men; the ladies *could not* face the terrible cold and sleet.

Mr. Moody took his position punctually at noon, and announced the hymn, "Sweet hour of prayer," followed by multitudes of classified requests for prayer, the classification giving a somewhat *bizarre* character to the requests, as, "*Eleven* sisters ask prayer for brothers." Mr. Moody resumed his previous noon discourse, "On the inquiry·room," instancing various faulty methods of dealing with inquirers, particularly condemning the statement often made, "Believe that you are saved, and you are saved;" and pointing out that saving faith must be faith on the Son of God. He passed on to consider right methods of dealing with anxious ones, giving many valuable hints and texts as he went along, speaking earnestly against mere discussion on his way. When Mr. Moody concluded, Mr. Sankey sang "Nothing but leaves." Has Mr. Sankey been listening to critics? The writer heard him sing the same hymn with far more effect in Glasgow at the convention. In London there was more of the *artist*, but it seemed less of the earnest gospel singer. In Glasgow, it was *solemnizing* and *thrilling!* In London, it was *very nice!* (*N.B.*—These are not the writer's criticisms, but the opinions expressed.) After singing, a gentleman (name unknown) spoke earnestly of the way and the need of working for Jesus. He was followed by another, who told a touching story of how the lost are found in London. A tract distributor offered a man a tract on Waterloo Bridge; jt·was declined with the remark, "I shall be in hell before night;" the words were heard and answered, "No, you will not, for I'm going to heaven, and will stick to you all day." They left the bridge together, the hungry man was supplied with food and taken to a place of worship. There he fell asleep. "Perhaps he has been walking all night," said his friend; "let him sleep!" Service over, he was conveyed home to supper, inquiring concerning all this kindness, "*What's up?*" He was fed, tended, reasoned with, instructed, and brought to the way of heaven instead of being in hell, as he had said.

REVIEW OF THE DAYS.

So ends the first three· of Mr. Moody's noon prayer-meetings and the first three nights of work in London. And it is simple truth to state, that such meetings were never held before in London, if ever they were in the world's history. In *three days* of noon and evening service, about *eighty thousand* have listened to the glorious gospel of the blessed God. Well might Mr. Moody express his thankfulness to God—the encouragement he had received and felt, and his deep sense of the sympathy and help extended to him and his colleague in their great work. Well might he dissolve in broken accents and tears of entreaty for a rich blessing on himself and those who, laboring with him, will share his eternal rest and reward. Surely, when bankers and rich merchants, and ministers holding high official positions, are content to be doorkeepers, it must be said, "We never saw it after this fashion," and this was exactly the case at the door of Exeter Hall yesterday.

On Friday evening, *March* 12, the last meeting of Messrs. Moody and Sankey for the week was held in the Agricultural Hall. The audience exceeded in number every night but the first. As far as the eye could reach there was the same dense multitudes above and below.

Mr. Moody gave notice that on Saturday evening there would be addresses by various ministers; Sunday services as before mentioned; Monday evening Rev. J. Spurgeon would preach, while Mr. Moody met anxious inquirers. On Tuesday, Wednesday, Thursday, and Friday there would be two daily services, at 3 and 8 P.M., when the same address would be repeated daily. Tickets would be issued for all these meetings. He then read Matt. vi., from 19th verse to end. Mr. Sankey read from Mark x. of Bartimeus, and sang "Jesus of Nazareth passeth by." The effect of this melody was simply wonderful in its stilling power on the audience; the sibillation of the "s" at the end of the line could be distinctly heard in a quiet as of death. Mr. Moody took for his text Isa. lv. 6, stating, for two evenings he had dwelt on man seeking God, but now he would speak of God seeking man; yet recommending earnestness in seeking God by many touching incidents and suggestions. This, among others, he thought "the dying thief might have had a praying mother." He also turned to the ministers around him and asked, "Did they believe that God was present, and willing to save?" and was instantly answered by an audible "Yes." A tearful, impassioned appeal followed to all classes to seek the Lord, and He would assuredly be found. Silent prayer succeeded, and Mr. Sankey sang "Almost persuaded." Then the audience were dismissed, and all anxious, and all workers, were invited to remain, an invitation that was accepted *by several thousands!* The whole space under the arched room was occupied by seekers and workers, while the responses to earnest prayers rolled around like the deep tones of the great sea waves at night. The Lord was there. Inquirer after inquirer made themselves manifest, until there were scores in the inquiry-room, and scores remaining in the hall speaking with the workers there. In the inquiry-room were seekers and workers in every direction, and very many found peace in believing. One fine young man fell to the lot of the writer, and it was emphatically good to watch the dawning of divine truth on the mind, as shown in the intelligent face—to see the look of anxiety and fear give place to the knowledge and love of God—to watch the birth of the soul to eternal life bring forth that look of brightness on the face which is never seen from any other cause. One young lady said "she was so happy, she seemed to tread on air;" and in instance after instance the testimony grew and multiplied, till we could only rejoice in believing that numbers were born again—not of corruptible seed, but of the incorruptible, which liveth and abideth for ever. Then the long, happy evening closed by Mr. Moody calling the workers together, and giving some brotherly advice and counsel concerning the details of work in the inquiry-room.

Oh for the time of blessing! Oh for the rain upon the weary! Oh for the coming, in mightiest power, of the loving Spirit and the King our Brother, among the ruined and lost—among the weary and burdened laborers on this rough and stony ground! Our Father, hear and answer Thy children's heart-cry, for Jesus' sake!

I. WORKERS.

Sunday, March 14.—On Sunday morning, March 14, the usual unbroken quiet of Islington experienced a striking change. From every direction, solitaires, couples, and bands of well-dressed people were hastening to the Agricultural Hall. Many parties of singers had arranged to meet in their different localities, and marched with songs to their destination. Sunday-school teachers resident in the line of march near to the Hall had invited their fellow-laborers to breakfast at a very unusual hour; while the vendors of hymns and papers round the Hall took their usual week-day positions, and transacted a large amount of buying and selling, to which multitudes made strong and indignant objection. Pouring in at all the entrances to the Hall, there was speedily convened such a gathering of its Christian workers as London had never seen. It was a complete re-union. Friends, whom the exigencies of work had separated for years, met and clasped hands once more; young men grown old in service met with others in like condition, whom they had labored with in years of strength; and comely matrons' faces were recognized as those of former girls in Sunday-schools. Long before *all* old friends could be recognized and greeted, the time for the service arrived, and the Evangelists stood face to face with many thousands of the Christian workers of the great Metropolis for the first time.

Cool, prompt, and business-like as ever, Mr. Moody announced the first song would be "Hold the fort," which, being recognized as peculiarly appropriate to the occasion, was sung with a vigor that left nothing to be desired. Earnest prayer followed, and then the hymn, "Stand up for Jesus." Mr. Moody read part of Isaiah vi., ending with, "Here am I; send me!" and called upon Mr. Sankey to sing the melody known by that title; which he did with a little difficulty, *perhaps* occasioned by the sharp morning air, or perhaps by having been not a little overworked recently. Then the congregation sang, "I love to tell the story," and Mr. Moody's address was given.

The text was Dan. xii. 3, "They that be wise shall shine," etc.; and Mr. Moody proceeded to say: We all like to shine, and had better *own up!* But who shall shine? The wise! and thus the glorious privilege of eternal splendor was held forth to all engaged in Christian work! But personal conversion must precede the conversion of others by us. And here he narrated a striking instance of a Sunday-school superintendent who was not converted, but finding this to be so, went honestly to his minister and offered to resign. The minister suggested a more excellent way — that the superintendent should first turn to the Lord at once, and then continue his labors This was done; he turned to the willing Saviour, and then became the means of the conversion of the teachers and a great revival in the school. It was the duty of each Christian — *not* duty, but privilege (away with mere *duty!* we did not talk of *duty* to wives and mothers, and why in religion?)—to speak to some person daily. For twelve years there had scarcely been a day in which he had not done this. Seek out friends, and bring them into the current, that they might get a blessing and pass it on. We must also get into sympathy with the unsaved. When he was laboring in the school at Chicago, a teacher, who was going away to die, came to him in bitter trouble about her unconverted class. He felt his strength too far gone to visit them; they were unsaved, and he was leaving them—going away, for ever. Mr. Moody procured a carriage, and they went together day after day for ten days, until the teacher had seen all, pleaded with all, and won them all for Jesus. The tearful eyes, the pale face, and the deep sympathy had triumphed for Christ! Then they all met him on the platform, and the wave of his hand from the carriage was a last, long farewell. The effect produced by this narration was very deep. Sobs and tears were almost universal. The ministers on the platform were wiping both eyes and glasses, and some were literally scooping away the tears with their hands. Strong men were weeping like children, and the speaker himself wept abundantly as he remembered and depicted the touching scene. Yes, he continued, we must get in sympathy—make their case ours, their troubles and sorrows ours, and then we shall have prevailing power. He spoke of a poor mother, whose child had been drowned in procuring drift-wood from the river, and whom he visited along with his little daughter. "If that was me," said my child, "wouldn't you *feel bad*, father? Don't you feel bad for the poor mother?" This unlocked the springs of sympathy, and I did feel bad for her. I found a grave for the poor child, and afterwards bought ground for a Sunday-school lot, to bury a hundred of our poor little scholars. In the midst of a most striking scene of weeping, such as that hall had never seen before, the address concluded, and Mr. Moody *attempted* to pray. So deeply was he moved, that he was compelled to pause in his prayer, amid dead silence, to recover himself, and be able to proceed. Then we sang, "Work, for the night is coming," and the benediction ended the first workers' meeting.

2. WOMEN.

On Sunday afternoon, at three, the first special meeting for women was held. The service commenced by singing "The Great Physician," after which prayer was offered, followed by the hymn, "I hear Thy welcome voice." Mr. Moody read Ps. lvii., and Mr. Sankey sang "The ninety and nine." Where all the singing is so good, it is hard to particularize; but this seems to be one of his own favorites, and is most certainly a favorite with the people. Then all joined in singing "Free from the law," and Mr. Moody commenced his discourse from Gen. iii. 9: "Where art thou?" Was ever such a gathering of women only, convened before, simply to hear the gospel of the grace of God? There were, at the lowest computation, about 17,000 women present; and the power of the Spirit was clearly there: tears and sobs and repressed cries, anxious faces, low, earnest words and entreaties for mercy were all around, as the discourse proceeded from point to point. God was the preacher of this sermon, said Mr. Moody; and though the first audience was small, the sermon has come rolling down the ages, and many, I hope, are asking themselves this question now. I am speaking to professors, to backsliders, and to those who never made profession, but all equally lost. There are three steps to ruin—neglecting, refusing, despising the good news of God. The discourse concluded, Mr. Moody offered earnest prayer; silent prayer followed, and then the soft, persuasive strain, "Come home," from Mr. Sankey, arose upon the meeting, the choir singing the chorus. Then all sang the hymn, "Lord, I hear of showers of blessing," and the meeting closed to allow inquirers to gather. Such a number accepted the invitation that the large inquiry-room could not contain them, and many were spoken to in the bitter cold without the room.

3. MEN.

The evening service was simply a repetition of the afternoon, but for men only, instead of women. Thousands of women, nevertheless, accompanied their male friends in hope of admission, but were disappointed—they could not be admitted. Nevertheless, the building was filled to its utmost capacity, and the doors were closed nearly an hour before the service commenced. The would-be infidel orator of London is in the habit of saying that "Religion is an affair of priests and women." Never again will he be able to repeat that taunt, after the meeting on Sunday evening last, *when nearly* 15,000 *men of London* were held breathless by the simple preaching and singing of the gospel of Christ. Before the address was delivered, Mr. Sankey sang "Jesus of Nazareth passeth by;" himself singing the verses, and the vast multitude joining in singing the last line in each verse, thus producing the effect of one of the mightiest choruses ever sung on earth. After the address the inquiry-room was opened, while the meeting in the hall continued with praise and prayer.

So great had been the effect produced, so large was the number of inquirers who were not "priests" or "women," that there were not enough workers present to deal with them. Nor can this be wondered at. Christians had been entreated and enjoined to stay away, that the unconverted might have all the room; and this request was too literally obeyed. It may also be noted that most of Mr. Moody's best helpers have much work of their own on the Lord's Day, which cannot be neglected even for the inquiry-room. With all the will to help, churches, chapels, and missions must not be left untended through the temptation of the attractive and pleasant work provided at Islington.

The noon prayer-meeting on Monday, March 15, was densely crowded—hall, galleries, and platform presenting an unbroken mass of believers in prayer, quite as well able to judge for themselves as any philosopher was able to judge for them, and having that which no unbeliever could have—experience—to guide them. After singing the hymns "Over there" and "Wondrous love," Mr. Moody read part of Isa. xii., and then proceeded to devote the meeting principally to accounts of the Lord's work. He had received accounts from Liverpool, that "the real depth of the work had just commenced, that it was better now than ever." At the Glasgow noon meeting, convened at that time, the prayers would be devoted to the work in London. The Earl of Cavan read a letter from Glasgow concerning the work in the Metropolis, and offered prayer in accordance with the letter. Mr. Quinton Hogg made a touching and earnest appeal for workers in the inquiry-room, asking if it was right that one worker should have ten inquirers. The Rev. R. W. Dale gave an account of the results of the work in Birmingham, dwelling largely upon the inquiry-room, asserting, "You know nothing of the work until you go there." He also spoke of the need of more cheerful singing in church worship, which was met by loud applause and clapping of hands, which Mr. Moody immediately and very decidedly repressed. Lord Radstock gave good news from Russia, and the meeting closed as usual.

On Monday evening the Rev. J. A. Spurgeon delivered the address to a greatly thinned audience, while Mr. Moody attended to the inquiry-room; Mr. Sankey also appearing on the platform, with the Rev. R. W. Dale and many other ministers and gentlemen. Mr. Billing, the chairman of the committee, invited the anxious and inquirers to St. Mary's Hall—an invitation that was immediately and largely accepted —the audience meanwhile singing "I am so glad that our Father in Heaven." Mr. Spurgeon offered prayer, followed by "Jesus, Lover of my soul," and the reading of Isa. lv., with running comments. Prayer by Rev. R. W. Dale, and "What shall the harvest be?" sung by Mr. Sankey. The address by Mr. Spurgeon was founded on "That spiritual rock . . . Christ." At the close of the service it was made known that all the workers were required in the inquiry-room, and there was no after meeting in the hall.

THE FIRST AFTERNOON MEETING.

On Tuesday afternoon, March 16, at three o'clock, the first afternoon meeting was held. There were about seven or eight thousand persons present, comprising those who were not occupied in their daily callings, with not a few lads and girls among them. The service commenced with "Rejoice, and be glad,' to a well-known Primitive Methodist tune, which was heartily sung by all present. Prayer followed, and Mr. Sankey sang "There's a light in the valley,"—not at all an easy melody, or likely to become popular. Mr. Moody read from John iii., and the meeting sang "Rock of Ages." Mr. Moody took for text, "Except a man be born again," saying they had better get the *text* than the *sermon*; there was *life* in the text. Men were not *baptized* but *born* into the kingdom of God. They must be born again, regenerated. Regeneration was not going to church—Satan went to church; not reading the Bible, not praying—Saul prayed daily before conversion; not attending the Lord's Supper—Judas did that, but was not regenerated; nor was it trying your best, as a woman told here in the inquiry-room, and was answered, "That's the way down to the pit!" Nor was it baptism; if regeneration could come by baptism, he would leave off preaching and take to baptizing—if he could save by baptism, he would get a bucket and baptize all whom he could come near. Has not God a right to save in His own time and way, and on His own terms? If you could save yourself, on your own terms, you could not make them so easy as God has made them. No man could save himself; God must save him. Under the law it was "*Do and live;*" under grace it was, *Live and do.* The address concluded with the story of the wounded soldier, who sent for Mr. Moody "to help him to die!" who was brought to peace in believing by the repetition of John iii. 14, "As Moses lifted up the serpent," etc. After the address the audience sang "There is life for a look," and the service ended. There were more workers than inquirers in the room at the close of the meeting.

THE EVENING GATHERING.

At the meeting in the evening, the address was repeated to the largest gathering yet crowded into the Hall. The demand for "more seats" was responded to as far as possible; but when the last seat was taken, many thousands were excluded who would willingly have heard the gospel. The hymns were varied from those sung in the afternoon, but the reading and the address were the same. Prayer was offered. Mr. Sankey sang the hymn called "Mary Magdalene." The meeting concluded with audible and silent prayer, and the hymn, "I hear Thy welcome voice." The crowd that retired to the inquiry-room was so large that hundreds sought admission in vain, and some were spoken to outside St. Mary's Hall. Whether those within were inquirers or workers could not be distinctly ascertained, as admittance could not be given.

The address at the noon prayer-meeting on Tuesday and Wednesday was on the subject of "Prayer," founded on "Ask, and ye shall receive," etc. The three words described three classes—those who *ask*, who *seek*, and who *knock*. Some *ask*, and don't give God time to answer. He heard of a little boy who asked for his father's *razor*, and when denied, cried because his father didn't love him: some *asked God for razors!* Many who thoroughly understood praying for others did not know how to pray for themselves, as Moses, Elijah, and Paul. The sweetest thing he had ever learned was to let God choose for him. God gave Christ without asking; what would He give on asking? A dozen knocking Christians would bring a mighty blessing on London. After the address Mr. Sankey sang, "Knocking, knocking," and the meeting was thrown open for prayer.

The meetings at the Hall in the afternoon and evening of March 17 were most encouraging. There was a very large gathering in the afternoon, and in the evening there was not a vacant seat to be seen. Before singing, Mr. Sankey led in earnest prayer for a blessing, and then sang with solemn feeling the hymn, "O Christ, what burdens bowed Thy head." Mr. Moody read from Gal. v., with a few comments, and we sang "There is a fountain." Mr. Moody's address was from John iii. 14, continuing the subject of the previous day. He commenced by asserting the greatest sin was unbelief; that no man then present would be lost unless he refused and despised the remedy God had provided. He instanced the supposed case of a man in consumption, near death and hopeless, visited by a friend who had been cured, and who had brought the remedy that had cured him—as *a free gift.* If the consumptive died, he died because he had refused the remedy. Mr. Moody proceeded to instance various ways in which the remedy was declined and refused; illustrating as he went in broadly dramatic fashion, which caught and held the thousands of enthralled listeners. "Suppose," said he, "a young man dying of serpent-bite. His mother waiting, watching, hearing of the serpent just set up, and putting her arms, strengthened by love, round her dying, perhaps her only son, and dragging him to the door of the tent. There she sees him look and live; strength returns to his limbs, color to his cheek, and joy to his heart. So is it when the sinner looks to Jesus—looks his sin and sorrow away together. Mr. Moody next pictured one bitten who would not look until he could understand the philosophy of cure by looking, and who died in his unbelief. The address concluded by the narration of an incident concerning a Jew, who came to the Chicago prayer-meeting just when it was ended. He had met the text, "It is appointed unto men once to die," etc., and it had broken him down; and he had heard of the prayer-meeting, and came there to be taught. Mr. Moody spoke to him of Jesus, but the Jew scowled in unbelief, and asked to be taught of the God of Abraham. Mr. Moody agreed to pray with him to Abraham's God for enlightenment concerning Jesus. The Jew prostrated himself with his forehead to the ground, and prayed earnestly for light and direction; and as he sought he found. "Oh, I would," continued Mr. Moody, "I could say something to move you, that would rouse the whole of you! One young man who was here last week, and declared his intention to attend the inquiry-room this week, is now in his coffin. Don't leave unsaved!" While "Almost persuaded" was solemnly sung, numbers left for the inquiry-room, whither we will now follow.

THE INQUIRY-ROOM—ST. MARY'S HALL.

St. Mary's Hall is a large concert-room, with chairs on the floor fronting the platform, and a deep gallery round the sides and end of the hall. Mr. Moody divided the inquirers, leaving the women on the basement, and sending the men into the gallery, and directed the workers to divide in the same way. All round the gallery were men in twos and threes, to the number of two or three hundred—each couple or three separated from their neighbors, and earnestly engaged in their own work, without taking any notice of those near and around. Here was a couple discussing a difficulty in the way. There another couple earnestly reading passages of God's word. Next was one pleading earnestly with another. Next one whose work was done, as the close, loving hand-clasp showed. Many were striving together in prayer, two by two. Here a worker earnestly asking for the light to come. There another pressing the inquirer to pray for himself, and others praying earnestly together. The writer had the pleasure of speaking with three in succession. The first was a young man who had made long, wearying endeavor to work out salvation; he had been *trying hard* to come to Jesus, but neither work nor trial had brought the assurance of faith. To one so much in earnest it was most pleasant to show salvation as the *gift* of God, and a little patience was richly rewarded by the dawning of the light. Then said he, "*I see it now; please to leave me alone with God!*" Most reverently and willingly this was done, and the second was spoken to; he also promised to accept the gift, and left to kneel before the Lord in seeking, as he was compelled to go.

The third had long had a form of godliness, but neither its power nor hope—he was just a sleeping nominal church member, who did not wish to be disturbed. He had wandered into the inquiry-room, thinking it was public, and he should hear an address. Unable to deal satisfactorily with him, the attention of another brother was called to him, and we passed on round the gallery. On returning, this one was praying earnestly, the second was gone, and the face of the first showed better than any words that he had lost his burden. Passing below to leave, a lady who was talking to three working girls claimed help, as help had been claimed in the case above. We held conversation, and speedily all three declared themselves on the Lord's side; and the bright, earnest young faces glowed with the thought of the gift received, and the "covenant unto death" with Jesus. As we saw, so we heard of many to whom light and peace came; nor was it the least impressive to mark how willingly help was given and received, how entirely absent were evidences of self and self-seeking. Conversions all around, an atmosphere of prayer and the word of God, the subdued hum of conversation with each other, and converse with the Father through the Son, gave a sense of "nearness of access," of personal presence, of a very present and loving help, that was as sweet as it was solemn. Verily it "was good to be there!" It was just eleven o'clock when, after three hours of delightful service, "the labor was done, and the laborers gone home."

Thursday, 18th.—There was the usual crowded hall at the noon prayer-meeting on Thursday, March 18. Mr. Moody spoke on "Prayer," specially the disciples' prayer, commonly called the "Lord's Prayer;" but, said he, the Lord's Prayer is found in John xvii. The principal point was the forgiveness in order to be forgiven. When he spoke of a woman whom he had exhorted to forgiveness, but who would not, he told she could not be saved until she forgave her foe. "Then," she replied, "I'll never be saved, for I'll never forgive her;" and she went mad! He spoke also of two girls who were impressed, who had been at variance, but forgave each other, and were themselves forgiven. It must not be, as some said, I can forgive, but cannot forget; but must be, as God does, both forgive and forget. He spoke also of believing we received what we desired. Speaking of an ophan boy who had been adopted into a family, and was asked if he could pray, and responded by praying as he had been taught by his dead parents, and adding, "Please make these as kind to me as my own father and mother were, won't you, Lord? of course you will!" After the address Mr. Sankey sang, and several brethren engaged in prayer. One of these ended by repeating the *disciples'* prayer, in which the whole gathering joined, producing a most striking effect; for as the subdued voices rose and fell, it was with a thrilling grandeur of sound, resembling heavy artillery heard far away.

The service in the Hall in the afternoon and evening showed clearly how the wave of attraction is rising higher and higher; though, perhaps, the unusual mildness and beauty of the day might have allowed many to attend who had hitherto been windbound. The afternoon service commenced with "Wondrous love," prayer, "Stand up for Jesus," and reading of part of 1 Cor. xv. by Mr. Moody. Then, by special request, Mr. Sankey sang the "Ninety and nine." Mr. Moody's address was on the word "Gospel," or "good news." The gospel was angelic news, and it was sung before it was preached. It was the knowledge of the life and death of the Son of God *for us!* It was the sight of Jesus; at which sight down went Paul into the dust, when he *drank* so deep a *draught* of conviction that he *couldn't eat* for three days! Every man likes his enemies out of his way, and the gospel took our three great enemies—sin, death, and judgment—out of our way for ever. For though we might die, death had nothing; the sting of death was buried in the bosom of the Son of God. The frontier men on the prairies, when they were on fire, set fire to the part near them, and when it was burnt bare stood upon it, and so saved their lives. There's one mountain-peak the fire of God's wrath has swept over, and now it is safe for ever, and that is Mount Calvary. Then he told of a father and son who were at enmity for years, but were brought together by the dying wife and mother, but only reconciled over her dead body; so the sinner was reconciled to God over the dead body of the Lord Jesus. Mr. Sankey sang "Come home," and the meeting was adjourned to the inquiry-room, whither many retired.

At the repeated service in the evening, at eight o'clock, the commencing hymns were the 100th Psalm and "Rock of Ages.' The lesson of the afternoon was also repeated, and Mr. Sankey sang "I love to tell the story," the audience joining in the chorus. The great hall was crowded to the utmost limit, the attention was most profound; and when the address closed, Mr. Moody announced that the inquiry meeting would be held in the gallery, and the prayer-meeting in St. Mary's Hall. Unhappily, this change of plan had not been made known to the stewards; they could not hear Mr. Moody in consequence of the people moving away, and the confusion that resulted was dreadful — thousands of people pressing in different ways, the centre of the crush circling round and round in vain efforts to escape, while the attendants shouted confused and contradictory direction. The unravelling was found by the sheer force of pressure sweeping the crowd into the wide street in front of the building. It is only needful to remember what happened at the Surrey Gardens when Mr. Spurgeon preached, and to think what might have been, to feel and express deep and devout thankfulness that matters were no worse.

Friday, 19th. — On Friday, March 19, Mr. Moody presided for the last time for the week at the noon prayer-meeting. The weather was bad, the audience greatly thinned, and Mr. Moody appeared to be suffering either bodily or mentally, very different from his usual happy self-possession. In his address on "Prayer," some things were said which had been far better omitted, being impossible to harmonize with his own exhortations to unity in work, so often repeated. Altogether, the meeting was far beneath that of the day previous in numbers, spirit, and power.

The subject of the address on Friday afternoon and evening was "Salvation." There was a very large attendance in the afternoon, but in the evening it seemed as if a human form was planted in every possible place. In front of the platform, and on every step of it, there were people crowded as closely as possible. The first hymn was "Wondrous love;" prayer followed; "I hear the Saviour say" came next; and then reading and brief exposition of Ps. xl. and part of Acts xvi. Mr. Moody remarked on Paul and Silas, with backs bleeding, feet in the stocks, and no supper, praising God—suggested that our praise, if there, would have been, "Hark! from the tombs a doleful sound." Mr.

Sankey sang, "Yet there is room;" and Mr. Moody gave the notices for the next week and commenced his address. He seemed a little straitened at first, but soon recovered all his wonted fire, and delivered a red-hot discourse on seeking and finding salvation. In speaking of leaving all earthly trust, he mentioned a miller who, in a boat asleep, came near the jaws of death by the mill-dam; he found a twig which could only stay his deathward progress; he therefore held on, and shouted for help with all his power. A friend heard and *let down* a rope (the help must come from above), but the miller could not grasp the rope until he let go the twig. This he did, and was rescued. Mr. Moody next narrated the well-known story of his own conversion by the ministry of the teacher in Boston, and his after meeting and influencing for Christ the teacher's son. A most earnest appeal ended the discourse, and the service concluded with prayer. The after-meeting was held in two inquiry-rooms; one in the gallery over the platform, where a large number of inquirers gathered, and many more than double the number of workers; and a few were also in St. Mary's Hall, with the same large preponderance of workers.

Now unto Him that is able to do exceeding abundantly above all that we can ask or think, according to the power that worketh in us, unto Him be glory in the Church, by Christ Jesus, thoughout all ages, world without end. Amen.

MEETINGS ON SUNDAY, 21st.

The morning service at the Agricultural Hall, which was for Christian workers only, was one of the most satisfactory meetings that have been held in London, there being about 16,000 persons present, all of whom were either Sunday-school teachers or persons employed in similar Christian work. The afternoon service, which was for women only, was attended by about 14,000. The evening service was for men only. There were about 19,000 present, as against fully 20,000 the few preceding evenings. Mr. Sankey sang the favorite hymn, " Jesus of Nazareth passeth by;" Mr. Moody taking for his text the words, " He that believeth on me shall be saved." The evening service being for men only, many who were not aware of the restriction came with their wives, whilst others from a distance came with young women. It was in vain that they pleaded that they had come many miles by train; the orders were peremptory, and no women were admitted. The committee, to accommodate the large number of women who had come, threw open St. Mary's Hall, and held a service there, conducted by Mr. Leithes, of Liverpool. At this service there were about 2,000 women present, many of whose companions were in Agricultural Hall.

III.

THE FIRST MONTH IN LONDON.

THE *Christian World* of April 6th thus summarizes the first month's work:

To-day the American Evangelists, whose names are on every lip, enter upon the second month of their London campaign. They have all but completed the series of meetings at the Agricultural Hall, in Islington, designed more especially for the benefit of the people dwelling in the great northern region of the metropolis; and now they are about to enter on the daily occupation of a building specially erected for their accommodation at the East End. From week to week we have furnished our readers with full reports of the proceedings. In this way, the public have been enabled to obtain a comprehensive, and we believe accurate, view of a series of meetings that certainly stand without a parallel in the religious annals of England. We may not be able to say, with a respected contemporary, that Mr. Moody is the modern Wycliffe — a name we should rather assign, if we used it all, to a great English preacher who has been proclaiming the Gospel to multitudes in London every week for more than twenty-one years. Neither are we prepared to coincide with the magnanimous assertion of a Wesleyan Methodist journal, that this movement puts the revival which was wrought by Whitfield and Wesley into the shade, in respect, at least, to the numbers brought under the sound of the Gospel. These are statements, as it seems to us, which would require to be greatly qualified before they could be accepted by thoughtful men. Yet, without going the length of our too exuberant friends, we can testify that the success of the gatherings over which Mr. Moody presides has been simply marvelous, and in its way quite unexampled, either within the memory of living men, or in all that has been recorded by the pen of the English historian of the Christian

Church. Whatever may be the view he takes of the work, as to its true spiritual significance and value, every candid onlooker must acknowledge that the present is a phenomenon which cannot be too carefully scanned, or too fully described by the contemporary journalist. It will unquestionably claim for itself a chapter of no inconsiderable magnitude in the book that deals with the religious history of England in the last quarter of the nineteenth century. Some little service to the future, as well as to the present-day reader may, therefore, be rendered by an attempt to gather up the salient points in the story of the first month spent by Messrs. Moody and Sankey in London.

And first of all we have to note the sustained, and it would even seem growing, interest which the public take in the meetings. Every day at noon Exeter Hall has been well filled; often it has been crowded, and there is no symptom of any falling off in the attendance, while it may be confidently expected that when the prayer-meeting is transferred, as it will be on Monday next, to Her Majesty's Opera House, the audience will be as great as that building is able to contain. That the interest in the primary purpose of the noon-gathering has not declined is made manifest by many pleasing tokens. Not the least eloquent of these was the statement made by Mr. Moody on Wednesday last, that the requests for prayer received that morning numbered no fewer than 180. The reports of spiritual work achieved in connection with the movement, not only in London, but also in the provinces, have been multiplying daily; and these form a feature of the proceedings at Exeter Hall which does much to keep alive the interest and to intensify the fervor of the assembly. Then there has been the appearance of new speakers from day to day—witnesses to the reality of the revival in Scotland, Ireland, and provincial towns of England. When the meeting is thrown open to volunteers, the result has not always been edifying; but Mr. Moody, as a shrewd and ready-witted president, keeps the most of the time well occupied with a swift and flowing succession of song, prayer, and exhortation, so that the hour seems to all present to be only too short, and is obviously most refreshing to their spirits. Mr. Moody is, perhaps, seen at his best at Exeter Hall. Some of his short addresses there have been gems of pithy exposition; and his occasional quaint bits of self-defence, and frequent touches of mingled humor and pathos, have been remarkably effective. People from the country have formed a distinctly perceptible element in the congregation; and we cannot doubt that these, along with the city brethren, have derived useful hints from Mr. Moody's method for the conduct of prayer-meetings in their own places of worship. In this way, we think it likely that a great deal of good may be done.

The three afternoon meetings held at Sanger's (formerly Astley's) Amphitheatre, were among the most successful of all the gatherings, and are said to have been the most fruitful in spiritual results. The place could not hold all the people who flocked to them; and a proportionately larger number of the "lapsed masses" were to be seen in these South-side gatherings than in the assemblies at the Agricultural Hall. The two afternoon Bible readings—the first held in the Conference Hall at Mildmay-park, and the second at Exeter Hall, and to both of which admission was procured only by ticket—were crammed, and they seemed to be greatly enjoyed.

As for the great meetings, those held every night (with the exception of Saturday) at the Agricultural Hall, and thrice on Sunday in the same enormous edifice, they have continued to attract an average attendance of at least eleven or twelve thousand down to the very last. On the two nights when the address was not given by Mr. Moody there was a great falling off in the congregation. On Good Friday the *Times* "felt bound" to express its "strong conviction that the interest of the meetings was rapidly falling off;" but the facts do not sustain this view. The largest congregations have assembled within the last ten days; and these have included all ranks and classes of society. Royalty itself, in the person of her Royal Highness the Duchess of Teck, has expressed its intention to come since the leading journal proclaimed the turning of the tide. On one evening there were at least sixty clergymen of the Establishment present, with Dean Stanley occupying a conspicuous seat on the platform; and on the night of Good Friday the evangelical Earl of Shaftesbury sat on the same chair which a few evenings before had been occupied by the Broad Church Dean. Lord Shaftesbury, at the close of the service, paid a visit, along with his daughters, to the inquiry-room. In respect to the numbers of the Agricultural Hall congregation, the floor of the building is capable of seating 9,000 persons; the raised platform for the choir and ministers, 250; the eastern side gallery, 900; the western side gallery, 1,000; the upper raised gallery in front of the platform, 1,350; the balcony in front, 850; and the upper western balcony, 350. Even on moderate computation, it would seem, that about 350,000 must have been the total of the numbers present at the Agricultural Hall services during the month; though it must be borne in mind that very many persons were frequent, and not a few constant, attenders. It would probably be a liberal allowance if we were to say that 200,000 separate individuals were present. The arrangements made by the committee for the comfort of the congregation and the preservation of order have, from first to last, been admirable.

With respect to the inquiry-rooms, they have been largely attended every night by Christian friends, clerical and lay; and the penitents pressing in for spiritual advice have, on many occasions, numbered several hundreds. But there has been no more excitement there than in the public service; indeed, the proceedings have been more subdued, and a quiet, solemn earnestness has characterized all that has been done in connection with this part of the work. Several gentlemen taking part in it have testified to the good accomplished; and Mr. Sankey in particular, who is active in the inquiry-room, describes the work of which he was witness on Sunday week, and on every succeeding night, as being in the highest degree encouraging. Many Christian workers, though not so many as Mr. Moody desires to see, have scattered themselves among the great audience at the ordinary services, for the purpose of speaking a word to their unconverted neighbors; and a case has been mentioned in which the young ladies of a certain seminary have, in this way, been instrumental in leading twenty individuals to the Saviour. With this we may bracket the case of a lady who took her ten servants to one of the services, and who reports that seven of these have been, in consequence, converted to God. Mr. Moody has detailed instances of persons brought to a knowledge of God in the inquiry-room one night, and appearing on the next with friends whom they desired to see sharing the peace which they had secured. Since the second Sunday a young men's meeting has been held every night at St. Mary's Hall, immediately after the public service; and latterly this feature has come more conspicuously into view, and been more pressingly urged upon the attention of the class referred to by Mr. Moody, who is ambitious of securing a band of at least a thousand to assist him in his work.

IV.

THE following discriminating, candid, and exhaustive review of the work of the Evangelists, is from the pen of the Rev. R. W. DALE, the successor of John Angel James, at Birmingham. It appeared in the March number of the *Congregationalist*, and has since been widely circulated in pamphlet form. It is here reprinted in full, and is well worthy a careful perusal.

An article which appeared in the *Congregationalist* for December, 1872, under the title "Have we Forgotten Christ?" closed with the following words: "Already there are signs that the power of Christ is ready to reveal itself again. In every part of the country, the despondency which has been occasioned by the depressed condition of the spiritual life in Christian people themselves, and the inconsiderable success of the Gospel among those who are outside, is giving place to courage and hope. Are we ready to receive the returning Christ? Many have prayed Him to come back, or rather to reveal His presence, which has never really been withdrawn from us. Have we learnt how sorely we need Him? Are we prepared to fall at His feet, and to confess that 'apart' from Him we 'can do nothing'? If we meet Him as we should, there are the strongest reasons to believe that He is about to baptize us afresh with the Holy Ghost and with fire."

During my absence in the East, the *Congregationalist* contained a series of articles on "Religious Revivals," written before I left England; and I had so deep a conviction that a great manifestation of the power of God was at hand, that I returned with a strong hope that I should find Church after Church in different parts of the country, bright with a new joy, and on fire with a new zeal. The hope was not fulfilled, and yet it was not altogether disappointed. At Derby, at Ipswich, and in some other places, there was already the dawn of a new day; and in many directions the darkness was beginning to melt, and those who had been long watching for the morning were growing more and more confident that the night was nearly gone.

In what form the new spiritual movement would come, or by what agencies, it seemed impossible to predict. In the series of articles to which I have referred, it was earnestly maintained that "if in our own times God comes to us in the greatness of His power and in triumphant love, His coming may not be manifested in precisely the same forms as in any of the great Religious Revivals of former days, and may not produce the same effects."* The reformation of the monasticism, and the great religious movement associated with it, extending from the close of the eleventh century far into the thirteenth; the Waldensian revival, which covered a part of the same period; the very remarkable outburst of religious life in the Low Countries in the fifteenth century; the Protestant reformation of the sixteenth century; English Puritanism; English Methodism, —were singularly unlike each other; but they were all the results of fresh communications to the Church of the life and light and power of the Holy Ghost. In one case there was the earnest and vehement preaching of Christian morality; in another there was a clearer apprehension of those spiritual truths which touch, and perhaps cross, the boundaries of Mysticism; in another there was a revolt against a priesthood that had separated the Church from God, and a rediscovery of the doctrine of Justification by Faith; in another a strong assertion of the necessity of the new birth. The men who, under God, did the work, differed greatly; they were monks; they were common people; they were popular orators; they were scholars. Some of them wrote books, others preached sermons. Some had remarkable powers of organization, and have stamped their names on great and permanent ecclesiastical institutions; others left the new life to take form according to its own laws, or to quicken the existing organization of the Church.

I thought it possible that in our own time the power of God might be specially manifested among children and young people. Nor has this expectation proved altogether unfounded. In several parts of England there has sprung up a beautiful and happy religious life among children,

* *Congregationalist*, Jan. 1873, page 2.

which is the promise of very large results, if we remember with devoutness and faith the words of Christ: "Suffer the little children to come unto Me."

But I certainly did not suppose that several of the great towns of the three kingdoms were to witness a remarkable religious movement, originated by two American strangers, one of them a man who had been trained for his work by his experience as a Sunday-school superintendent, and the other with a fine baritone voice and playing an American organ.

A few years ago I had read, week after week, with great interest, the reports in the Chicago *Advance* of Mr. Moody's addresses at the noon-day prayer-meeting in that city; but I had never heard of him as an evangelist. Indeed, until he came to England he had never taken an evangelistic journey.

It is not my purpose to attempt any general view of what these two guests of ours have done—or rather of what God has done through them—since they have been on this side of the Atlantic. They began their work, I believe, in York; but in York they had very little success. Their first great impression was made in Newcastle. In Glasgow, Edinburgh, and Dundee, the impression was still greater; in Dublin and Belfast greater still. At Manchester and Sheffield they collected vast crowds of people, and there is reason to believe that in both places a very considerable number of persons were led to repent of sin and to confess the authority and mercy of Christ.

During the last fortnight of the month of January they were in Birmingham. Their first meeting was held on Sunday morning, January 17th, at 8 o'clock, in the Town Hall. The meeting was for "Christian workers," and the admission was by ticket. The morning was cheerless, damp and raw; but the great building was crowded in every part. In the afternoon they held an open service in the Hall, and thousands went away unable to get in. The great test, however, of the measure of the expectation which they had excited came in the evening. Last October twelvemonth, when Mr. Bright addressed his constituents after his return to the Cabinet, he spoke in Bingley Hall, a building used for the annual cattle show, and as a drill hall for the volunteers. Various estimates were made of the number of people that listened to Mr. Bright on that occasion; it seems probable that most of them fell far short of the truth. At Mr. Bright's meeting in October, 1873, there were no seats on the floor of the hall, and without seats there is now reason to believe that the hall will hold between twenty and twenty-five thousand people; it was crowded in every part. For the recent religious meetings, the "Moody and Sankey Committee" hired upwards of nine thousand chairs. On the very first Sunday evening, long before eight o'clock, when the service commenced, not only were all the chairs occupied, but several thousand people were standing, and thousands more could not gain admission. It is difficult to estimate accurately the real magnitude of such a crowd; but I am inclined to think there were thirteen thousand people present. Every night through the first week the Hall was thronged in the same way, and there were vast crowds outside.

On Sunday morning, January 24th, it was filled with people who obtained admission by tickets, and who before they received their tickets declared that they were not in the habit of attending any place of worship. In the afternoon of the same day it was filled with women, and a second service was held in the Town Hall for the overflow. In the evening it was filled with men. There was a break on the Monday of the second week, when Mr. Moody had an engagement at Manchester, to meet those who professed to have received Christ during his visit to that city. Mr. Bright spoke in the Hall that night, and it was most inconveniently crowded; but some of the police were of opinion that on several of the following evenings the crowd that filled the Hall for the religious services was denser than that which filled it for the political demonstration. Night after night, long before the hour of service, long rows of carriages stood in the street, filled with persons who hoped that when the crowd about the doors had thinned, they might be able to find standing room just inside, and thousands streamed away because they found they had come too late to have a chance of pressing in.

In addition to the evening service, there was a prayer-meeting every morning at twelve o'clock, at which Mr. Moody gave an address of twenty or twenty-five minutes' length, and Mr. Sankey sang. The meeting was held at first in the Town Hall, which was generally quite full; on the last four days it was held in Bingley Hall, and the attendance varied from four to six thousand. At three o'clock, after the first day or two, Mr. Moody gave a "Bible lecture;" he began in Carr's Lane Chapel, which was soon found to be too small. It was then transferred to Bingley Hall, and the attendance varied from five to seven thousand.

How is all this to be accounted for?

"You advertised the Americans well," it has been said, "by holding special prayer-meetings every day for three weeks before they came—prayer-meetings in which all the Evangelical Non-conformists and some of the Evangelical clergy united." Well, no doubt the prayer-meetings were a kind of "advertisement" of the services, and assisted to attract large numbers on the first few days.

It is said again: "The local newspapers helped you. One of them published a series of articles on Mr. Moody and Mr. Sankey before they came, describing the impression they had produced in Scotland and Ireland. The *Morning News* generally gave several columns day after day to reports of the services; the *Daily Post*, though prevented by pressure on its space from reporting the services at equal length, gave great prominence to them; and even the local Conservative organ, the *Daily Gazette*, always had enough about 'Messrs. Moody and Sankey' to attract attention." Granted: the Birmingham newspapers helped us greatly.

It is also true that the local Committee advertised the services most efficiently. The walls of the town were covered with their placards, and these were constantly renewed. Further, it must be acknowledged that when once it was known that Bingley Hall had been filled to hear the strangers, a certain measure of popular excitement and curiosity was created, which made it almost certain that the hall would be filled again.

I have had some experience, however, of popular agitation. I think I know pretty well what is likely to be effected by newspaper articles and advertisements; and these do not seem to me to explain the interest which the services created from the first. They explain still less the deepening of the interest from day to day; they do not explain at all the effects which I believe have been produced.

Some people have said that it is easy to get crowds of women to 'hysterial" religious services. But although the morning and afternoon meetings were largely attended by women, I believe that the majority of the evening congregation always consisted of men, and of men of all kinds—rough lads of seventeen or eighteen, workingmen, clerks, tradesmen, and manufacturers. I happen to have on my desk a list of persons that came into Carr's Lane Lectureroom one evening to tell me that they had "found Christ" during the fortnight that Mr. Moody and Mr. Sankey were here; out of twenty-one on the list, eleven are men. I have another list of persons who came to me the same evening who had been quickened to earnest religious anxiety, but were not yet at rest; out of thirteen, eight are men. I believe that these lists imperfectly represent the proportion of men to women among those who were impressed by the services, for I generally find that men are slower to express religious decision than women.

Nor were the services at all "hysterical;" the first sign of hysterical excitement was instantly repressed by Mr. Moody, and although I attended a very large number of the meetings, I saw nothing of the kind again. It was very curious, too, that although the crowds were so enormous, very few women fainted. I do not remember more than three or four cases.

The most plausible explanation that I have heard from an "outsider" was suggested to me by a Unitarian friend, who said that since all the Evangelical Non-conformists and some Evangelical Church people united to make the meetings a success, it was inevitable that many thousands of people should come together. But it so happens that of all the towns in the kingdom, of which I know anything, Birmingham is the least curious to listen to strangers, whatever their reputation and on whatever subject they may have to speak. The Birmingham people are very loyal to their own leaders, and seem to care very little about men who come from a distance. The Evangelical Non-conformists are no exception to this rule.

How, I ask again, is the great interest of the people in these services to be accounted for? The truest, simplest, and most complete reply to the question which I can give is, that the power of God was manifested in an extraordinary degree in connection with them; but there were concurrent circumstances which deserve notice.

(1) As I have said, I attribute very much to the attention and expectation excited by the preliminary prayer-meetings;

I attribute still more to the articles in the local newspapers, describing the impressions which had been produced by Mr. Moody and Mr. Sankey in other parts of the kingdom. I also attribute very much to the reports of "revival work" which have appeared for many months in such newspapers as *The Christian World* and *The Christian*—reports which have convinced large numbers of religious persons that the services of our American visitors have originated a religious movement more remarkable than any we have seen in England since the middle of the last century. Thirty thousand copies* of *The Christian*, containing an account of the services at Manchester, were distributed in the congregations of the town a week or two before Mr. Moody and Mr. Sankey came to us.

(2) I attribute very much to a fact which is perhaps not sufficiently recognized by any of us. There are, I believe, a very large number of persons—many of them regularly attending public worship, many of them never crossing the threshold of church or chapel—who have had deep religious impressions, which have not issued in a clear decision to serve Christ, but which have left a dull aching of heart for God. The sense of dissatisfaction with their condition never wholly leaves them; it sometimes makes them very restless. But when they listen to the preaching of most of us, they feel as if we were moving in regions which are inaccessible to them. If they come to our places of worship, they come without any hope of receiving help. Many of them, having found that we do not help them, never come at all. When such people heard that within a very few months thousands of men and women had declared that, while listening to Mr. Moody and Mr. Sankey, they had passed from religious indifference or despondency into the clear light of God, they began to think that for them too there might be hope. I think it probable that many of the "converts" will be found to have belonged to this forgotten class.

(3) There must be large numbers of persons in Birmingham who have relatives and friends in the towns that the American Evangelists had visited before coming to us; and I have no doubt that mothers, brothers, sisters, cousins, old school-fellows, and old shop-mates wrote urgent letters to

---

* I think this was the number, but am not quite certain.

them, entreating them to attend the services. At one meeting for "inquirers" I met a young man who seemed quite careless about religious thought and duty, and I asked him how it was that he remained to that meeting. He told me he had promised his friends "to go to the Moody and Sankey meetings;" and he seemed to suppose that to remain to the inquirers' meeting was part of the process to which he was pledged to submit himself.

(4) After the first day or two, the services were "advertised" in a very much more efficient manner than by newspapers or placards : every evening, at the "after-meeting," a considerable number of persons received Christ as their "Prince and Saviour," and, judging from those with whom I conversed, most of them went home with overflowing joy. I had seen occasional instances before of instant transition from religious anxiety to the clear and triumphant consciousness of restoration to God; but what struck me in the gallery of Bingley Hall was the fact that this instant transition took place with nearly every person with whom I talked. They had come up into the gallery anxious, restless, feeling after God in the darkness, and when, after a conversation of a quarter of an hour or twenty minutes, they went away, their faces were filled with light, and they left me not only at peace with God but filled with joy. I have seen the sunrise from the top of Helvellyn and the top of the Righi, and there is something very glorious in it; but to see the light of heaven suddenly strike on man after man in the course of one evening is very much more thrilling. These people carried their new joy with them to their homes and their workshops. It could not be hid. On the Sunday after Mr. Moody and Mr. Sankey had left us, I invited those members of my own congregation to meet me who had come to Christ during the services of the preceding fortnight. A few who were still out at sea longing to make their way to quiet water came with them. Nothing was easier than to tell the difference between the two classes; I think I could have separated them into two divisions without asking a question and with scarcely a mistake. Those who were still "inquirers," if they did not look anxious and troubled, looked like other people; the "converts" were bright with their new joy. It is as yet too early to obtain any general information about the extent of the influence which I have attributed to the converts themselves; but among the names that I have on several lists of persons that I saw myself, I find the names of two clerks who sat side by side at the same desk, three pairs of brothers and sisters, three husbands with their wives; and four brothers—rough, working men — all of whom have been awakened to religious thought by Mr. Moody's addresses.*

(5) Nearly all the "living" and active members of the various Evangelical Churches hoped that the services would achieve great results; and many Christian people whose religious life was depressed and sad, trusted that they might find their way to the light.

(6) Direct efforts were made to induce those who had not been at any of the meetings to come to them. In one manufactory in which 600 people are employed, I believe that there was an attempt to induce all who were not in the habit of attending public worship to go to the special meeting that was held for that class of persons. Hand-bills were distributed from house to house in the poorer parts of the town. Very many persons of all ranks, who had become interested in the services, urgently pressed their friends to go with them to hear the American strangers.

(7) The services themselves were attractive.

Mr. Sankey's solos evidently touched very many hearts; and the effect produced by the manner in which the vast audiences united in such songs as "Hold the fort, for I am coming," and "Safe in the arms of Jesus," and "The Great Physician now is near," was sometimes very thrilling. The "songs" have been sharply criticised. It is very easy to criticise them; it might be more profitable to consider why it is that both the music and the words are so popular and effective. About their popularity there can be no doubt. There were sometimes ten or twelve thousand people in Bingley Hall for more than an hour before the services began. With intervals of a few minutes they occupied themselves with the more popular hymns and melodies; and the delight with which they sang them was obvious. Passing along the streets I

---

* Some of these are not persons with whom I had conversation at the "after-meetings," but are persons who have given their names to me as wishing to enter Carr's Lane Church.

hear men whistling "Safe in the arms of Jesus." I have long held the conviction, and often expressed it, that the reformation in our Psalmody which has been going on for the last five-and-twenty years, though it was very necessary, and though in some particulars it has been very admirable, is, in some respects, unsatisfactory.

The tunes which were sung by Nonconformist congregations thirty years ago were often vulgar, but they were real tunes, easily learnt, easily remembered; and they haunted people during the week. Most of them were destitute of artistic merit, but the people liked them, and they were the natural expression of their emotion. Many of the new tunes are not "tunes" at all. They are not vulgar, but they are uninteresting. They differ from their predecessors very much as the dullness of a "respectable" dinner-party differs from the merriment of a picnic at which the people are just a little unrefined, but at which they have resolved to enjoy themselves. I do not like either, but on the whole I prefer the picnic. The men who have composed or adapted the new tunes are for the most part organists, who know very much more about how to get solemn effects out of their instrument than how to give the people something to sing. Mr. Sankey's melodies — whatever their demerits — are caught by thousands of people of all kinds, cultivated and uncultivated, men, women, and children, and are sung "with a will."

I agree with those who say that we ought, if possible, to get really good music for God's service, but it must be on one condition: that we do not sacrifice "God's service" to the "good music." Our first business is to enable Christian congregations to give free and happy expression to their joy and trust in God's love, and their reverence for God's majesty: the promotion of their musical taste is a matter of only secondary importance. Moreover, my contention is that much of the new music differs from the old chiefly in one particular: there is not more musical genius in it, but less life. Let a scientific musician write tunes which lay hold of the imagination and heart of all kinds of men as powerfully as some of those which Mr. Sankey has brought together in his little book, and most of Mr. Sankey's melodies will soon be forgotten.

The same principles are applicable to the hymns. Critics have said that they are "childish," that they have no "literary merit," that there is something ridiculous in hearing a congregation of grown people singing with enthusiasm, "I am so glad that Jesus loves me." Well, the fact that hymns which are simple even to childishness are sung by grown people with so much earnestness, that hymns with no "literary merit" kindle new fire in the hearts of men and women who know something of Shakespeare, Milton, and Wordsworth, is surely worth investigating. Is it the "childishness" which accounts for their power? Is it the absence of "literary merit?" I think not. Give the people a collection of hymns characterized by equal fervor, expressing with the same directness the elementary convictions and the deepest emotions of the Christian heart, and if they have also the literary merit which is absent from many, at least, of Mr. Sankey's songs, they will become equally popular, and their popularity will be more enduring. But our hymn-books are too stiff and cold. People want to sing, not what they *think*, but what they *feel*; and if they are asked to sing hymns in which there is no glow of feeling, and in which the thought is perfectly commonplace, they will not sing at all. "I am so glad that Jesus loves me" is a childish way of expressing our joy in the love of Christ; but if hymn-writers will not help us to express it in a more masculine way, we must express it as best we can. How few hymns there are in our language which express thanksgiving for salvation in a popular and really lyrical form! how few which express exultation in the large freedom which is the inheritance of those in Christ! Again, it is of no use asking people to sing to God in a language remote from the language of their common life: hence one of the difficulties of writing a really good hymn. There is similar difficulty in writing good secular songs; we have an infinite number of songs which are musical in their language, and graceful in their thought, but which have never found their way to the heart of the nation; the number of songs which have really high literary merit and are also popular is perhaps smaller than the number of successful hymns. Mr. Binney's "Eternal Light" has the simplicity, fervor, and dignity which constitute a perfect hymn; but I am not sure whether its dignity does not impose a kind of strain upon very many minds, which though very good for them occasionally, interferes with their delight in singing it. There are, however, comparatively few hymns which combine the simplicity necessary both for the cultivated and uncultivated in acts of happy thanksgiving, praise, and worship, with elevation of thought and manner.

But it was not the singing only which made the services interesting: there was great animation and variety in them. In the evening they began with a hymn which the people sang together; but what would be the "order" of the service no one knew, and I suspect Mr. Moody did not know beforehand. Every man who is accustomed to conduct public meetings for any purpose can easily tell whether the people are interested: Mr. Moody has this instinctive perception in a remarkable degree.

After the first hymn somebody generally offered a short prayer; if it was clear that the heart of the audience went with the prayer, he would then read a chapter and make a few remarks upon it as he read; if not, he would ask Mr. Sankey to sing a solo, or a solo with a chorus in which the people joined, or else one of the most popular hymns. Then he would read the chapter, and perhaps have another hymn or offer a short prayer himself. Then would come another hymn, and then the sermon. Sometimes the sermon was followed by a solo from Mr. Sankey, sometimes by a hymn in which all united, sometimes by a prayer. Everything was determined by what was felt to be the actual mood of the moment. Generally the whole service was over in a little more than an hour and a quarter. Then came the "after meeting," of which I will say something presently.

Of Mr. Moody's own power I find it difficult to speak. It is so real, and yet so unlike the power of ordinary preachers, that I hardly know how to analyze it. Its reality is indisputable. Any man who can interest and impress an audience varying from three thousand to six thousand people for half an hour in the morning, and for three-quarters of an hour in the afternoon, and who can interest a third audience of thirteen or fifteen thousand people for three-quarters of an hour again in the evening, must have power of some kind. Of course, some people listened without caring much for what he said; but though I generally sat in a position which enabled me to see the kind of impression he produced, I rarely saw many faces which did not indicate the most active and earnest interest. The people were of all sorts, old and young, rich and poor, keen tradesmen, manufacturers, and merchants, and young ladies who had just left school, rough boys who knew more about dogs and pigeons than about books, and cultivated women. For a time I could not understand it—I am not sure that I understand it now. At the first meeting, Mr. Moody's address was simple, direct, kindly, and hopeful; it had a touch of humor and a touch of pathos; it was lit up with a story or two that filled most eyes with tears; but there seemed nothing in it very remarkable. Yet it *told*. A prayer-meeting with an address, at eight o'clock on a damp, cold January morning, was hardly the kind of thing —let me say it frankly—that I should generally regard as attractive; but I enjoyed it heartily; it seemed one of the happiest meetings I had ever attended; there was warmth and there was sunlight in it. At the evening meeting the same day, at Bingley Hall, I was still unable to make it out how it was that he had done so much in other parts of the kingdom. I listened with interest; everybody listened with interest; and I was conscious again of a certain warmth and brightness which made the service very pleasant, but I could not see that there was much to impress those that were careless about religious duty. The next morning at the prayer-meeting the address was more incisive and striking, and at the evening service I began to see that the stranger had a faculty for making the elementary truths of the Gospel intensely clear and vivid. But it still seemed most remarkable that he should have done so much, and on Tuesday I told Mr. Moody that the work was most plainly of God, for I could see no real relation between him and what he had done. He laughed cheerily, and said he should be very sorry if it were otherwise. I began to wonder whether what I had supposed to be a law of the Divine kingdom was perfectly uniform. I thought that there were scores of us who could preach as effectively as Mr. Moody, and who might therefore, with God's good help, be equally successful.

In the course of a day or two my mistake was corrected; but to the last there were sensible people who listened to him with a kind of interest and delight with which they never listen to very "distinguished" and eloquent preachers, and who yet thought that though Mr. Moody was "very simple and earnest," he had no particular power as a speaker. I do not intend to suggest any comparison between Mr. Moody and our great English orator, but I have met people who have talked in the same way about Mr. Bright, and who seem to think that to speak like Mr. Bright was possible to nearly everybody.

One of the elements of Mr. Moody's power consists in his perfect naturalness. He has something to say, and he says it— says it as simply and directly to thirteen thousand people as to thirteen. He has nothing of the impudence into which some speakers are betrayed when they try to be easy and unconventional; but he talks in a perfectly unconstrained and straightforward way, just as he would talk to half-a-dozen old friends at his fireside. The effect of this is very intelligible. You no more think of criticising him than you think of criticising a man that you meet in the street, and who tells you the shortest way to a railway station. I can criticise most preachers and speakers; I criticised Dr. Guthrie, though I was either laughing or crying the greater part of the time that I was listening to him; but somehow I did not think of criticising Mr. Moody until I had got home. Generally there seemed nothing to criticise; once or twice in the simplest and most inartistic manner, he said things which at the moment he said them I felt were of the kind to give a popular speaker a great triumph, but his whole manner threw me out of the critical attitude. Some men force you to be critical. It is impossible to take a single coin from them without ringing it on the table and looking to see whether it is properly "milled." From first to last, they provoke "watchful jealousy." It is clear that they are taking a great deal of trouble with their sentences; it is disrespectful not to examine their work. It is clear, too, that they are giving you their best thoughts, their best arguments, and their best illustrations, and they show them to you just as a collector of gems shows you his last triumphant acquisition. It is impossible—it is almost insulting—not to criticise. When a speech or sermon is plainly a work of art, criticism is inevitable. It is not necessary for anyone to paint pictures, to sing songs, or to deliver artistic addresses; but if a man insists on being an artist, and lets you know it, he forces upon you a critical examination of his performance.

Mr. Moody—so it seems to me—has an "art" of a very effective kind; but he is infinitely more than an artist, and therefore most people listen without criticising. This is an immense element of power. If our congregations came to hear us preach, instead of coming to hear *how* we preach, the effect of our sermons would be immeasurably great. Now and then Mr. Moody quoted a text in a very illegitimate sense; Now and then he advanced an argument which would not hold water; now and then he laid down principles which seemed untenable; and there was a momentary protest on the part of the critical faculty; but the protest was only momentarily. I was not thrown out of sympathy with him.

It is objected that he is too "familiar" with sacred things. Generally—not always —the objection comes from persons who are extremely *unfamiliar* with them. The fault that is charged against him—if it be a fault—is perhaps not too common in these days. There are not too many people who live, and move, and have their being in the fair provinces of Christian truth, and Christian hope, and Christian joy. Mr. Moody is, no doubt, very "familiar" with things about which he talks. He is like a man who keeps Sunday every day in the week; his mind does not put on Sunday clothes when he begins to speak about religion. Religious truth is the subject of his constant thought; he does not therefore assume the "Bible tone" when he begins to pray or preach. In one of Mr. Ruskin's books there is a very remarkable passage on ecclesiastical architecture, which has occurred to me very often while thinking of Mr. Moody and Mr. Sankey. Mr. Ruskin says that the great builders of the Middle Ages never thought of building a church in a different style from that in which they built a house. There was no "ecclesiastical" style of architecture. There were houses in every street with doors and windows and niches in the walls for saints, just like the doors and windows and niches of the cathedral. The cathedral was larger, the materials used in it were richer, the work was very much more elaborate; but when a man went to worship God he did not feel that he was in a building different in style from the common buildings about him. Mr. Ruskin does not discuss the question whether for religious reasons it is desirable to have an "ecclesiastical" style of architecture, but he insists that those who erected the great ecclesiastical buildings of the Middle Ages did not intend to produce the kind of feeling which these buildings produce upon ourselves. We feel when we are in Lincoln or Notre Dame, that we are in a building which is so distinctively religious that it would be almost profane to apply the style to common uses. This is because our houses are not built in the same style as the churches; but when those great churches were erected they were illustrations of the ordinary house architecture carried to perfection. This is Mr. Ruskin's theory; and he maintains that we can never have good church architecture until our house architecture is sufficiently noble to be used for church purposes.

Now the architecture—if I may so speak—of Mr. Moody's discourses is not ecclesiastical. The windows, and the doors, and the furniture, and the decorations are of the kind with which we are familiar in our every-day life. He does not tell stories because they are amusing; but if an amusing story helps him to make a truth clearer, or to expose a common mistake, he does not refuse to tell it merely because it is amusing. The common things of common life are about him all the time he is speaking. He uses the words of the home and the street: the plainer they are the better he likes them. The gowns and bands which some of our preachers wear are the symbols of the special costume in which they think it proper to array religious truth. Mr. Moody does without gown or bands, and speaks to men as he would speak to them at a meeting of the "United Kingdom Alliance," or at a political meeting during a contested election. He has given himself to God, all that he has, all that he is, and he uses every faculty and resource of his nature to prevail upon men to hate sin and to trust and love Christ. To him nothing is common or unclean. He has humor, and he uses it; he has passion, and he uses it; he can tell racy anecdotes, and he tells them; he can make people cry as well as laugh, and he does it.

Some people say that he is "irreverent." If he is, I must have been singularly fortunate, for I have never heard him say anything which justifies the charge. But what people seem to mean is that he does not regard with religious respect everyone that is mentioned in the Bible. Why should he? When he said that Bartimæus, after getting his sight, was eager to go home and to "see what kind of a looking woman he had for a wife, for you know that as yet he had never seen *Mrs*. Bartimæus," some people who saw the report in the newspapers thought this was a proof of the irreverence of which he is said to be guilty. But I do not know that there is any reason for speaking reverently either of Bartimæus or his wife. As a matter of taste, most of us would prefer to describe the woman as "the wife" of the blind man; but why the "Mrs." should be thought irreverent it is difficult to understand. Reverence is due to God alone, and to Him in whom God is manifest in the flesh; of God, of our Lord Jesus Christ, there was never a word which was not inspired by fervent love, perfect trust, and devout worship. Of great saints, good men will speak with affection and respect; and it was thus that Mr. Moody spoke of them.

There was something in his way of telling Scripture narratives from which preachers may learn very much. The Oriental drapery was stripped off, and he told the stories as though they had happened in Chicago just before he had left home, or in Birmingham an hour or two before the service began. At times this gave the stories a certain air of grotesqueness, but it made the moral element in them intensely real. We are in the habit of making a double demand on our hearers; we ask them, first, to reproduce, by a strong effort of imagination, the Oriental circumstances of the narratives, and we then ask them to apprehend the human passions and follies and virtues which the narratives illustrate. I believe that they get so interested in the mere drapery that the substantial facts are often missed; or else the enduring human element looks so strange in its unfamiliar costume that its power is lost. I have heard men say that of late years the scenery and the dresses at the great theatres are wonderfully improved, but that the acting is very inferior to what it once was. Mr. Moody cares nothing for the scenery and the dresses. If he were a "manager" he might bring Julius Cæsar on to the stage in the uniform of an American general, and Hamlet might put on his "Ulster" when he was going out to meet the ghost, but he would insist on making the plot and passion of the play intensely and vividly real.*

Of the aspect of the truth on which he dwells it is not necessary to say much. His great topic is the infinite love and power of Christ. That Christ wants to save

---

* To prevent misunderstanding it may be well to say I do not intend to suggest that all preachers ought to strip off the "Oriental drapery" from the Bible stories. Can we not keep the proper "drapery," and yet make the stories real?

men, and can do it, is the substance of nearly all his discourses. I asked him, after one of the morning services, whether he never used the element of terror in his preaching? He said that he did sometimes, but that "a man's heart ought to be very tender" when speaking about the doom of the impenitent; that the manner in which some preachers threatened unbelievers with the wrath to come, as though they had a kind of satisfaction in thinking of the sufferings of the lost, was to him very shocking. He added that in the course of his visit to a town he generally preached one sermon on hell and one on heaven. That night he preached on the text, "Son, remember!" I greatly regret that I happened to be absent; I should like to have heard how he dealt with this difficult subject. As the readers of the *Congregationalist* know, I believe that in modern preaching there is too little said about the awful words of our Lord concerning the destiny of those who resist His authority and reject His salvation. The unwillingness of most of us to speak of this terrible subject ought to suggest very earnest self-examination. Christ's love for men, which was infinitely more tender than ours, did not prevent Him from speaking of "the worm that dieth not, and the fire that is not quenched," and it is surely presumptuous of us to assume that we are prevented from speaking of future punishment by the depth of our sympathy with the Divine mercy.

The possibility of "instantaneous conversion" was one of the points on which he insisted incessantly. I think I should prefer to speak of the certainty of Christ's immediate response to a frank trust in His love, and a frank submission to His authority. These, however, are only two ways of presenting the same truth; and the vigor and earnestness with which he charged his hearers to obtain *at once* the pardon of sin and power to break away from a sinful life, were extremely effective.

Almost invariably the preaching was followed by an "after meeting." Cards of admission to the meetings for inquirers had been distributed among the ministers who co-operated with the movement, to be given by them to ladies and gentlemen to whom they could entrust the duty of conversing with persons agitated by religious anxiety and needing sympathy and advice. The intention of this arrangement was to prevent "inquirers" from being left in the hands of unwise and incompetent people.

How many of these "cards" were distributed I do not know; in my own church I gave away between a dozen and a score, and it was pleasant to me to see many of my friends at their work night after night. The arrangement broke down. The number of persons who remained for the "after meeting" was so large that a general appeal had to be made again and again to Christian people in the congregation to give their help. Some responded who had more enthusiasm than good sense. But, notwithstanding this, the results of the "after meeting" were extraordinary. I have already spoken of the number of persons with whom I conversed myself, to whom, while I was conversing with them, the light came which springs from the discovery of God's love and power, and from the acceptance of His will as the law of life. Testimony after testimony has reached me from "converts" to whom the same light came while conversing with others. "I went up into the gallery," said one young man to me, a day or two ago, "and Mr. Sankey walked up and down with me, and talked to me as though he had been my own father; and I found Christ."

The preaching without the "after meeting" would not have accomplished one-fifth of the results. It was in the quiet, unexciting talk with individuals that the impressions produced by Mr. Moody's addresses issued in a happy trust in Christ, and a clear decision to live a Christian life. The galleries were a beautiful sight. Mr. Moody's quaint directions were almost universally followed: "Let the young men talk to the young men, the maidens to the maidens, the elder women to the elder women, and the elder men to the elder men." Cultivated young ladies were sitting or standing with girls of their own age, sometimes with two or three together, whose eager faces indicated the earnestness of their desire to understand how they were to lay hold of the great blessing which they seemed to be touching but could not grasp. Young men were talking to lads—some of their own social position, others with black hands and rough clothes, which were suggestive of gun-making and rolling mills and brass foundries. Ladies of refinement were trying to make the truth clear to women whose worn faces and poor dress told of the hardships of their daily life. Men of business, local politicians, were at the same work with men of forty and fifty years of age. And there was the brightness of hope and faith in the tone and manner and bearing of nearly all of them. Christian people who want to know the real nature of the work of our American brethren, and to catch its spirit, should take care to spend a few hours at the "after meeting." If they go twice, they will find it hard to keep away.

Separate arrangements were made for those of the young men who preferred an after meeting of their own. A Presbyterian church in the neighborhood of the Hall was thrown open for them, and the attendance was generally very large.

Mr. Moody does not approve of the publication in newspapers of the number of persons who have declared that they have been led to begin a Christian life as the result of these services, and I therefore do not feel at liberty to publish in these pages the information on this point which is in my possession. A week after he had left us he returned to hold a farewell meeting for "converts" and "inquirers." Ministers sat at the office of the Young Men's Christian Association to receive applications for tickets from both these classes of persons. In every case I believe that there was personal conversation with the applicants. Their names and addresses were registered, and the congregations with which they were already connected, or with which they intended to connect themselves. One hundred and twenty names have been sent to me of persons who are already attendants at Carr's Lane, or who mean to attend there. These include eighty-five professed "converts," and thirty-five persons who have been awakened to religious earnestness, but who cannot say that they have rest of heart in Christ. The large majority of them, so far as I have been able at present to analyze the list, are working people, and most of them young men and women. In some cases the young men told me that they had been in the habit of swearing and using bad language up to the night when the truth came to them. "And never since then?" I have asked. They smiled, as though I had asked a very unnecessary question, and answered, "Never, sir." And when I talked to them about their conduct at home to their parents, and about their temper, it still seemed that I was going over ground that they had already gone over for themselves: "Things don't put me about now, sir, as they used," was the answer of a rough boy of seventeen or eighteen. I heard through a friend, that a manufacturer, who had a violent temper, and who had been accustomed to swear a great deal at his men, was suddenly so much changed that the men noticed it, and, of course, inferred that he had been to "Moody;" for a whole week they tried, "for the fun of it," to get him to swear at them again, but failed. I heard of another case that was very sad. A poor girl came to one of the meetings and was deeply impressed; when she got home, her father, who was half drunk, insisted on knowing where she had been, and when she told him, he was in a great rage and violently abused her. She bore this quietly, and went to bed. The neighbors, however, got to know it, and the next morning, as she went to work, they hooted at her and chafed her in the street. When she reached the shop where she is employed, her shopmates began to tease her and annoy her; she bore it a long time, but at last gave way and turned upon them in a burst of passion, and poured out on them a torrent of curses. The deepest remorse came upon the poor girl, and she thought that it was impossible for her to be recovered from her fall. I have no doubt that the Christian lady who is caring for her told her of one who, though he denied Christ with oaths and curses, was forgiven, and restored to all the honors and joys of his Apostleship.

The effect of this work has extended beyond those who were present at the services; and very much of the good that has been effected is never likely to be known. Since I began to write this paper, a son of one of the members of my own Church, a lad of seventeen, came to me and said he wished to enter the Church. I talked to him for a few minutes, and took for granted that Mr. Moody's services had led him to religious decision. He had all the brightness and joyousness which I have come to regard as characteristic of the typical "Moody convert." I asked him which of the services had had the greatest effect on him, and he said that his business engagements had prevented him from going to any of them. "How was it, then," I asked, "that you came to trust in Christ?" "Well, sir," he said, "I could not go to the meetings, but I heard a great deal of what these two gentlemen were doing, and I came to the conclusion that they could not be doing it themselves, but that God must be doing it; and then I came to see that I could look to God myself and get all the good."

Some of the most remarkable results of the visit of our American friends are to be found, perhaps, among those who have been long members of Christian Churches. I hardly know how to describe the change which has passed over them. It is like the change which comes upon a landscape when clouds which have been hanging over it for hours suddenly vanish, and the sunlight seems to fill both heaven and earth. There is a joyousness, and an elasticity of spirit, and a hopefulness, which have completely transformed them; and the transformation shows itself in the unostentatious eagerness with which they are taking up Christian work.

If I thought it worth while, I could speak of some things in this work which are not to my taste, and some things which my judgment disapproves. But before Mr. Moody and Mr. Sankey came to Birmingham, I had arrived at the conclusion that what was said of the early evangelists at Antioch was the truest account of the work of these American evangelists in Scotland and Ireland, "The hand of the Lord was with them: and a great number believed and turned unto the Lord." This conviction has been deepened and confirmed by all that I have seen of them. When Whitfield and Wesley were renewing the religious life of England, there were learned, orthodox, and devout ministers who were distressed by "The Decay of the Dissenting Interest," and the low state of religion throughout the country; there were ministers who had written pamphlets on these subjects in the hope of reawakening in the Christian Churches of that time the faith and zeal of earlier and better days, but who regarded Whitfield and Wesley with a distrust like that with which Mr. Moody and Mr. Sankey are now regarded by some excellent people. The very objections which are urged against Mr. Moody and Mr. Sankey were urged against the leaders of the great Evangelical revival which saved England from sinking into atheism. The result was inevitable; these ministers and their churches missed the blessing for which they had been longing and praying. When "the power of God" is with men who preach what we acknowledge to be the great truths of the Gospel, it is surely our clear duty to cooperate with them heartily and frankly. If in their methods, and if in their very conception of Christian truth and the Christian life, there are some things which we cannot accept, these may surely be borne with and even forgotten. Those men especially who are in the habit of insisting on "breadth" of sympathy with all in whom there is genuine Christian earnestness, and who are always saying that rigid accuracy in doctrinal definitions is of inferior importance to a living faith in Christ, ought to be able to rise above the kind of objections which seem likely to alienate some of them from this work.

It is possible that in some places our American visitors may not achieve the kind of success which has hitherto followed them. Before they came to Birmingham I felt very doubtful whether they would accomplish here what they had accomplished in Dublin and Belfast. I believe they will accomplish very little in any place where they are not sustained by the hearty sympathy of Christian people, and where Christian Churches do not earnestly entreat God to manifest in connection with their work the transcendent greatness of His power and love. There were people among whom our Lord Himself " could do no mighty works, because of their unbelief."

PROGRESS AND CONCLUSION

OF THE

WORK IN ENGLAND.

The following extracts from the *Christian World* and *The Christian* will serve to exhibit the spirit and success of the movement through the four months preceding the departure of the Evangelists to the United States.

I.

The American Evangelists are now nearing the end of their second month in London. During the greater part of April services had been conducted daily in each of the four divisions of the metropolis. Messrs. Moody and Sankey have divided their labors almost equally between the East and the West ends—officiating at Her Majesty's Opera House in the Haymarket at the daily noon prayer-meeting, and also at an afternoon Bible reading, while in the evening they have generally been present at the service in the Bow-road Hall. On two evenings of each week they have returned to the Agricultural Hall in Islington. The first week after their departure from that hall the services there were conducted by Rev. William Taylor, of California; but the attendance instantly dropped from 12,000 to 2,000, and sank to as low as 1,000 before the week was done. In the second week Mr. Taylor was succeeded by the Rev. W. H. M. Aitken (Episcopalian), of Liverpool, who secured much larger congregations, there being occasionally as many as 5,000 and 6,000 present to hear him; and at the Victoria Theatre, on the South side, Mr. Taylor held daily meetings, where his labors would appear to be better appreciated than they were at Islington. The prayer-meeting at the Opera House has not been so well attended, on the whole, as that at Exeter Hall; but the Bible readings have attracted great congregations, these including many members of fashionable society, led by Her Royal Highness the Princess of Wales, who was present on Thursday, April 15. In an article on "The American Revivalists in England," the New York *Independent* says: "We presume that the aristocracy and the literati will scarce hear of the movement that is about them. It is an after generation that builds the monuments of the prophets. Bunyan got no words of honor from the Duke of Bedford, whose descendant has lately set up his statue." Several months before these words were written, Mr. Moody had sojourned as a guest within the walls of Dunrobin Castle, the northern seat of the Duke of Sutherland; and weeks before he had dined with the Lord Chancellor of England, at Bournemouth. At his first meeting in the Agricultural Hall he was assisted by a peer of the realm, and other noblemen took part in subsequent gatherings, while Lord Cairns, the Earl of Shaftesbury, and many other members of the aristocracy formed part of his audience. The favor with which his labors are regarded by a large section of the nobility has been still more conspicuously displayed since the opening of the services in the Haymarket, and especially since the visit paid by the Princess of Wales. Standing somewhat in the same relation to Mr. Moody that the Countess of Huntingdon did to Whitefield, her Grace the Duchess of Sutherland has been well-nigh a daily attender, accompanied sometimes by her daughter and Lady Constance Leveson-Gower. Twice last week the Duke and Duchess of St. Albans were seen in the royal box, the Prince Teck has also been present, and so have the Duke and Duchess of Marlborough, the Countess of Gainsborough, Lady Dudley, Lord and Lady Rendlesham (the latter a daughter of the late popular Earl of Eglington), and many more of the "upper ten thousand." To crown all, it is alleged, not only that Lord Dudley interested himself in securing the Opera House for the American Evangelists, but that his lordship was encouraged to do this by no less a personage than the Heir-Apparent. Dr. Donald Fraser and Mr. Newman Hall have preached to excellent congregations at the Opera House; but when Mr. Moody's place at the Bow-road Hall was taken by the Rev. Mr. Howie, a Free Church minister from Glasgow, and a powerful preacher, the congregations instantly melted away. The young men's nightly meeting at St. Mary's Hall, Islington, was conducted until the end of last week by Mr. Henry Drummond, a nephew of the founder of the Stirling Tract Enterprise; and a pleasant feature of the work at the East-end has been the giving of a comfortable meal on the Sundays to many hundreds of poor people, brought together by young men visitors, assisted by some devoted ladies from Glasgow.

THE LAST DAYS.

II.

NOON MEETINGS, VICTORIA THEATRE.

Tuesday, July 6.—To-day's meeting was of very varied and absorbing interest.

Mr. Denny presented 254 requests for prayer for relations, friends and places, including thirty by persons for themselves.

Mr. Moody's address was on "Obedience." He showed how this was the rock on which man originally fell, and it was by obedience only that man was restored. He contrasted the two Sauls of the Bible. The Saul of the Old Testament was disobedient to God's command, and he lost his crown, his kingdom, his family, the friendship and counsel of Samuel, God's favor, and, at last, his life. The Saul of the New Testament, on the other hand, by obedience, gained more than the disobedient Saul ever lost. What a happy city London would be if everyone in it would obey God! From the exclusion of Moses from the promised land, on account of his disobeying God's command, he read a solemn lesson to Christians to beware of this sin of disobedience, and also addressed some words of warning to parents concerning the effects on their children of neglecting to train them to obey God's commands.

A young convert, on the platform, gave some testimony that was very indistinctly heard by the audience. Mr. Sankey prayed that he might have courage to confess Christ and work for Him. He also offered prayer for friends who had come to the meetings from the country.

Rev. T. Richardson read a letter from a man who was converted in Bow-road Hall a fortnight ago, and who was now rejoicing in the conversion of his wife through his instrumentality. Mr. Richardson had read the letter, he said, at his own church last Sunday, and this was the means of stirring up one of his congregation to speak to his wife about salvation, which ended also in her conversion. The husband himself related the circumstance to Mr. Richardson on Sunday evening, at the after-meeting, and there, before the assembled congregation, the family altar in that household was set up. Mr. Richardson urged converted husbands to speak to their wives. "Believe on the Lord Jesus Christ, and thou shalt be saved *and thy house.*"

A gentleman requested special prayer for India and for the Lodiana Mission. He said the proportion of laborers in that vast country was something like what it would be if we had *eight* foreigners laboring for the conversion of London. At Mr. Moody's request, silent prayer was offered for India.

Dr. Ziemann, of Manchester, had a very thrilling story to tell of the wonderful workings of God in Manchester during the past six months.

In speaking of the difficulties encountered in getting a site for the tent, he said that, when all their efforts had failed, they laid the matter before the Lord at one of the noon prayer-meetings. That afternoon they received a letter from some solicitors, stating that if a large sum of money were given they might get a temporary site. Dr. Ziemann sent a messenger to the solicitors, with instructions to get the ground at the best possible terms. He returned with the joyful tidings that he had got a site. When he went to the solicitors, the owner of the ground happened to be there. Hearing what the messenger said, he asked what was to be done with the tent. The reply was given that it was wanting for meetings. Revival meetings? Yes. The owner rejoined that he had just come from London, where he had attended Mr. Moody's meetings, and had liked them very much. He added, "If you are going to have Moody's meetings, you are quite welcome to the place without money and without price. I have another place in the town; if you like, you can have it too, and I will level it without any expense." Last Sunday the tent was opened, and not only was it crowded three times, but hundreds of people could not get in, and there were a great many inquirers.

Dr. Ziemann also told the following striking circumstance. One evening, some months ago, he was going to the meeting in the Museum, when he saw a man standing at the door, apparently hesitating as to whether he should enter. He was invited in and took his seat in a corner. Dr. Ziemann was guided to speak about men professing themselves to be wise and becoming fools. He had been visiting the prison, and if any one wanted to know the truth of these words, let them go to the prison, and ask of those who are there. The man in the corner burst out, "That's true, that's true!" At first it was thought he was going to make a disturbance, but they soon saw he was in earnest. The fact was, he had come that evening direct from Millbank, where he had been in prison seven years. He had come to Manchester with his little bundle, and the Lord led him to that meeting, where he was converted that night. The returned convict became a son of God. Next morning he said he was going home to his mother, whom he had not seen for long years, and instead of taking with him his ticket of leave, he was going with his Bible to tell her he was now a saved and happy man.

Mr. Sankey sang, to gratify the wish of a little invalid girl who was present, "Knocking, knocking."

The Rev. Newman Hall related a very beautiful incident of a stray, starving lamb he found in his travels on the Westmoreland hills, and which he, after ineffectual attempts to find its mother, handed over to the tender care of the strong, stalwart farmer with whom he was staying. Some time after, on revisiting the place, he found the half-starved lamb had become one of the stoutest and fattest of the flock. His description of the lamb and its bleating voice, that seemed to say, "Pity me, help me, save me," was very touching, and the application of the story most appropriate.

Wednesday, July 7.—At the meeting this day, Mr. Sankey sang two fresh solos, one of them asking in the refrain, "Who is on the Lord's side?" and the other, a very sweet and soothing song, commencing—

"I need Thee every hour,
My gracious Lord,
No tender voice like Thine
Can peace afford."

Mr. Henry Varley gave the address from the story of the man with the withered hand whom Christ cured on the Sabbath day. Mr. Varley, in condemning the conduct of the Pharisees, who "watched" the Saviour that they might find occasion against Him, was very severe on those hearers who go to church or meeting, not to catch souls for Christ, but to catch some stray word of the preacher that might afford them the opportunity for criticism. He also dwelt on the power of Christ, as set forth in this passage, to give a present blessing to all who really feel their need and seek relief.

Mr. Paton gave some interesting extracts from letters, and the bulk of the second half-hour was spent in earnest prayer that the closing services of the Evangelists might be signally instrumental in the conversion of sinners.

Thursday, July 8.—Mr. Moody's address to-day was an exposition of the Christian life as portrayed in Scripture under the figure of water. He quoted a number of passages setting forth the fulness of blessing treasured up in Christ, who is both the bread and the water of life. One reason why we do not get more of the water of life is because we do not thirst for it. It would be a good thing to have a meeting for dissatisfied Christians—dissatisfied, not with Christ, but with themselves.

Mr. Sankey spoke of having recently been up the Thames with some friends, and he noticed as the tide flowed in how it covered up a great many unseemly places, and when it was full, how the water came up to the very brink of the land. There was the beautiful landscape to be seen, with the flowers along the shore, and the whole scene looked so sweet and lovely, I thought how impossible it would be for man ever to make this tide to flow in. So when the water of life flows into our hearts, how the things that are unseemly are covered up, and how it cleanses and beautifies our natures. He thanked God for the tide of blessing that had been flowing through the land and into the hearts of God's children. How many hearts had been flooded with God's love, and were to-day feeling how gracious the Lord is. Do not let us try to make the tide flow, but let us lie low at the feet of Jesus, and say, "Come, Lord Jesus, into this heart of mine." Mr. Sankey then sang one of his new hymns—

"It passeth knowledge, that dear love of Thine."

There is a certain stateliness and majesty in the melody, especially in the *crescendo* passage in the fourth line, that interprets well the reverential tone of the hymn.

Friday, July 9.—The last meeting in

this theatre where the familiar voices of our brethren should be heard. The spacious building was crowded by twelve o'clock—stage, boxes, pit, upper circle and gallery. We were reminded of the closing meeting at the Opera House. On the stage were a large number of those who have been prominent workers at these evangelistic services.

The meeting opened with the hymn, "Tell me the old, old story," one of the most precious, and, at the same time, one of the most popular in Mr. Sankey's collection.

Among the requests for prayer, classified and read by Mr. Denny, there were many of touching significance. A large list of requests for praise, for various blessings experienced through the services, were also presented.

Mr. Moody said that, when he first came to this country, they had a little meeting one day at twelve o'clock, at 165 Aldersgate street, and he was led to take for his subject "The gift of the Holy Spirit." That subject had been agitating his mind for a number of years. He believed it was the privilege of every child of God to be filled with the Spirit; if they were not, it was their fault. If he should ask all to rise up in the meeting who were filled with the Spirit, he did not think any one would dare to stand up; they were all living beneath their privileges. In Matthew xv. 18 it says, "Be not drunk with wine wherein is excess, but be *filled* with the Spirit." A great many Christians were satisfied with the bare life they received at Calvary. That was the work of the Holy Ghost, but there was such a thing as receiving the gift of the Holy Ghost for service. He believed this gift the Church had lost sight of and mislaid. He quoted various passages to show further that the Apostles and early disciples received the gift of the Holy Ghost on several occasions, both before and after Christ's death. Many Christians, he said, seemed to think that they could receive power at one time that would last them all their life. But there was such a thing as a constant anointing. He believed it was right to pray that the Holy Ghost should come on us. The grace that God gave them in order to do his work in Liverpool did not suffice for the work in London. And if they were to attempt to go back to America, and work for God there on the grace God had given them here, they would very soon break down. If they were *filled* with the Spirit, there would be no room for the devil, for self, for pride, or darkness. It would be all light, and peace, and joy—the fruits of the Holy Ghost. May the prayer of every heart to-day be to be filled with the Holy Ghost. If we were filled with the Spirit, we would have a hundred times more influence in this dark world than we have. Instead of getting up and talking so that people were glad when we sat down, we would have something fresh to say. One hour's work would tell more than a whole day of service without the filling of the Spirit. He thought Christians would not lose much by going into the desert for a few days, and get there a fresh anointing for the work that lay before them. The Church wanted nothing so much as Holy Ghost power. Some people could not get on without having men's hands laid on them. If he was sure who were the undoubted successors of the Apostles, he might not object to having their hands laid on him; but they could all receive the Holy Ghost from the hands of Jesus Christ. Let us pray that the Holy Ghost may fill every one of us and qualify us for God's service. He engaged in very earnest prayer that thousands might be brought to Christ through their closing services, and thanked God for all the blessings and successful services of the past.

Mr. Sankey sang the appropriate hymn, "I need Thee every hour;" and followed in prayer.

Dr. Andrew Bonar, of Glasgow, quoted the verse, "To Him that is able to do exceeding abundantly above all that we can ask or think." He said the showers that had brought this year's harvest to maturity would not do for next year. So it was with us in our Christian life; even if we have received the Holy Ghost, we must come to be filled with it again and again. That was the root of the whole matter. In the Old Testament we have some very remarkable lessons as to God's people needing constant renewal of strength. There was Jonathan, who, with his armor-bearer, could put a whole army to flight, and yet, when one man came out and challenged the army of Israel, Jonathan did not dare to meet him. And David himself, who slew a lion and gained victory over the giant of Gath, in the latter years of his life was almost killed by a brother of Goliath's. He came upon him with his sword, and if it had not been for one of David's captains, he would have been killed by the brother of the giant he had slain. So we need fresh grace. Let us go back to the verse, "To Him that is able to do exceeding abundantly, above all that we can ask or think." Let us be ashamed of our little faith and expectations. It would be an interesting study to look up all the places in the Bible where God's people asked for something and God gave them a great deal. The prodigal son was a notable instance. He wanted to be a servant, but there were the best robe, the ring, the shoe, the fatted calf—everything. "To Him that is able to do exceeding abundantly above all that we can ask or think, to him be glory in the Church, world without end, amen."

After silent petition, Mr. Varley offered the closing prayer, and Messrs. Moody and Sankey's meetings in Victoria Theatre ended with the Doxology and the Benediction.

III.

CAMBERWELL-GREEN HALL.

We find it difficult to realize that we are called upon to chronicle the closing services of Messrs. Moody and Sankey in this country. By their self-denying labors for the evangelization of our fellow-countrymen and women, they have endeared themselves to every honest, loving Christian heart, and their long stay of two years in our midst has almost made us look upon them as two of ourselves; we have well-nigh lost sight of the fact that they came to us from across the sea, and we have, as a consequence, not been disposed to dwell upon the other fact that they must needs return to their own kindred and their own land. We find it hard to use the language of Job, and say, "The Lord gave; the Lord taketh away, and blessed be the name of the Lord." But facts are stubborn things, however unwelcome they may be; and the fact that, before these lines are read, Messrs. Moody and Sankey's public labors in Great Britain will have ended, only reminds us of the onward march of time, and the coming end of all things. Well, we suppose we must accept the inevitable, and in the midst of our sincere sorrow at being severed from those whose names have become, and will ever remain, household words in our mouths, and who have been used of the Lord to do such great things for us, we will seek to comfort ourselves with the thought that the parting is not a final one, and also with the blessed assurance that the Lord of the Harvest remains with us, however His servants may come and go. Still, no words of ours can tell the heartfelt sorrow that clouds this parting hour, and we will not seek to conceal it. When we have sometimes felt bodily weariness in the congenial task of attending so many meetings, and telling out to our readers the goodness of God as we have witnessed His saving power in the great congregations, day after day and week after week, we have been re-invigorated by the remembrance of how our American brethren have stood the strain of two years' incessant toil without the thought almost of personal rest or ease. What an example of persistent, devoted, loving service these friends have set us. It is one, unhappily, that we are slow to imitate. Would it were otherwise!

But we must cut short these reflections in order to give our readers some brief account of the services that have closed up the four months mission in London. As may be supposed, there has been intense anxiety shown to attend the services of the last week. Many of our country brethren and sisters having hoped against hope that Messrs. Moody and Sankey would pay their respective districts a visit, and having seen this hope flicker and vanish, as a last resource have come to London specially to attend the meetings. Many, if not most, have come just to see and hear our two friends; but we are assured that not a few of them have had the eyes of their understanding opened, and have seen Jesus as the Saviour they needed. We have heard of some who have come up from their homes with a heavy heart and sin-burdened conscience, and have returned to their homes justified and rejoicing in a Saviour's love and pardon. We hope there have been many such from among the crowds of strangers who have every evening found their way to Camberwell Hall.

It has been asserted that the meetings in the South have proved, to all appearance, as fruitful as in any other part of London. The last week has, we believe, sustained the promise of those that went before. Invariably at the close both of Mr. Moody's Bible readings and Gospel addresses, the searching power of the Word has been made manifest in large numbers who stood up for prayer, and afterwards flocked into the inquiry-rooms, to seek and receive counsel and direction from those who, having found Christ to be precious to them, could say, "Come thou with us and we will do thee good." We have heard of many promising and decided cases of conversion, but the Lord alone can register the number of those who have been saved at these closing week-day meetings by faith that is in Christ Jesus.

Time and space would fail us to tell of the deep impressions made by the singing of Mr. Sankey, and how hearts that were frozen and sealed against the Saviour's love have been melted into submission by some tender message of mercy wafted to them on the unseen wings of sweet song, the way being thus paved for the reception of the Gospel more fully delivered by the voice of the preacher. When the history of this movement comes to be written by competent hands, we doubt not full justice will be done to the part Mr. Sankey has, in God's good providence, been enabled to play in this most blessed work of pointing sinners to Christ. In his hymn-book he has left us a legacy, the value of which, we believe, will never be exhausted, as long as there is a single singing pilgrim left in this vale of shadow and of tears. During the closing days of the week, Mr. Sankey has afforded us the opportunity of hearing many songs till now unknown to the Christians of this land. Amongst those recently composed by himself, the one that has gained the firmest place in popular favor is, "I am praying for you," and we think deservedly so. Mr. Sankey, we believe, has used it largely as a letter-leaflet, and we cannot think of a better. We hope this hymn, now popularized by his sweet melody, may speak for the Master to many a heart.

Mr. Moody's closing Bible reading on Friday afternoon was on the word "Able," and specially suited to confirm the faith and courage of the young converts, though useful for all. Towards the end, Mr. Moody referred to the fact that it was his last Bible lecture in London, and thanked the people for coming out day after day. Nothing had encouraged him so much during the past two years as having the people come to those Bible readings. He had seen so many Christians with a better knowledge of Scripture than himself come, that he had wondered they came at all; but it had encouraged him to study the Word of God more. He had one request to make, one favor to ask—that they would pray God to bless them both, that they might know more and more of His love, and more and more of His blessed truth. He had been for two years constantly unable to study much, and he felt great leanness of soul. He felt as if he would like to go into the desert, and when he was gone he hoped they would pray for him. He did not know what they would have done if it had not been for the prayers of God's people; for if they had come to criticize, as some had come to do, the whole mission must have failed. Of one thing their friends in this country might rest assured, they would be praying for them; they had become very dear to each other. Let them pray God to increase their love for Him, and their passion for souls. He finally made a very affectionate appeal to all who were out of Christ to come to Him without delay. The whole meeting was deeply moved, and many shed tears as Mr. Moody spoke of their departure. A number rose to be prayed for, and afterwards were conversed with in the inquiry-rooms.

The evening address was on the Ark, and gave Mr. Moody another opportunity of entreating sinners to flee from the wrath to come, by taking refuge in the Ark of Safety —Christ Jesus.

The gentleman who offered the closing prayer committed our brethren to God's care while on their homeward voyage, and prayed that great blessings might attend their labors in their native land.

Sunday, July 11.—LAST DAY.—How shall we write of it? Fresh from the meeting in the evening, we feel how inadequate are human words to portray that most marvelous close of a no less marvelous season of revival throughout our land.

The doors for the morning service were opened at half-past six, and by seven o'clock the Hall was comfortably full. An hour yet intervened before the commencement of the service, and the time was profitably and pleasantly occupied with a service of song from the familiar book. About half-past seven it was announced that there were thousands outside, some of them from a great distance, and if the audience would kindly sit more closely, a few hundreds more might be got in. The request was good-naturedly complied with at once, and room made for a few more. For the great numbers who were unable to get admission, an overflow-meeting was held in the Presbyterian church close by, and was addressed by Dr. A. Bonar, of Glasgow.

Shortly before eight o'clock Mr. Sankey appeared, and delighted the audience with a few solos. Before singing "I am praying for you," he said he hoped in the days to come, they would not forget to pray for Mr. Moody and himself when they were gone.

In consequence of the crowd at the gates, Mr. Moody could not gain an entrance, and had to be conveyed through a private house opening from the back upon the site of the building. This delayed the opening of the service till about ten minutes past eight o'clock—a thing altogether unusual at these gatherings, as one of the most noticeable features of them has been the punctuality observed by our brethren.

Mr. Sankey having sung "Only an Armor-bearer," the audience swelling out in the chorus, very earnest prayer was offered, making special reference to the occasion.

Mr. Moody then delivered his well-known address on "Daniel," beginning with the secret of his wonderful success, which he attributed to his being able to say "No" at the right moment. He sketched the eventful career of this man, "beloved of God," through the reigns of Nebuchadnezzar, Belshazzar, and Darius, showing how he was delivered from all the many snares laid for him by his enemies, because he was faithful to God and His commandments. The history of Daniel in the telling of it rouses Mr. Moody's enthusiasm, which he succeeds in a large degree in imparting to the audience, and many thousands of hearts were stirred by this closing address to Christian workers. Before parting we sang with Mr. Sankey "Dare to be a Daniel."

The afternoon service for women was a deeply interesting one, both in itself and from the fact of its being the last of them. The hall was crammed in every corner. The opening hymn was "Yet there is room," very appropriate to the occasion.

All through his mission in Great Britain, Mr. Moody has striven to make the Gospel so plain as to be understood by the meanest comprehension. He has avoided collateral issues, and eschewed theological discussions —and held to the proclamation of the good news of salvation through faith in a crucified and risen Saviour. One of his favorite texts has been the question of the gaoler, "What must I do to be saved?" and this he chose for his final gospel addresses to London audiences. Many people, he said, still disbelieved in sudden conversion, and he proceeded to draw from the treasury of Holy Scripture numerous illustrations to show that the new birth is, of necessity, an instantaneous act, and not a gradual change. He quoted the Ark, the salvation of Lot from Sodom, the preservation of the children of Israel in Egypt by sprinkling the blood on their doors, the cities of refuge, and others, as well as illustrations from history and from daily life. At the close he spoke, with much emotion, of how he had tried in all possible ways to allure sinners to Christ, and entreated those present not to go out of the building without receiving Christ as their Saviour. They might never hear his and Mr. Sankey's voice again on earth, but he hoped there would not be one missing at the last great meeting. Many rose in response to his pressing appeal at the close, and the inquiry-rooms were afterwards the scene of much earnest conversation and prayer with the crowds of anxious sisters. The evening meeting for men was almost filled before the last of the inquirers and workers had left the building.

The last meeting of all will, we think, be reckoned, by those who have attended the London meetings throughout, the best of all. It was as closely packed with men as could be : how many were left outside we cannot tell. A meeting for them was held in the Camberwell-Green Hall. Mr. Sankey took his seat at the instrument about half an hour before the time, and while he was singing for Jesus to the eager crowd of listeners, Mr. Moody and a few friends were in the little waiting-room below, supplicating God for a Pentecostal blessing on this parting service. And their prayer was answered of a truth. We have not witnessed such a wondrous scene during any of the many gatherings these last four months; the only approach to it was one Sunday afternoon at a women's meeting in the Opera House.

Several of Mr. Moody's American friends were present to witness the crowning service of this mission.

Mr. Moody took for his subject, as in the afternoon, that all-important query, "What must I do to be saved?" and the bulk of his discourse was essentially a repetition of that delivered to the women. But the power and presence of the Holy Ghost to apply the spoken word was far more wonderfully manifest.

As he drew to a close, Mr. Moody became very earnest and urgent in his appeals to the vast and intensely interested audience to accept Christ. "Just let me pause here," he said; "ask yourselves whether you ought not to receive the Lord Jesus Christ now. Who is there in this assembly will receive the gift of God and be saved?" After a brief pause, a voice came from the

CONCLUSION OF THE WORK IN ENGLAND.

left-hand gallery, somewhat feebly, "I will." It was speedily followed by others from all parts of the house. "Well," continued Mr. Moody, "Thank God for that. I am just passing round the cup of salvation; who else will take it?" "I will," "I will," "I will," "I will," "I will," came resounding on every hand. "That's right, my boy," replied the speaker to a little fellow down in front of him, whose "I will" came up to the platform with the rest. "Will the Christians keep praying? Men do not speak out like this unless God is at work. Who else will accept the gift?" Again a perfect volume of "I wills." "Would it not be a glorious thing if every man here would take it to-night? Is there another?" "I will." "Another?" "I will." "That is right; speak out. If you are willing to have God's gift, just say so." Then there came a louder response from a manly voice in a distant part of the Hall, followed by the shrill tones of a little boy, and many other "I wills" came to our enraptured ears in close succession. They came so thick, we could not note them down.

And then Mr. Moody said: "The time has come for us to close the two years and three weeks we have been trying to labor for Christ among you. This is the last time I shall have the unspeakable privilege of preaching the Gospel in this country at this time. I want to say that it has been the two best years of my life. (" Have another week, Mr. Moody," shouted a stentorian voice from the crowd below.) My friends, you can all be saved this night, if you will believe on the Lord Jesus Christ. If I stayed another week I do not know what more I could say. I have brought Christ before you; I have told you of His beauty. It is true I have done it with stammering tongue. I have never spoken of Him as I would like; but I have done the best I could. And now, in this closing hour, I want once more to press Him upon your acceptance. I do not want to close this meeting until I see you all safe in the ark, safe behind the walls of the city of refuge. How many are to-night willing to stand up before God and man, and say by that act that they will join us in our journey to heaven? You that are willing to take Christ now, would you just rise. (A mighty army of men rose to their feet at once.) Why not three thousand? The God of Pentecost still lives." Numbers more stood, until one could scarce distinguish between those sitting and those standing. Then Mr. Moody led in prayer, with a faltering voice, often choking with suppressed emotion. He besought the power of the Holy Ghost to fall upon those who had risen and those who had not, and that great multitudes might be saved.

"And now," he said, "we will sing, 'Safe in the arms of Jesus.'" While it was being sung, the inquiry-rooms were filled with seeking ones, and our hearts were rejoiced to see many come tearfully, others calmly and trustfully, declaring their trust in Jesus as their Saviour and everlasting refuge. It was a season of ingathering never to be forgotten by those who were present, and was a fitting close to the labors of Messrs. Moody and Sankey in Great Britain.

IV.

CLOSING MEETING AT BOW-ROAD HALL.

It is good to be *always* zealously affected in a good cause, and it was really a brave farewell that the good people of the East accorded to the services at the Bow-road Hall. The place was comparatively full by 7.15, and thronged by the time of commencing. The choir sang as if they appreciated the importance of their last opportunity, and the heart and delicacy of the rendering have not been surpassed in any section of London. Then in the intervals rose the sweet, weird notes of the Jubilee Singers, telling of the "Mansions of bliss," and of "peace" in the "valley of the shadow of death."

The solemnity of the occasion would have inspired a less sensitive and far-seeing man than the Rev. Hay Aitken, and we were more thankful than surprised at the thrilling address that he delivered on Luke xix. 37-44. He bade those who had been blessed of God, either personally or relatively, to join in the loud "Hosanna;" and called on all prepared to do so, to "stand on their feet, and bless the Lord." A very large number responded, singing as they did so, "Praise God, from whom all blessings flow." It was a never-to-be-forgotten moment, and gave a slight idea of what the harvest shall be in the "sweet by-and-by." Then followed terribly weighty words on the "day of visitation" which has come to our land. The women's meeting was crowded with inquirers up to a late hour, and the last words heard in the hall were the thanksgivings of new-born souls, and the praises of grateful workers.

And thus closes the brief existence of this hall built in faith that God would fill it with His glory. He has done it, and to Him be the praise.

V.
FAREWELL AND THANKSGIVING MEETING.

NOT the least interesting of the long and wonderful series of meetings held in London in connection with our beloved brethren Moody and Sankey was the farewell at Mildmay Conference Hall on Monday afternoon. The area was crowded with ministers and laymen, and the three galleries with a mixed audience of ladies and gentlemen. Dr. Bonar's hymn, "Rejoice and be glad, the Redeemer has come," was sung as the key-note. Then the Rev. C. D. Marston offered prayer; and Mr. Stone, chairman of the Central Committee, spoke out the thanks of himself and his brethren for the unremitting grace with which God has blessed this movement.

Mr. Moody said we were met to give thanks to God, and not to honor men, and very emphatically laid it down that nothing should be said about the instruments.

Dr. Andrew Bonar gave a deeply interesting address on the man who lay at the Beautiful Gate of the Temple, which we must reserve for our next. He then gave the most unqualified testimony to the wonderful results of the work in Glasgow. All the ministers are agreed that at least 7,000 have been added to the Church membership in that city. The results to the poorest of the people are manifest in the tent work every Sabbath-day.

Rev. R. C. Billings said that it was premature to speak of the results of the work at the Agricultural Hall, and added that one of the most cheering has been the increased union of Christians.

The hymn "Only an armor-bearer" was now sung, and then Rev. Archibald Brown noted some of the features of the work: religious stagnation swept away; a longing desire to hear the Gospel created and developed; God has taught His people to get up early on Sunday morning to hear how to study the Bible; an intense desire for the conversion of souls; such prayers offered as we used not to hear; God has shown that the Gospel of Christ is the power to move the masses; the gauntlet thrown down by philosophy God has taken up, and we are seeing the fruits at the East-end.

Dr. Donald Fraser bore testimony to the fruitfulness of the meetings at the Opera House, and dwelt specially on the necessity of now feeding the flock of God, and giving Bible instruction to the higher as well as to the poorer classes.

Rev. T. Richardson gave some most interesting statistics resulting from the meetings at the East-end; Rev. R. Taylor and Rev. Mr. Flindt gave most thrilling accounts of the work in the South; and Rev. Mr. Newton added to what Mr. Billings had already said about the North of London.

Dr. Jobson rejoiced in this work as an open and public rebuke of scepticism; it shows the power of Christianity to save the souls of men, and has checked the flood of worldliness flowing in with the increasing wealth of the nation; it has shown the importance of lay operation in the work of the Lord; and he reported that in all sessions and assemblies of members of his own denomination this work was never referred to but with sympathy and interest.

"Rescue the perishing" was now sung, and Mr. Moody, with broken utterance, led in fervent and humble thanksgiving in prayer.

Rev. Marcus Rainsford, dwelling on the greatness of the work, said that the masses had been more influenced than the ministers. He proceeded to tell a most graphic story of the way in which a costermonger showed the Gospel to an inquiring comrade, and some cases which came within his own experience. This work is not to stop. Our dear brethren are going away, but God is not going away. Why should we not all be preachers of the Gospel—each man, and woman, and child—in his position? If God Almighty will just pour his Spirit upon us, we may have 10,000 Moodys and 10,000 Sankeys to welcome them when they came back among us.

Rev. J. P. Chown said: While you meet, representing London, not only London, but the country also, is represented in this blessing on London. The Master is saying, "Thou shalt see greater things than these." If we can't be the Peters and preach great sermons, we may be the Andrews to bring Peters to Christ. The great want of the present day is the lifting up of a personal Christ. We want to be all Christ's—Christ living in us. Live so that Christian must be written with a capital letter.

Rev. W. Hay Chapman thanked God that these beloved brethren had been sustained through these long weeks of service; and added, that had there been still more union and sympathy among ministers, there would have been a still greater work.

Mr. H. Varley, feeling that there was a certain amount of weakness, yet believed there was a greater spirit of unity than ever before. He hoped the young converts would be cared for. Some old sheep don't believe in lambs; they never were lambs.

Rev. W. H. M. H. Aitken had had some work to do in gleaning after the American brethren; had been preaching at the East-end with the Jubilee Singers as coadjutors, and the Hall, instead of being thinly attended, was crowded. He was pleased to observe how reverentially the Jubilee Singers entered into the work, and much blessing resulted from their singing. He repeated the opinion of a publican at Liverpool, that if Moody and Sankey had remained there five months, instead of one, half the public-houses in the town would have been closed, and mentioned some most cheering facts connected with his own congregation.

Mr. Smithson, the indefatigable secretary of Mr. Moody's Dublin Committee, gave very interesting facts with respect to Ireland.

After some emphatic testimony from Lord Shaftesbury, Mr. Sankey sang "The ninety and nine," and the gathering shortly afterwards dispersed.

VI.

THE FAREWELL MEETING AT LONDON.

(Second Article.)

LAST week we were able to give only a very condensed sketch of the farewell and thanksgiving service convened by Messrs. Moody and Sankey at Mildmay Conference Hall, on Monday week. We now give the most important of the numerous addresses delivered by representative London clergymen and others on the occasion.

It may interest our readers to know that of the 700 and odd ministers who were present at this memorable gathering, there were 188 belonging to the Church of England, 154 Congregationalists, 85 Baptists, 81 Wesleyan Methodists, 39 Presbyterians, 8 foreign pastors, 8 United Methodists, 7 Primitive Methodists, 3 Plymouth Brethren, 2 Countess of Huntingdon's Connection, 2 Society of Friends, 3 Free Church of England, 1 Bible Christian, and upwards of 20 not known. These figures we take from the official statement supplied at the meeting, and they significantly show the catholic and unsectarian character of Messrs. Moody and Sankey's services, as well as the universal esteem with which our evangelist brethren are regarded by all sections of the Church of Christ in this country. A large number of influential laymen and Christian workers were also present, among the best known of them being Lord Shaftesbury, Lord Cavan, Mr. Cowper-Temple, M.P., Mr. Alderman M'Arthur, M.P., Mr. Samuel Morley, M.P., etc.

As already stated, we only give those of the addresses containing interesting facts and statistics relating to the movement.

Rev. R. D. Wilson, of Craven Chapel, said a new spiritual glow had come into the hearts of many during the last four months. They had learned, too, that their cherished traditions had no more sanctity or authority about them than the new things, which startled some of them at first, but with which they had now become most blessedly familiar. It was too soon to speak of the results as a whole, but within the last three days he had met no less than twelve or thirteen distinct cases of conversion in consequence of the ministrations of the evangelists. He read the following extract from a letter he had received: "I feel it my duty and inexpressible pleasure to tell you that I and one of my brothers were converted at one of Mr. Moody's meetings last week. Could you know my inner life for the past ten years, you would indeed say I have been plucked like a brand from the burning. I cannot cease to marvel at the greatness of my salvation." The mother of that young lady, said the speaker, had come to him yesterday, and stated that for twenty-five years, with few exceptions, she had regularly attended the service of the sanctuary, but the happiest day in her Christian experience was the previous Sunday, when she sat with her converted daughter on her right hand and a converted son on her left. As the speaker told this affecting little story, we felt certain that the tear of joy gathered in many an eye, only we could not see for the mist that came across our own. He went on to say, that we had never known what it was to "sing the Gospel" of Jesus Christ till our two brethren came. We could now understand how the sweetest tones could become the highest sort of Christian eloquence, in declaring to men the Way of Life. He would so far disobey the rule that no reference was to be made to the two evangelists, as to assure them that they would carry home to their American country the warmest love and heartiest esteem of the ministers and Christian people of this country. At this remark the pent-up feelings of the audience could no longer be restrained, and they burst out into loud and prolonged applause. We were extremely glad that the natural emotions of the congregation for once refused to be smothered by that false and frigid idea of decorum which obtains too much in our religious assemblies, and prevents the legitimate expression of the deepest feelings of the heart. But this is a digression. Mr. Wilson continued: "We shall not forget, when the Atlantic lies between their home and ours, at our family altar, at the place of secret meeting with our God, in our prayer-meetings, and in our Sabbath assemblies, to pray that God's richest blessing may rest upon them there. And it will be a glad day for us all, if ever that day comes, when we shall hear from the other side of the Western Main the intelligence that they are coming again. Until then we shall continue to pray that, when God sees meet that they should come, they may come in the fulness of the blessing of the Gospel of Christ."

Rev. Thomas Richardson, of St. Benet's, Stepney, said the effect of the meetings in the East-end had been to make his church and congregation "enlarge the place of their tent, and stretch forth the curtains of their habitation." He would rather wait for a year before he gave his testimony as to results, as there were many reasons why they should not now begin to count. But he had no doubt that thousands of souls would be recorded in their various chapels and churches all over London by next year. His district visitors had sent in to him formal returns, showing that of 1,008 families in his parish, 672, or two families out of every three, had attended the services at Bow-road Hall. Further, he had two direct testimonies that the attendances at the theatres of East London had sensibly diminished. Some of the officials of these theatres had given up the profession, and he had only to-day had an interview with one who was starting a different course. He had something too to say about the influence of the movement on the dock laborers. He had received testimony from several of the large docks that the men did not swear so much since Messrs. Moody and Sankey came: praise God for that. Besides, drinking was not so prevalent amongst the dockmen, and that was the kind of work that the world believed in. He had been privileged to attend every service in Bow-road Hall, and he would thank God to all eternity for it. He had seen the power of sympathy —that sympathy which brought Christ down to die for sinners. Sinners had felt its power, so they had stood up and declared they wanted to be saved. He had had the privilege of conversing personally with 450 anxious souls; his wife had spoken to 150, and his curate had spoken to 100. There were thus 700 souls whose names and addresses they knew, and to whom they had written. Formerly, he had an after-meeting once a month; now he had one every Sunday evening, and not a Sunday passed without some souls being gathered in. The direct results of the meetings were seen in his church, his wife's Bible-class, his young men's meetings, and among his district visitors. He urged the general adoption of the after-meeting, as being the key to the success of the services, and added that if the Spirit led him to adopt Mr. Moody's style of preaching he was going to do it.

Rev. Robert Taylor, of Norwood, gave some intensely interesting facts respecting what had transpired in the inquiry-room at Camberwell-green Hall. He had to do what Mr. Moody called "police work" there, and in this capacity he was able to take a general view of the inquirers who, night after night, thronged the rooms. One or two things had struck him. First, the large number of old people who came as inquirers, and who went away as very young Christians. He was afraid that, in their anxiety to shut up and shut in the young, they had been in danger of shutting out the old. They had fallen into the unbelief of Nicodemus, who said, "How can a man be born when he is old?" But many blessed births of the old had been seen in the inquiry-room at Camberwell. He was also struck with the amazing variety of opinion —religious opinion and no opinion—represented. One evening he gave up his seat in the hall to a distinguished literary man, who lately wrote that "there was a Power above us that, at least, we know to be working for righteousness." One evening, in the inquiry-room, he met a young woman, and asked if she was anxious. Yes, to know if there was a God. Did she not believe it? Well, the sum of her belief was that "there was something above us." He could tell of a wife, deserted by her husband, who had been in such utter misery and agony that she had twice contemplated going to London-bridge to commit suicide. In that inquiry-room she was brought to faith in Jesus Christ and peace with God through the preaching and singing. Afterwards she prayed so beautifully for her husband that the lady who conversed with her was deeply touched as she listened. She did not pray that he might be restored to *her*—now she did not care so much about that,—but that God would bring him to *Himself*, and that they might be re-united in heaven. He could tell of several Roman Catholics brought to simple faith and sweet peace in Jesus. He could tell of a man who for twenty minutes hid his face from the lady who spoke to him, so deep was his distress and shame. He afterwards told her how he was standing at St. Giles', and tossed up whether he should go the theatre or the meeting. It was, "Heads the theatre, tails Moody and Sankey." It was tails. He went to the meeting, was led to go into the inquiry-room, and, as he described it in a letter to the lady who was the means of bringing him into light, "She fought manfully with him for the Lord Jesus," and he went home a rejoicing believer. These were but few specimens of hundreds of cases he could quote, and when friends said to him the night before, with sad hearts, they were so sorry the meetings were over, he could only reply, "Yes, and I am so glad the work is so gloriously begun."

Rev. G. Flindt, of Denmark-hill, also spoke of the work in the inquiry-room at Camberwell. He said that one result of the services had been to increase the local congregations. In his own church they had on several occasions not had standing room during the visit of the evangelists. He had learned this lesson: that, if the ministry is to be useful, a personal Christ must be lifted up. A man in the inquiry-room had said to him, "It seems as if that man (referring to Mr. Moody) had his Friend quite close to him, and he was talking about him." Only eternity would reveal the good that had been done in the South of London. The night before, at the closing service, there were scores of anxious ones who came asking if it was possible to get a grip of the hands of the evangelists, and thank them for what had been told them about the Lord Jesus Christ; and tears of gladness flowed down many a furrowed cheek, when they were asked to go home and tell God all about it, and thank Him for the messengers He had sent. One remarkable circumstance in connection with the Camberwell services had been in the attendance of a number of medical students from the various hospitals. Some of the medical men in the neighborhood had found time and opportunity to invite them to their houses to dine, and afterwards had spoken to them about salvation. If only half a dozen young students were brought to Christ, what might not be the result when they were attending the sick-beds of those who should be committed to their care? The South London Committee were 100 strong, and they were going, by God's grace, to work shoulder to shoulder and hand to hand in this blessed work.

Rev. Marcus Rainsford said he felt we were living in days which many had looked and longed for, but had not seen. He thought that God had been working much more with the masses than the ministers. For his own part, he had learned much since Messrs. Moody and Sankey came to London. Many prejudices had been broken down, many difficulties removed, and many a lesson learned that he would never forget. He had been taught by a costermonger how to preach the Gospel. He was talking to a costermonger one evening, and trying to show him the great salvation, when a bright-looking young fellow came up and quietly put him aside, saying, "Sir, I found Christ last week; I think I can talk to this man better than you." 'Well, let us hear what you have to say.' "I never heard such lingo," said Mr. Rainsford. "Now, Joe, s'pose it was all up wi' yer; mother starvin', wife starvin', children starvin', and the mackerel nowhere. S'pose I see yer lookin' very pale, and sad, and miserable; and, says I, 'Joe, here's a fat half for you.'" (I wondered what that was, but the other seemed to know all about it.) "I give it yer with all my heart; be off and do your work. Away you go to Billingsgate and spend the fat half." (It means half a sovereign, and a sixpence means a "thin half.") "You get the mackerel, and bring it home; you get the money, and you bring home some bread; yes, there it be at home; now what would you say?" "I would say, 'Thank you; God bless you!'" "Well, say that to Christ, for He didn't give you the fat half, but the whole." And that was the Gospel as ably and spiritually preached, and as blessedly preached, as the Archbishop of Canterbury could preach it. After some further striking experiences, he expressed a hope that the work would go on after our brethren had left, and that many would be found to imitate their example in telling of Jesus to all around.

The Earl of Shaftesbury said, nothing but the positive command of Mr. Moody would have induced him to come forward on the present occasion and say but a few words in the presence of so many ministers of the Gospel. But as Mr. Moody had asked him to speak of what had occurred during the past four months, he did so with the deepest sense of gratitude to Almighty God that he had raised up a man with such a message and to be delivered in such a manner. And though Mr. Moody said they were not to praise him or his friend Mr. Sankey, yet if they praised God for sending them such men as these, they did no more than express their admiration of the instruments that He had raised up while they gave Him all the glory. He had been conversant for many years with the people of this metropolis, and he might tell them that wherever he went he found the traces of these men, of the impression they had made, of the feeling they had produced, and of the stamp that he hoped would be indelible on many of the people. He could speak that as the truth as to many parts of London, and the lowest parts of London. Only a few days ago he received a letter from a friend of his, a man whose whole life was given to going among the most wretched and the most abandoned of the populous city of Manchester, and who spoke of the good that had been effected there by the preaching of Moody and Sankey. A correspondent in Sheffield had also written him that he could not satisfy in any degree the wants of the people; that they were calling out for tracts and something that should keep up the appetite that had been created. He said, "For God's sake, send me tracts by thousands and millions!" Even if Messrs. Moody and Sankey had done nothing more than to teach the people to sing as they did with energy and expression, such hymns as "Hold the fort, for I am coming," they would have conferred an inestimable blessing.

Mr. Sankey then sang the hymn commencing, "There were ninety and nine that safely lay," after which he said that when they got to their own country they would often sing this hymn again, and they trusted that God's blessing would accompany the singing of it. They asked their friends here to pray for them, and that the Lord would continue to bless them. They would be glad to hear from their friends here, and they trusted to hear that the work was going on.

Mr. Moody said he would ask them to spend a few moments in silent prayer, but, before they did so, he begged to thank the ministers for the sympathy they had shown them in the past two years. They had had nothing but kindness shown them. He also wished publicly to thank the Committee, and also the stewards, who had manifested towards them nothing but kindness. He had also to thank the reporters for the press. He knew that he had made mistakes, but they had not reported his mistakes or his failings. In fact, they had all been kind. He also wished to thank the police for the considerate manner in which they had performed their duty. He had one favor to ask of them—he would not ask them to pass a resolution, for their hearts were worth more than a resolution—he asked them to pray for them, and to continue to pray for them as they had done for the last two years. He now asked them to pray for a short time in silence.

The congregation then bowed their heads, and, after the lapse of two or three minutes, audible prayer was offered by some one on the platform, after which Messrs. Moody and Sankey hastily retired, in order to escape the painful ordeal of bidding so many of their friends a formal good-bye.

VII.

IN NORTH WALES.

During the two years sojourn of the American Evangelists in England, Scotland, and Ireland, they have so assiduously devoted themselves to the unfolding of God's wonderful working in grace towards a lost world, that they have been left no time for any survey of His scarcely less wonderful working in nature, which any of these three kingdoms can show. They have passed the best part of three summer seasons in our midst; but while other laborers in the Lord's vineyard were taking their now universal "holiday," Messrs. Moody and Sankey were hard at work, entreating men and women to be reconciled to God. Now that their regular labors have ceased, a brief week or two has been devoted to well-earned recreation ere they return to their native land. As our readers know, Mr. Sankey traversed part of Switzerland, but Mr. Moody has not gone so far afield, but has contented himself with a short tour among the Welsh mountains, in company with Rev. Mr. Aitken, and Mr. Balfour, of Liverpool, at whose country residence he has been a guest. Like Mr. Sankey in Switzerland, he has had to pay the penalty of a wide-spread popularity. His fame has preceded him wherever he has gone, and thrice he has been induced to break silence, and give Gospel addresses to the immense multitudes that have gathered to hear him.

GREAT GATHERING AT WREXHAM.

Of the great meetings at Wrexham on the 1st inst., and at Rossett the following day, I can speak from observation. Reaching the former place about eleven o'clock on the evening of Saturday, 31st ult., the first thing that my eye caught, by the light of a street-lamp, was a poster announcing that Mr. Moody would preach in Wrexham the next day (Sunday), at six o'clock P. M. The town seemed in a bustle even at that late hour, and was palpably very full, as I had some little difficulty in securing a resting-place for the night.

At half-past seven o'clock next (Sunday) morning, a united prayer-meeting was held in the Corn Exchange, to ask God's blessing on the services of the day. I was not aware of it, and consequently did not attend, but I understand that upwards of 500 persons were present, including several of the ministers of the town, and that the proceedings were marked by a spirit of much devotion and earnestness. Early in the day, I encountered in the street an old man busy reading Mr. Sankey's little hymn-book, who stated that he had walked seven miles in the morning to be at the prayer-meeting, in the expectation of hearing Mr. Moody there. He also informed me that one of Mr. Moody's "followers" was to preach at St. Mark's Church in the forenoon. I found out afterwards that Mr. Aitken was meant, and the description was not so much amiss, as Mr. Aitken has of late acted somewhat as lieutenant to the great evangelist, and accompanied him on his short Welsh tour. At eleven o'clock St. Mark's Church was densely crowded. I am not greatly skilled in such matters, but it seemed to me that the mode of conducting the services was what is known in these days as "High." It looked strange, and contrasted strongly with the simple, unadorned services of Messrs. Moody and Sankey, which have formed the staple of my church-going for six months past, to see a choir of thirty men and boys slowly marching up the centre aisle, followed by the Vicar and Mr. Aitken in parti-colored robes. Everything was "intoned," to the "Amens" of the congregation. Any formality, however, in the early part of the service was soon forgotten in the powerful and heart-searching sermon of Mr. Aitken, on the words, "Come, for all things are *now* ready."

Prayer-meetings were held in the afternoon in various of the Sunday-schools, and at an early hour a large crowd had collected in the beast-market, where Mr. Moody was to speak. This triangular space in one of the outskirts of the town is about an acre and a half in extent, and a rough but substantial platform was erected for Mr. Moody, the ministers and choir, etc., at one side of it, in front of the National School-house, which was intended for use as an inquiry-room. As it turned out, it was not brought into requisition for that purpose. Long before six o'clock (the hour fixed for beginning the service) the market was crowded closely from end to end, except at the extreme ends, where it was neither possible to see or hear. The people must have flocked from all the surrounding districts, as it was freely asserted that 20,000 persons were present ; and judging from the vast sea of heads that was presented to view from the platform, and the area covered by the closely-packed multitude, I am not disposed largely to reduce this estimate. Mr. Sankey's hymns were sung for some time, led by an efficient local choir, and, as a variation, a Welsh verse was sung to a beautifully plaintive air, which carried me away to a Highland churchyard, where the Gaelic-speaking population meet periodically to observe the Lord's Supper. It was joined in more heartily even than Mr. Sankey's hymns, and testified to the affection of the Welsh people for their native language.

When Mr. Moody arrived, and saw the immense congregation, he expressed his doubts as to whether he could make them all hear, and suggested that an adjournment should be made to an adjoining field, where a more central position could be got for the speaker. As all the preparations were made, however, he agreed to go on with the service, though, as it turned out, his opinion was the correct one.

"The old, old story,
Of Jesus and his love,"

was the opening hymn, and then Mr. Moody plunged into his discourse from the text, "The Son of Man is come to seek and to save that which was lost." He pitched his voice in a high key, and labored hard to make his audience hear, but it was evident from the moving of the outside portion of the crowd that he was only partially successful. He was very earnest, and the sermon, which we had before heard him give many times, seemed as fresh as at first, many of the illustrations being told with thrilling effect. After speaking for about twenty minutes, his voice seemed to be fast giving way under the great strain to which he was subjecting it, and he wisely closed the discourse. After prayer he announced that a service would be held in the field in half an hour, when Mr. Aitken would preach.

To the field multitudes flocked accordingly, though a good many went away, the second gathering being a good deal smaller than that in the market-place, but still large enough to form an imposing congregation. Mr. Moody having recovered his voice, proceeded thither also, and himself resumed addressing the people, being heard freely by all. He asked and answered the question, "What must I do to be saved?" and his words were carried home with such power that a number held up their hands at the close as desirous of being prayed for. An invitation was extended to those who were really anxious for salvation to go to the Public Hall; and upwards of thirty responded, and were conversed with by Mr. Moody and others, and subsequently addressed by Mr. Aitken. The record of such as were saved through the day's services is on high, but, at all events, the good seed was unsparingly and faithfully sown; to fructify in eternal life, we hope and believe, to many of our Welsh countrymen and women. The weather was beautifully fine, and this added much to the comfort and picturesqueness of the out-door proceedings.

SERVICES AT ROSSETT.

Next day, the 2d inst., the pretty little village of Rossett, about five miles north of Wrexham, was invaded by a multitude of people such as, I suppose, it has never witnessed before, and probably never will again. Mr. Moody was advertised to lay the foundation-stone of a new chapel belonging to the Calvinistic Methodist body at 3.45 P. M., and afterwards to give a Gospel address at 4.30 in the park surrounding Mr. Balfour's residence at Mount Allyn. Early in the day the people began to converge on the spot where the ceremony was to be performed, coming from all points of the compass, and by all sorts of conveyance. The day was dry and warm; the roads were very dusty. A crowd clustered around the intended site of the chapel, and some hymns were given out, but being on the edge of a narrow roadway, only a small proportion of the assembled concourse could get near.

It will not be necessary to give the details of the ceremony of laying the foundation and memorial-stones, which was performed with great good-humor and anxious care by Mr. Moody and Mrs. Balfour, inconveniently hemmed in as they were by a curious crowd of onlookers.

The ceremony being over, Mr. Moody and the friends who accompanied him drove to Mr. Balfour's fine and extensive park, followed by a multitude of people, which, when completely assembled, must have numbered some 10,000. Here, standing in the carriage, Mr. Moody once more proclaimed the glad tidings of salvation from the story of Nicodemus and his conversation with Christ on the new birth. Mr. Moody was listened to with the most devout attention, and at the close of a long and most earnest address some dozen hands were lifted in answer to his appeal. Mr. Aitken supplied Mr. Sankey's place, in his much-regretted absence, and led the singing as well as offered prayer. Under the circumstances an inquiry-meeting was hardly possible, and the benediction being pronounced, the great crowd slowly melted away, much gratified, I doubt not, at having seen and heard the eminent evangelist, but above all, I hope, more desirous of profiting by "the words of this life" which had been so earnestly, though simply, proclaimed. Mr. Moody cannot fail to carry with him to his own country a fragrant remembrance of the short season spent in Wales, and the honor so universally accorded to him, both for his own and for his Master's sake.

VIII.

IN LIVERPOOL.—FAREWELL TO GREAT BRITAIN.

In few of the towns visited by Messrs. Moody and Sankey during the past two years has the work of their hands prospered more than in Liverpool. We call to mind with thanksgiving the wonderful scenes we were wont to witness in the Victoria Hall during the month of February last — the "eager, anxious throng" that waited patiently outside the doors for hours, despite the intense frost and cold that then prevailed; the multitudes who stood on their feet requesting prayer at Mr. Moody's call; and the long array of anxious seekers who filed into the large inquiry-room behind the platform, filling it to the doors, so that the great hall itself had to be brought into requisition for the same gracious, happy work of pointing the sin-sick, weary souls to Him who alone can give rest. After our brethren left Liverpool for the metropolis, the work of revival was sustained with more than usual vigor and completeness of organization, and amid tokens of the most unquestionable success. The work among the young men of Liverpool, and the efforts that have sprung out of the services held in the Circus, have been simply wonderful. The past six months have been very eventful ones in the spiritual history of thousands in Liverpool. In view of this, and seeing that Liverpool was the point of the evangelists' departure from our shores, it was fitting that their final words of farewell to Great Britain and Ireland should be uttered here. There was, no doubt, a local interest attaching to the meetings on Tuesday and Wednesday of last week, but the occasion was almost a national — I had almost said an international — one. There were friends present from many parts of the country, and Mr. Moody's long and important farewell address will be read with interest, and, we trust, profit, all over the world, as it was heard by those who were present. It need scarcely be said that the announcement of the farewell services gave great satisfaction, and the demand for tickets for the meetings on Tuesday, the 3d inst., was far in excess of the supply.

I arrived in Liverpool on Monday evening. The long train which carried us to Chester was so crowded that I, with about forty others, was fain to clamber into the guard's van, and take refuge among the heaps of luggage, notwithstanding that the journey had to be performed with scarcely any light and less ventilation. From the landing-stage I proceeded straight to the Circus, which I found well filled with the after-meeting, presided over by that devoted servant of the Lord, Major Cole. One of Mr. Moody's oldest and best friends, Mr. J. V. Farwell, of Chicago, was giving testimony for Christ in his own unpretentious, but most telling way. Others followed, and before the gathering dispersed there was yet a third meeting for conversation with the anxious, of whom there seemed to be not a few.

The noon prayer-meeting on Tuesday at Victoria Hall was very well attended—more largely, I suppose, than usual—and was a deeply interesting one. Messrs. Moody and Sankey were not present. The address and the prayers were felt to be a very appropriate introduction to the memorable gatherings that followed in the after part of the day. It was announced that the Hall must be emptied, otherwise a large number would have remained for the two hours that intervened before the afternoon meeting, in order to secure their places.

AFTERNOON CONFERENCE.

About a quarter to three o'clock, Messrs. Moody and Sankey, with many well-known friends, emerged from the trap-door in the middle of the platform, and were greeted by an audience that crowded every nook and corner of the great building. The heartiness of the welcome found vent in a universal clapping of hands, which, however, Mr. Moody speedily stopped by a wave of his hand. Some kind friends had placed very beautiful bouquets of flowers on Mr. Sankey's organ—a custom which is, I believe, common in American churches, and, within certain limits, a very pleasant and unobjectionable one. It would be invidious to mention the names of those on the platform without giving all, but it will be sufficient to say that they included most of the local friends of the movement, and many from London, Manchester, etc., and also from Scotland.

In the course of the opening exercises, Mr. Sankey sang alone two verses of the fourteenth hymn, "Tell me the old, old story,"—and here my opinion must conflict with that of the correspondent of a contemporary last week, who asserts that "the *timbre* of Mr. Sankey's voice is perceptibly impaired." I was struck with the clearness and rich quality of Mr. Sankey's tones, and the circumstance was to me most gratifying. It would be no cause for wonder if, after the constant strain of two years' daily public singing in all states of this changeable climate, his voice should be affected, but as far as I am capable of judging, it seemed on this occasion to ring out with more power and expression than ever. And considering the great work that we all hope lies before our brother in his own, and, it may be, other lands, I repeat that this is matter of genuine satisfaction and thankfulness.

Mr. Moody read part of the 105th and 107th Psalms, and from the 12th Chapter of Isaiah, and said he wished to give two key-notes for the addresses that should follow; his turn to speak would come in the evening. The first thing was to praise God for what he had done—to praise God and not man. He would call to order any one who attempted to praise man; we want to get man out of sight, and exalt Christ. The next thing was to "advance." What were they going to do? His faith had grown since he had been in this country, and he did not know why Christians should not go out and possess the land.

The first address was by the Rev. Mr. Aitken, who said he thought they could not meet in that hall without feeling that the departure of their dear friends for America very greatly enhanced the personal responsibility of all who called themselves Christians. The blessing which God had been pleased to shower upon His work in various parts of the land had put them on a vantage-ground, for they occupied a better position now than they ever occupied before in this land. He did not believe that the Church of Christ had ever occupied a better position in this land than it did at the present moment; and if that was so, their responsibility must be proportionately heavy. And if they allowed themselves to lose their vantage-ground and slip back into the dull routine of the past, they would have themselves to blame. The question before them was a very practical one, and it was, How were they to push on the advantage? If they were really to avail themselves of the opportunity, they must expect further successes. He was apprehensive of Christian people allowing themselves to think that the period of reaction had come —that they had been having such great encouragement that for a little time they must rest on their oars. If they placed themselves in this attitude, they would have themselves to thank for it, if God turned the heavens above into brass, and made the earth as iron beneath their feet; therefore he felt it incumbent upon him to sound this note of warning. He thought that their attitude should be this: That they should thank God, and then rush on against the foe with fresh determination, believing that the victory was only commencing, and that inasmuch as God had given them a position of advantage, they must push it on and fight the battle out, until God in His own good time placed the crown of victory on their brow.

How was this to be done? God expected every one of them to come forward with the gospel of grace in their hearts; and if they realized their personal responsibility and went into the battle fully determined to win souls for Christ, England would very soon feel the results of their efforts. He desired to warn them against this season of revival being followed by a period of reaction. Before the present work closed, he thought that ministers of Christ and also lay people, especially those who occupied influential positions, should ask themselves solemnly what were the permanent lessons which had been brought before them in this great movement. Mr. Moody had given himself up to the work of evangelization, and he (Mr. Aitken) could not help believing that the Church of Christ from a very remote period had practically ignored the evangelist's office. They had their local pastors, but he thought that the evangelist was more likely to be powerful in a locality where he was not permanently fixed than in his own country. If they were desirous to see God's work still carried on on a large scale, those whom God had in a large measure gifted with the power of the evangelist should consider whether they could give themselves entirely to the work. He had done so, but he utterly disclaimed all credit on that score. He did not think he should have had the courage to take that step, but domestic circumstances had rendered it imperatively necessary that he should leave his flock in Liverpool. He had, however, long been convinced of this truth, that if a man was to be a practical evangelist, he must give himself over to the work; and he called upon God's people to take this matter into serious consideration, and say that the great work which had been undertaken must be followed up in all our towns and villages, for he believed that even the villages needed it more than the towns.

A great responsibility also rested on the ministers of Christ. In almost all the places where the wave of blessing had passed, there would be a large number of young converts who had given themselves over to God and wanted something to do. Their duty at this moment was to set all those young Christians to work. There were a great many ministers who fell into the mistake of trying to do all the work themselves. What was wanted to be done was to find specific spiritual work for those who had given themselves to God, and encourage them; and he wished to point out that unless this was done they must be the last persons to find fault with those extravagances which otherwise must develop themselves. If, instead of young converts being taken by the hand, they were left in the rear and not given any kind of encouragement, the result would be that they would either draw themselves up in their shells altogether, or rush into the opposite extreme.

It seemed to him that now was the golden opportunity; and unless they got their young Christians to work they would have to regret it to the end of their days. If after the departure of their American brethren they resolved to have a holiday time of it, then good-bye to their usefulness, and God's blight would rest upon them instead of God's blessing; whereas if they put themselves into God's hand, depend upon it this wave of blessing which had swept over the land was but the beginning of good things. He closed his stirring and practical address in the words of Wesley, which, he said, used to be sung at the close of his conferences:

> "A rill, a stream, a torrent flows,
> But send the mighty flood;
> Awake the nations, shake the earth,
> Till all proclaim Thee God."

The Rev. A. N. Somerville spoke next, and it is not too much to say that the meeting was fairly electrified, as "the old man eloquent" poured out the wealth of his declamation and illustration in a perfect torrent of burning words, accompanied by highly dramatic and expressive gestures. He said Messrs. Moody and Sankey did not want them to occupy time by throwing their arms round their necks and kissing them, but they had given them the motto, and that was to "advance." What, he asked, is our great encouragement? "All power is given unto Me in heaven and earth; go ye therefore and teach (or disciple) all nations." Just before Christ ascended He said, "Ye shall receive power after that the Holy Ghost is come upon you, and ye shall be witnesses unto Me in Judea and Jerusalem, and unto the uttermost ends of the earth." Why did the Lord Jesus tell us He had received all power? That He might confer power upon us. Mr. Somerville recounted the exploits of the mighty men in the days of the Judges, upon whom the power of God fell and proceeded. The day has come when it will not do for us to remain within this little isle. Larger efforts must be made to proclaim Christ's name throughout the world. We read that Alexander the Great, while a young man (he died before he was thirty-two), crossed the Hellespont with only 35,000 infantry and 5,000 horsemen. He had provisions and money to last them only one month, yet they went forth and took possession of the world. What! Is Alexander the Great to be always spoken of as the only man who can do the like of this? Is Jesus Christ not strong? Why should we not gather round Him, and in the power of His Spirit take possession of the world? We must not only send out men to engage in this blessed work, but the whole Church must by prayer and sympathy, by the voice of encouragement, and by liberal support, work together as one man for this great end. When I was in India, I felt that wherever I went I was borne up by the sympathy and prayers of many dear friends in my own city of Glasgow, in Edinburgh, in London, and in many parts of the world besides, and I was strong through their sympathy. If a man is sympathized with and encouraged in that way, he will do twice as much as he would do otherwise. Mr. Somerville illustrated the power of sympathy by telling how Alexander the Great was traversing a desert with his followers, who were suffering greatly from thirst. Some one brought him a little water in a helmet, and as he was about to partake of the precious refreshment, he looked towards his followers, and seeing their sufferings, he refused to drink. His men were roused to action by the sympathy thus shown by their leader; they put their spurs to their horses, and sped on to a place where relief could be found. Speaking of the necessity of humility in Christian work, he quoted a beautifully apt simile, in the use of which he seems to excel. He said the Rhine before it reached Basle received no fewer than 1,200 tributaries. How was this? It was *by keeping at its lowest level*. If it had not, these streams would have flowed somewhere else. He roused the audience to such a pitch of excitement, that when he sat down they burst into applause, which no attempt was made to suppress.

Mr. Sankey then sang "My Prayer," a beautiful hymn of consecration. He prefaced it by saying that he would be able to go out and work better if we had the blessing of which the hymn told.

Dr. Barnardo then gave an address, in the course of which he said the question was frequently asked, "How shall we reach the masses?" He knew only of one answer: "Go and preach Christ to them." That must be the bait; but there must be something more than that. Not only must they preach Christ in His boundless love to a dying world, but there must be the hook —such an application of the truth as should enter men's hearts and draw them to the Saviour. What was the great prerequisite to success? It was given in the two words of our Saviour, "Follow me." That was the secret of successful service: there was no royal road; their brother, Mr. Moody, had no knack in it. God help them to follow Christ, that they may be truly fishers of men.

Mr. Stalker, of Edinburgh, said he felt that the past two years had been years of great importance to the whole country, and would be remembered for many years to come as great years. One thing that had made them interesting and memorable was that religion had been made respected among the young men of the country. Young men had been apt to look down upon evangelical religion; but in the part he came from they dared not do that now, because in all classes of the community the very backbone of these young men had been won to Christ, and they were bearing themselves so in the ordinary business of life that it was impossible for those around them not to respect them. He never thought of this movement without his mind wandering away into the future; and he thought not only of the number of men who had been saved, but of the young men who were devoted to Christ going on in their various spheres—in the family, in social intercourse, in business, at the University, in their shops, as clerks, and in all the different walks in life—distinguishing themselves, and showing that their Christianity, instead of keeping them back, was helping them on; that their spiritual regeneration had been at the same time moral and intellectual regeneration; and that they were determined to be men in all the departments of life. He read often with pity the remarks made by some, of the weakness of those who took part in this movement. At the University of Edinburgh, last April, there were only six or seven men who secured first class honors, and three of these were head and shoulders in this work. Only one man got what was called a "double first," and that man he had heard addressing these revival meetings. That was the kind of revival of religion they were having now; and he thanked God for it with all his heart, for their preaching to young men was far more effective if they could show them that their religion was making them get on well in business, and do their business well, and come to the front in the ordinary walks of life. Let them seek to serve God by doing their work thoroughly, and at the same time standing on that vantage ground, exhort all their brethren to get that which had made men of them.

Dr. Ziemann, of Manchester, then briefly addressed the convention, pointing out the evangelical work that was being carried on in Manchester, and said that many young men in that city had taken part in the movement, and were determined to carry it on to victory. He asked those whom he addressed to continue the fight, and let their watchword be "Victory."

THE EVENING MEETING.

In a touching chapter of family biography, that Mr. Moody was wont to use for illustration, he spoke of Thanksgiving Day in America, the observance of which brings all the scattered members of the household together. We were strongly reminded of this by the great gathering of friends who had come from far-off parts of the country to be present at the farewell services and departure next day. Many who had taken an active part in the London services came down expressly to say good-bye to their Evangelist Brethren, and not a few had come from across the border, as well as the other English towns.

We have seldom had to record anything

else but crowded meetings in connection with Messrs. Moody and Sankey's services, but this last evening meeting of all must have been—if that were possible—more crowded than any. As one of the local papers of next day puts it: "Every inch of space where a person could sit, or stand, or crouch, was occupied." As a consequence, the heat was exceedingly great, and in order to obtain more ventilation, some resorted to the dangerous expedient of breaking the windows. As the hall was filled to overflowing long before seven o'clock, the time was profitably used in singing the well-known hymns, in prayer, and in the delivery of short addresses by various friends.

Mr. Sankey sang, "I am praying for you," and before doing so, he said: "When we are gone from among you, we hope that you will remember to pray for us, as we will surely remember to pray for you. Pray God that He may use us in our own dear land as He has used us here, and even more abundantly. May the blessing of God rest upon the singing of this hymn to-night."

Mr. Moody then commenced his address, and spoke for more than an hour, but to the very last there was the most rapt attention. By some means the gas could not be lit, and as the fading twilight deepened into darkness, the scene became intensely solemn, as Mr. Moody's earnest and sometimes faltering words fell on the hushed and eagerly attentive multitude. At the close of his address he offered fervent prayer. He besought God's blessing on England and America, on the work among the young men, and on the ministers, his utterances anon being stayed by his evident emotion.

Mr. Sankey's voice found expression for the last time in the farewell hymn which he has sung at many of the towns visited, though not in London. As Mr. Sankey sang it, by the light of a candle, to the justly popular tune of "Home, Sweet Home," the audience was much moved. It was the last time many of them will probably hear Mr. Sankey's voice, and we are sure none of those present will be able to forget it.

Before the meeting was dismissed, Mr. Drysdale, the Secretary of the Liverpool Committee, announced that the hall would be opened for continuous services in October next, and that Mr. Aitken had agreed to take the first week, commencing Sunday, October 3.

Mr. Moody stated during his address, and much to the satisfaction of many, that a service, chiefly for young men, but open to all, would be held next morning at half-past seven o'clock.

On Tuesday evening, after the general meeting, a few friends gathered at the Compton Hotel, and two or three hours were spent in an informal conversation on the subject always uppermost in Mr. Moody's thoughts—the best way to benefit young men—to conserve and utilize in the way of righteousness, for the glory of God and the good of men, the young manhood of Great Britain, America, and the world. Happily, Mr. Shipton, the Secretary of the Young Men's Christian Association, was present, and he entered heartily into the feeling of all present that the time and opportunity had come in the providence of God, for that Association to take a more decided stand than it has ever yet done, as an aggressive organization for the evangelization of young men. It was felt that the various branches throughout the country may become much more closely and intimately allied in Christian work than hitherto, and that as so very much depends on the secretaries, as to the tone and conduct of each branch, it was felt to be of the first importance that those who occupied a position of such influence and trust should be men qualified for their office by capacity and education, as well as by Christian character and consistency, and that they should be adequately remunerated. Mr. Moody remarked concerning this, that it is easier to raise £500 a year for a well-qualified man than £100 for an inferior one; for people will give liberally and willingly in the one case, but grudgingly or not at all in the other.

The conversation turned upon Manchester, and the amount of influence for good which one man may exercise was remarkably illustrated by the effect of Dr. Ziemann's devotedness and energy upon the noon prayer-meetings, and the evangelistic work generally, since Messrs. Moody and Sankey's departure from that city; and it was felt by all that men like Mr. Henry Drummond, whose conduct of the young men's meetings had so well supplemented the work of our American brethren, are needed in order that the Young Men's Christian Associations throughout our land should develop the power and usefulness which lie latent in them. Mr. Moody also mentioned the case of the Young Men's Christian Association at Washington, which some years ago was in a declining condition, but soon became one of the most important and successful branches in all the United States, under the leadership of its present secretary, who was induced to give up his position as a Wesleyan minister, in order to devote himself to this no less honorable and influential ministry. When it was proposed to offer him the post of secretary, some objected that he was a Northern man, and would be unpopular in a Southern city; others said that, being a Methodist, he would be objectionable to those of other denominations, but the fact that a man's personal fitness outweighs all difficulties as to his circumstances and surroundings, was signally proved in this case, for soon this excellent young man was surrounded by a numerous band of young men as devoted as himself, who were so much attached to him, that when he was wanted for another sphere they absolutely refused to let him go.

We believe that if one thing more than another will induce Mr. Moody to return to Great Britain, it will be the desire to weld together its Christian young men into a band of fellow-laborers, that, by the operation of the Spirit of God, shall be in the midst of many peoples as a dew from the Lord, and as a lion among the beasts of the forest. And to tell the truth, we expect that it will not be years before we see our brethren again among us.

THE LAST SERVICE IN ENGLAND

was held by Mr. Moody on the morning of their departure, so that we may say they left our shores "with their harness on their backs." Mr. Sankey was not present.

The hall was opened at seven o'clock, by which time considerable crowds had gathered at all the doors, and before Mr. Moody made his appearance at twenty minutes past seven there were some 5,000 or 6,000 persons in the hall. After praise and prayer, Mr. Moody read part of the first Chapter of Joshua and twenty-sixth Chapter of Leviticus. He proceeded to give a short address to the young men, the first part of which was an earnest plea for a systematic study of the Bible and Bible characters, and for union with some organized body of Christians. He also urged on them the necessity of having some definite work to do, and not to attempt too many things at one time. "I have been wonderfully cheered," he continued, "during the past months by the tidings coming from Liverpool. I want to say from the depths of my heart, God bless you, young men. The eyes of Christendom are upon you. Perhaps there has not been a place where the work has been so deep and thorough as the work here among the young men. I believe it was in answer to the prayers that went up for it when we were here six months ago. And now, as we cross the Atlantic, it will cheer us as tidings come that the young men are still advancing. Do not fold your arms, and say, 'We will have a good time next fall.' God is just as ready to work in August as in July. If some have gone out of town on their holidays, the work should not stop; I think it is the best time to work when many are away. Every man ought to be worth the five or six that are away. Then the work will go on. The great revival at Pentecost was in the hot weather, and also in a very hot country. People think there cannot be any interest in the warm months; but if the prayer goes up to the throne, God does not look to see what month it is. He is as ready to bless in one month as in another. Let me give you the watchword we had yesterday afternoon — 'Advance.' I hope there will be a fresh interest awakened in Liverpool as there has been in Manchester. I do not know of anything that has encouraged me more than to hear of the work going on in Manchester for the last six weeks. I hope Liverpool and Manchester will shake hands in carrying on the work, and let the lies of those sceptics who say it is only 'a nine days' wonder' be driven back. I cannot talk longer. I say from the depths of my heart, I love you; God bless you, and may the power of God come upon you this morning afresh."

In his prayer that followed, Mr. Moody made special reference to the work being carried on by Major Cole, and he afterwards made some remarks, urging all to support and help on these labors.

After the hymn, "Free from the law," had been sung, Mr. Alexander Balfour said, "I do not know whether I am the proper person on behalf of this audience to say good-bye to our dear friend, Mr. Moody, and our absent friend, Mr. Sankey; but I feel that there must be some mouthpiece to say to them what we really do feel. We thank them from the bottom of our hearts and souls for what they have come here and done. Unless Mr. Moody had been a man like a cannon ball for hardness of material, for directness of aim, and for strength of will, he could never have done what he has been privileged by God to do. His wisdom has been conspicuous in discovering this —that our young men in Liverpool and elsewhere in this country have been greatly neglected, and in choosing them to be, for the future, not merely the recipients of God's grace, but the distributors of it. I do feel that Mr. Moody, in having given so much attention to our young men, has really done the right thing. Many know that Liverpool has been a curse to young men. They have come here and been led astray into all kinds of mischief and wickedness. How many broken hearts are there in this country because of the mischief done to young men in Liverpool ! On behalf of the mothers and sisters of this country, I want to give Mr. Moody the most heartfelt vote of thanks that it is in my power to convey ; and on behalf of thousands who shall be influenced by the young men in Liverpool, I want to convey to him the tribute of gratitude for what he has done. As President of the Young Men's Christian Association, I want to say this : That it is our purpose as young men to go on with the work ; and, by God's grace, we shall not go back, but advance in our endeavor to do our duty before God and men."

Mr. Moody, in reply, simply said, " I will now shake hands with you all in the person of the President of the Association ;" and the meeting having been closed, he returned to the Compton Hotel, surrounded by a large crowd, which sang, " Hold the Fort," and the " Doxology" in the street in front of the hotel. Many of them lingered there during the hour and a half that elapsed before Mr. Moody, Mrs. Moody, and family, accompanied by a large number of friends, drove away to the landing-stage. They were followed by the enthusiastic cheers of the assembled multitude. Mr. Sankey stayed at the residence of a friend, and so escaped much of the popular attention that Mr. Moody had to undergo.

IX.
THE DEPARTURE.

A SPECIAL tender was provided for the conveyance of the evangelists and their party to the " Spain ;" and Mr. Sankey, who spent the night at Edge-lane, and most of the friends, went on board of it shortly before Mr. Moody. As Mr. Sankey passed across the landing-stage, upon which a large number of people had assembled, he was warmly cheered.

There was a large gathering of people in front of the Compton Hotel to see Mr. Moody leave. As Mr. Moody emerged from the hotel, a hearty cheer arose from the crowd, and people rushed to the door of the cab on each side to shake hands with him, and bid him good-bye. The cab was, however, immediately driven away to the stage amidst renewed and warm cheering. For some time prior to the hour at which the special tender was to leave for the "Spain," people began to assemble on the Prince's pier and the landing-stage, and when Mr. Moody arrived there were several thousands present. A wide strip of the stage was kept clear by the police for the party to walk to the tender, and as Mr. Moody went on board he was heartily cheered, which he acknowledged by bowing. When the company were all on board the tender steamed away. As it passed down the river the people upon the pier and the landing-stage cheered with increased heartiness, and waved their hats and handkerchiefs. Their example was imitated by the people on the ferry-boats moored at the stage or crossing the river ; and when the cheering had subsided the people on the stage struck up one of the wellknown hymns. The sorrowful countenances of many of the people showed that it was with no ordinary feelings of regret that they saw the evangelists going away.

The tender reached the "Spain" about an hour before the time for the ship to weigh anchor, and the interval was fully occupied in taking leave of the evangelists, and in receiving from them or conveying to them parting words of comfort and encouragement. Mr. Moody again urged those who have been his fellow-laborers in this and other districts to remain united, and to carry on the work with courage and determination ; whilst on the other hand, there were very numerous expressions of the hope that a success equal to that of the last two years may attend the evangelists' labors wherever and whenever they may be resumed. Many of the leave-takings, from their intense earnestness, were very affecting. Only when the "Spain's" anchor was being raised, and the tender was upon the point of starting, could many of the friends tear themselves away. As the last of the people "for the shore" were leaving the ship, those who were already on board the tender sang the hymn, "Safe in the arms of Jesus." As the "Spain" moved slowly down the river, the people in the tender, which was still alongside, cheered heartily, and the passengers on board the "Spain" replied with another cheer, and the waving of handkerchiefs and hats. As the "Spain" passed on ahead, the people in the tender sang the hymn, "Hold the Fort," and afterwards the hymn, "Work, for the night is coming." Mr. Moody and Mr. Sankey stood at the bulwarks of the "Spain" and bowed and waved their handkerchiefs until the two ships were out of sight of each other. Shortly before the tender reached Liverpool, prayer was offered up on board by Mr. R. Radcliff, and other gentlemen for the safe arrival of the evangelists at their destination, and for the subsequent success of their labors, whether carried on in England or America.

X.

"TWO AND TWO."

ADDRESS BY DR. A. A. BONAR, OF GLASGOW, AT THE MILDMAY FAREWELL MEETING, JULY 13.

WE must all have noticed that this is the manner of our Lord to carry on important works by more than one instrument; He likes to work by two. He sent out His disciples two and two in the days of His flesh. You read in the Acts of the Apostles that Paul is scarcely ever alone; it is Paul and Barnabus, or Paul and Silas, or Paul and Titus—always somebody helping.

And it is just so in this present work. The invitation to our beloved brethren in America was sent from two—one who you all knew so well here, Mr. Pennefather, and the other Mr. Bainbridge—both of them now in glory. Then when God was going to work here, He sent over two brethren, both now so well known amongst us.

I think it is good to notice this; and I notice it for another reason. It brings me to a passage of Scripture that will be interesting and profitable perhaps; if the Lord use it, it cannot but be profitable to us all. One of the most remarkable cases of conversion in all the Bible was wrought by two men in company with each other—Peter and John. In the 3d chapter of Acts we find what I look upon as the most singular instance of immediate conversion and free salvation that we have in the whole Word of God; and it is by the instrumentality of *two* of God's servants, as we have said. You recollect the lame man at the Beautiful Gate of the Temple. He had been lame for forty years, and did not wish to be anything else than lame. I want to show you how he was a grand monument of free grace.

He went to the Temple—carried, I should say, up the fifteen steps of the Corinthian gate,—and there he lay to beg. The Temple was his house of gain, never his place of worship. He did not go to the altar; he did not go in any way to praise God in that Temple. This man stood out all Christ's ministry. And there was a remarkable day, when he must have seen a tumult in the temple—the buyers and sellers driven out by the Saviour. And then the day of Pentecost came, when there were three thousand souls brought to Christ. He must have seen the 120 as they passed down the steps, with their countenances changed; but it produced no change in him. He was there to get money, for it was money he wanted.

And yet such is the sovereign grace and power of God through the Holy Ghost, that this man was wonderfully in a few minutes brought out of darkness into light. Peter and John, we are told in the chapter, were going into the Temple, and they saw this man carried up before them and laid down at the gate. And it is said, "He seeing Peter and John, asked an alms." I do not think he had been wont to do so, for he knew that Christ's disciples were very poor. But I suppose he had heard that the Christians had been throwing their money into a common stock, and thinking that Peter and John would have money to give, he asked an alms of them. They said to him, "Look on us," and Peter, as the spokesman, said, "Silver and gold have I none, but such as I have I give thee." He held out his hand as he said, "In the name of Jesus Christ of Nazareth, rise up and walk." He held out his hand, as much as to say, "Will you take a gift to-day, in the name of Jesus of Nazareth, whom you have despised?"

Now, you know, this was a time when the Holy Ghost was working, going through Jerusalem quickening souls. How the scales fell from the man's eyes in a moment! The man is healed; and not only so, but he is saved. He stood, and he never had stood before. He stood, and then he leaped, and then he walked;—he had never leaped or walk'ed before. Now he felt he was whole in body; his ankle-bones had received strength. And that he was whole in soul is quite evident from Peter's testimony in the 16th verse: "His name, through faith in His name, hath made this man strong; yea, the faith that is by Him hath given him this perfect soundness in the presence of you all!"

I wish you to notice that here was free salvation given to this lowest of sinners. No questions were asked about his former life; no payment was asked. Did you ever notice that on the day of Pentecost there was something evidently required; they were pricked in their heart for some hours? But here the man is not kept a minute waiting. He gets salvation like a flash of lightning: it is as sudden as it is free. Heaven flings an alms to this beggar, and the Holy Spirit enables him to close his hand upon the ransom-money. I do not know a more wonderful case of sudden, sure conversion than this, and no case in which you can so certainly say, "He paid nothing for it."

And so we ask sinners here—for it is not likely that all here are saved—we ask any unsaved one to-day, "Will you take the gift to-day in the name of Jesus of Nazareth?" Away to the Cross, and read what is written there—"Jesus of Nazareth." Will you grasp the Cross, and, as you look in His face, say, "I take Him?" This is the time when the Holy Ghost is quickening souls and moving the hearts of men.

How did the man that was made whole and saved act on that occasion? It is very beautiful to notice the words—"walking and leaping and praising God." The first thing he did was to praise God, and it is twice stated that he praised God as he walked on the Temple floor. Here was the lame man leaping and the tongue of the dumb singing, and I am sure there never was a song so sweet sung in the choir at Jerusalem as was the new song that this saved man sang that day in the ears of all the people.

That is not all. Some people seem to think that grace does not allow us to say anything at all about instruments—to ignore them. That is a misunderstanding of grace. It leads us to put them in their place, but not to ignore them. This man teaches us a lesson. Was he ashamed of Peter and John as instruments? Did he say, "I will praise the Lord; I will say nothing about these men?" No, he did not. In the narrative we have a most beautiful account of him. As the lame man, it is said, was walking in Solomon's porch, he *held Peter and John by the hand* while he praised God. He does not seem to have mentioned their names, but then he would not let them go, and all the while he sang his song of praise. He acknowledged the instruments, and praised God for them. Let us do the same to-day.

Do you want me to say a little about Glasgow before I close? We can give most satisfactory testimony about God's work in Scotland; but as I come from Glasgow, I want to speak of it. Here is one thing about it as to the ministers of the Gospel. If you want a man to believe, it is about one of the worst ways you could take to talk about faith. Speak about the object of faith. So we never talked about union; to talk about union is not the way to bring it about. We talked about Him who unites us all. We found ministers of all denominations that hold the Head, meeting together in union, and from that day to this we have worked in perfect harmony, asking no questions. We found this —and I had better mention it in case I forget—we are prepared to state, over and over again, that *at least* 7,000 souls were gathered in last year in Glasgow. All these are under the superintendence of our ministers. You in London have this disadvantage, that you are not likely to know, and take the converts under your care, as we can do in a smaller place. I wish to say this too: All these 7,000 know something about the shorter Catechism, so that we have confidence in them, that it is not feeling only.

Some brethren in Glasgow were not at all pleased with Mr. Moody and Mr. Sankey at first for not going to the lowest classes. *I think it would have been a great mistake if they had.* What we needed was that those in our churches who had but the name to live—yet had the name and were in most peril—should be reached first; and the blessing came to them first. We have found that there was not one minister who took an interest in this work who did not gain great additions to his congregation. There have been added to my own congregation over 100 souls, and there is scarcely one I have any doubt of.

When God's people had been greatly stirred up, the work among the lower classes began. And it is going on in a way that is most wonderful. If you are spending a Sabbath in Glasgow you could do this: you could go to the Green, and there see 2,000 outcasts every Sabbath morning get their breakfast, and then the Word preached; and you never saw a more attentive audience. At eleven o'clock you would find in the same tent at least 1,000 people, and in the afternoon perhaps 1,200. These ragged children are a most interesting congregation. In the evening the tent is overflowing to hear the preaching of the Word. Taking the numbers outside and in, you never have less than 3,000. Then there are meetings during the week, and all over the city smaller works are going on. The Lord is amazingly blessing us. If there is anyone who has stood aloof, I do not wonder he does not believe in the work. If he has only come once or twice, he might as well not have come at all. But go among them, and you will discover the reality of the work. You know Christ said about believing, "If any man will do His will, he shall know of the doctrine."

There is a solemn thought that is always occurring to me as well as many others— What is this amazing work to end in? In London we have been noticing that your newspapers have been far more favorable than ours were to the work. Why is all this? Is it not as though God has given the four angels charge to hold back the four winds for a little till He has sealed His servants in the forehead? Then comes the terrific storm that ushers in the coming of the Lord.

XI.

IMPRESSIONS OF THE RECENT REVIVAL IN EDINBURGH.

(A Letter to a Friend.)

LAST year, when I came to Edinburgh, I was agreeably surprised to find that the glowing reports I had read of the "work" in Edinburgh were true, and in no way overstated. This year these favorable opinions have been confirmed, and I think I may best convey to you the information I have collected in answering the following queries, presuming, of course, the existence of what we may call a revival. The queries I shall attempt to answer are these:

Is the revival movement still going on?

How are the professed converts standing?

Is there revived Christian life? and if so, how is it manifesting itself?

First, then, "Is the revival movement still going on?" My answer is, Yes. I have every reason to think so, because the meetings are still being held. These meetings are the Mid-day Prayer-meeting, the Saturday and Sunday Evangelistic, Cottage, Hospital, and Open-air Services, singing bands, etc. I have seen and attended most of these. The mid-day meeting is both well attended and joyous. I have been to several, and see little difference from those of last year, excepting in attendance, which is a little smaller; but, on Saturday, not any less. Of the other meetings the same may be said, and I am justified, therefore, in saying that the *revival is going on*.

The same simple, stirring Gospel addresses are being delivered. I attended the Saturday night's Evangelistic Service. This service, being the outgrowth of the revival, may be taken as an index of its success, both past and present.

When I went in (some few minutes after the hour), the speaker, an intelligent doctor of medicine, was speaking from the text "Deliver me from blood-guiltiness, O Lord." It was an exceptionally good Gospel address, and delivered in the quiet, intelligent way of a believing, educated Christian. Its effect was quiet, but quite marvelous to me, and I thanked God from my heart for the address.

A minister then applied by anecdote and exhortation what had been said. A working-man illustrated it by his own conversion; and a mere youth further found an illustration in his case; and so on.

But meetings are not only being held, but conversions real and lasting are occurring. In all the churches, too, I find the fruits of the awakened zeal of ministers, and numbers are being reaped in the steady growth of the membership of these churches.

The second query I am to answer is, "How are the professed converts standing?"

True conversion is a change from a *life* of ignorance to one of knowledge; from indifference to trust and obedience; from acts of sinfulness to holy living. It may be well to test the converts by this text, and ask, "Is there in their conduct and conversation such evidence of these things as we may reasonably expect?

This query has been prosecuted by such

experienced examiners as Dr. Horatius Bonar, the Rev. J. H. Wilson, of the Barclay Church, Professors Blaikie and Charteris, and the report of these gentlemen is, "Yes, —unequivocally Yes!" I asked a person —most likely to know—if there were signs of a growth in grace on the part of those converts (because I attach importance to this point), and he told me this was remarkably apparent in the case of most. I was satisfied, therefore, that those conversions were not the result of mere excitement, but of adequate knowledge of the merits and requirements of the Saviour.

My third query to answer is—" Is there revived Christian life, and how is it manifesting itself?"

We know the existence of a thing by its manifestation. If, therefore, we see manifestations of revived Christian life, we know there is such. It is manifesting itself in more thorough consecration.

In listening to the prayers at Monday's Mid-day Prayer-meeting, in thinking over the tenor of the chairman's remarks, in following the drift of the anecdotes and illustrations, I was forcibly struck with the desire almost everywhere observable, of the anxiety to be more free from dominant evil, and more entirely used for God's glory. But it is not only desire that those Christians have; it is work. Just run over the engagements for the week; just see the daily notices in the *Scotsman* newspaper; just listen to the long list of requests for prayer for some special effort to rescue the perishing; and you will know whether or not it is merely desire or work that those Edinburgh Christians want. And they are *not* each stifly following out his *own* plan— working only as he or she thinks best—but meeting in conference, and agreeing upon the adoption of any or every suggestion calculated to accomplish the end they have in view. And they know how frail they are, and also where they can get the strength they need, and are found in large and steady numbers daily at the throne of grace in the Mid-day Prayer-meeting.

I leave this God-gifted city as a man who is leaving a genial party in a warm and cosy room steps into the cold, and sleet, and blast of a drear November night.

But I am reminded by a friend, who is taking part in Mr. Moody's meetings in London, that the same God who is working so unmistakably in Edinburgh and London can do the same glorious things in South Shields. That He may do so is the prayer of your absent friend.

XII.

THE following is given as the number of meetings and aggregate attendance during the four months that Mr. Moody has been in London:

In Camberwell, sixty meetings, attended by 480,000 people; in Victoria, forty-five meetings, attended by 400,000; in the Opera House, sixty meetings, attended by 330,000; in Bow, sixty meetings, attended by 600,000; and in Agricultural Hall, sixty meetings, attended by 720,000. The amount of money expended for buildings, printing, stewards, etc., is $140,000. Messrs. Moody and Sankey have declined to receive any compensation from the committee. It is stated that a prominent business man has bought the Victoria Theatre, and intends to fit it up for religious work.

THE END.

SERMONS AND ADDRESSES

BY D. L. MOODY.

I.

GOD'S HUMAN INSTRUMENTS.

1 Cor. 1: 17 to end of chapter.

I WANT for a few minutes to call your attention to a truth that you will find in these verses that I have read. There are a great many Christians in London that are praying for God to revive His work. I have received letters from a great many, and the thing that I have to fear most in coming to London is that many might be leaning upon man or upon the arm of flesh, or upon the great meetings, and get their eyes off from the Lord. Now if there is going to be a work in London, God must do the work. It is not any new Gospel that London wants; it is not any new power. It is the same old power, the power of the Holy Ghost, and it is the same old story—nothing new. The world is running here and there after something new, and they come and hear the old, old story, and they say, "Well, it is not anything new after all." I want you to understand, if you have come here to-night expecting to hear something new, you will be disappointed. We have not come with any new Gospel, but are just going to preach the same old truths that these ministers before me have been preaching. And not only that, but we are come in weakness. There are hundreds of men in London that can preach a good deal better than we can, and if you are leaning upon man you will be disappointed; but if we lean upon God, and all our expectations are from Him, we will not be disappointed, "Cursed is the man that maketh the arm of flesh his trust." What we want is to cease from man, and get done with men, and look right straight away from man up to God. The world is seeking after wisdom, but they don't know God by wisdom. It is not the wisdom of the world. God's thoughts are not our thoughts, and God's ways are not our ways. Let us keep that constantly in mind. If God is going to work in London, He is going to work in His own way, and we must not mark out channels for the Holy Ghost to work in when He comes. He will work in His own way when He comes; and He will use the instruments that He pleases. God is a sovereign. He may take up this man; He may take up that man; He may take up that boy, and use him. God will choose His instruments, and God will work in His own way, and what we want is to get into our places as Gideon's army did, and let God work. Yon is a mountain, and God wants to thrash that mountain, and there lies a great bar of iron—ten thousand men could not lift it—and right by its side a little weak worm. The Almighty passes by that bar of iron, and takes up the little worm to thrash the mountain. That is what God has been trying to teach us six thousand years. He uses base things, contemptible things in the sight of the world. In this chapter Paul sums up the five things that God does use—foolish things, weak things, base things, despised things, and the things which are not. What for? "That no flesh should glory in His sight."

THE CALL OF NOAH AND OF MOSES.

When God was going to destroy the world, and wanted an ark built, He did not tell a nation to do it; He did not tell a great city to go and build it; He did not call forth hundreds of men, but one man, who was contemptible in the sight of the world. The world laughed at Noah and at his ark. They mocked him and made light of him. But that is God's way, not man's way. What is highly esteemed of man is abomination to God, and what is highly esteemed of God is abomination to man. God's thoughts are not our thoughts, and God's ways are not our ways. When God wanted to bring three millions of people out of Egypt, out of bondage, how did He do it? (An interruption here took place, in consequence of people attempting to enter at one of the side doors, and Mr. Moody called upon the people to rise and sing a hymn until the confusion ended). He then went on to say: We were talking about the weak things that God uses, and I was just going to say when God wanted to bring the children of Israel out of Egypt His way of delivering them was different from ours. We would have sent down there a mighty army. We would have called for an army with chariots and with weapons, or, if we were going to send a man down to plead with Pharoah, we should not have sent down that man who had been forty years on the backside of the desert; a man who really was not known. He had been so long out of Egypt that his name had been forgotten and his influence at the court was gone, if he ever had any; and he says himself that he was slow of speech, not an eloquent man. I suppose he was what we call a stuttering man—the last man we would have thought of sending down there. We would have picked up some great orator, some eloquent man, to lay it all before the king; but the Lord's ways are not our ways.

GOD'S BLANK CHEQUE.

And when Moses said, "If they ask me who sent me, what shall I tell them?" God said, "Say I AM sent me;" and, as some one has said, that was a blank cheque, and God told him to fill it out; and when they were in the desert and wanted water He filled out the cheque and drew water from the rock. When he wanted bread He filled out the cheque, and God gave him bread from heaven. Yes; *I Am* sent him, and God delivered three millions of bondmen. Pharoah looked down upon Him with scorn and contempt. "Who is God, that I should obey Him?" But he soon found out what the God of Moses was, and what we want is to be filled with the Spirit of God, and they will find out who our God is. It is of very little account who we are or what we are. All we want is, to be vessels fitted for the Master's use, and just willing to be worked in God's way, and to be fools for Christ's sake. That is what we want. There is not a man in the world of God whose name shines out upon the page of Divine history who was eminent in God's service, but who was considered the greatest fool in his day. I have not any doubt but that Enoch was considered the greatest fool in his day in the sight of the world. They looked upon him with scorn and contempt, but "he walked with God," and God thought so much of him that He said, "Come up higher;" and he is up there walking with God now. God liked his company. Noah was the laughing-stock of his day. Men made sport of him. He was the greatest fool, as the world would call it. He was willing to be a fool for God's sake, and God used him and blessed him; and if you and I are to be used by God we must be willing to be fools in the sight of the world.

Look at Joshua going round the walls of Jericho—a most absurd sight in the eyes of the world. How the London press would come down upon a scene like that—the idea of seven priests going round those walls blowing rams' horns. Fancy the Archbishop of Canterbury and some of your great potentates going right round London blowing rams' horns. Everybody would be disgusted, and say they must have gold trumpets at least, and not rams' horns. But that is not God's way. They went round those walls and compassed that city by faith, and by the grace of God they took it.

THE WORK OF SAMSON.

It was very foolish in the sight of the world, but God's ways are not our ways, and God's thoughts are not our thoughts, and what the Christians of London must learn—and we must learn it very quick if God is going to commence the work here—is that God is going to work in His own way; and it will be a very foolish way in the sight of the world. Look at that man Samson. Why, when the spirit was on him, how he did work. With the jawbone of an ass he slew a thousand men. We are not willing to work with the jawbone of an ass: we want some polished weapons; we want some weapons that the world won't have anything to say against; but Samson came down from the rock—he had been on the rock — and he took up the jawbone of an ass that he came across, and he went out and slew them right and left. And what we want is for every one of us to grab up the first jawbone of an ass that we come across, and not to wait to do some great thing, but to do anything that the Lord will permit us to do. Let the world mock and laugh on; when we are willing to be fools for Christ's sake, then God can use us. How absurd Gideon must have looked in his day, with his three hundred men. There were thirty thousand—too many; God could not use them. There were ten thousand—too many; God could not use them; and He got them down to three hundred, and they had empty pitchers. What queer weapons, were they not? It was God's way, however, and every man stood in his place, and God stood beside them, and the result was that they routed the whole of the army. To be sure London is a great city, but we are enough here to rout London, if God is on our side. Any man with God on his side, though a fool, must succeed. God sent Moses, and he did not fail. God sent Elijah, and he did not fail. Look how absurd it was. Elijah out there, fed by ravens—contemptible, unclean birds feeding Elijah; and then when God sent him somewhere else it was not to a palace, to a table laden with good things, but to a widow just ready to die, who had scarcely enough for a meal for herself and her boy beside. But that is God's way, though very absurd in the sight of the world. So it is all through Scripture, from beginning to end, and God is unchangeable. It is said we are living in an enlightened age; we may be, but to God it is the same, and He takes the contemptible things and the despised things of the world and uses them.

WHO IS JOHN BUNYAN?

When He wanted a book written to do some great good to the world, He did not call forth a philosopher, but some Bedford tinker (the devil got his match when he got hold of John Bunyan), and he took up his pen and wrote a book for the blessing of nations. The world looked down upon him with scorn. "Who is John Bunyan?" How they turned up their lips with scorn and contempt in his day; but he lives to-day. Many a man that stood high in society in Bunyan's days is forgotten now. We do not know who they were; but John Bunyan now lives, and will live so long as the world lasts. That despised tinker! I hope there are some Bedford tinkers here to-night that the Lord will bless, and send out to bless the world. God can take up the learned and the unlearned, and there is not a man here whom God cannot use if he is willing to be used. Some one has said there was not a man in all Saul's army but knew that God could use him to meet Goliath; but there was only one that believed that God would use him, and God used that one. And what we want is not to believe that God can use us, but that God will use us. Out went that young stripling, and he met the giant. Forty days he had come out and pointed his finger at and defied all Israel. Forty days was Saul trembling from head to foot, and he was a head and shoulders taller than any one else in his army; but he was not the man God had chosen to meet the giant. It was the little stripling, the youngest son of Jesse, the smallest and the weakest of the sons of Jesse; and God used him, and went with him, and God delivered the giant into his hands.

GREAT PREACHERS IN LONDON.

And God will use the weak ones here in London if they will only let Him. You have got great preachers here—I do not believe London ever knew a day when it had so many great and good preachers as at the present time. It is not that; it is not human power that London wants; it is Divine power; it is not the eloquence of man; it is the eloquence of heaven, the power of the Holy Ghost breaking men's hearts. That is what London wants. And the moment that you and I get ready to receive the baptism that comes from on high—that moment the power will come. "Why!" says Jonathan, "there is no restraint in the Lord; He can save by few as well as by many." Well, we are few, and what we want is just to let God work. I think John Wesley said, if he had one hundred men that loved no one but God, and feared nothing but sin, he would set up the kingdom of God on earth, and shake the gates of hell in twelve months. And I believe he would have done it. One hundred such men never lived at one time. Talk about Alexander making the world tremble with his army—talk about Napoleon making the world tremble with his army—why! the little tentmaker of Tarsus made the world tremble without any army at all. Saul of Tarsus! I would give more for such a man in London than for ten thousand of the men who are mixed up with the world. What we want is to be out-and-out on the Lord's side, brain and heart both on fire for the Lord. It is said of David's mighty men that they were right and left-handed. They were wholly consecrated; they could use their left or their right hands for the king.

That is what we want in London. Men who are right-handed and left-handed for the King of Glory. Men who can use their eyes, and tongues, and ears, and everything for the Lord Jesus. Then London will be moved, and it will take very little, thank God, to move this city from end to end. This is a great city, but we have a great God with great power.

NO HEART FOR THE WORK.

But this is not all. God uses human instruments. Sometimes it is a wonder to me that He does not take the work out of our hands and put it into the hands of angels, or some one able to do it. There are but few now that say, "Here am I, Lord; send me." The cry now is, "Send some one else." "Send the minister," says the elder; "don't send me." Or if he is not an elder, he says, "Don't send me; send the church officers, the church-wardens, but not me. I have not got the ability, the gifts, or the talents." Ah! honestly say you have not got the heart, for if the heart is loyal, God can use you. It is really all a matter of heart. It does not take God a great while to qualify a man for his work if he has the heart for it. He may not have many talents, but if he makes good use of what he has, God will soon increase his talents. Look at Elisha! There is another man. We would not have thought of Elisha to take the place of the wonderful prophet. We would have gone to Bethel, or Jericho, to the school of the prophets, and picked out a theological professor, or some great man. But Elijah finds a man in the fields behind twelve yoke of oxen, and Elisha slew his oxen, and consecrated everything to God, and started off with Elijah. And Elijah says one day, "What can I do for you?" "Well," says Elisha, "give me a double portion of your spirit!" "Well," said Elijah, "that is a great thing that you have asked; but if you see me when I am taken up you shall have it." Then they started from Gilgal, and Elijah says, "You stay here, and I will go down to Bethel and see how the prophets are." But Elisha says, "As the Lord liveth, and as thy soul liveth, you shall not go without me." And I can see the men arm-in-arm going to Bethel. And when they got there, "Now," says Elijah to Elisha, "you stay here and I will go to Jericho to see how the prophets are going on there." He was going to visit the theological seminaries.

"Well," says Elisha, "as the Lord liveth, and as thy soul liveth, you shall not go without me. And arm-in-arm they went to Jericho together. And when they got there, says Elijah, "You stay here and encourage these prophets and I will go over Jordan"—Jordan means death and judgment. "As the Lord liveth, and as thy soul liveth, you shall not go without me," says Elisha, and arm-in-arm they went to Jordan together. And Elijah took up his mantle and struck the waters, and God held back the waters in the palm of His hand. And they walked over dry shod. But it had been revealed to those prophets that Elijah was going to be taken away, so fifty of them went out to watch. By-and-by there came a chariot of fire, and Elijah stepped in and swept away home. And as he went up his mantle fell, and Elisha left his own mantle and took Elijah's mantle. Then he went back to Jordan, and he took Elijah's mantle and struck the waters, and came through dry shod. And when the fifty prophets saw him, they cried, "The spirit of Elijah doth rest on Elisha." So it was. And God qualified him to take Elijah's place.

THE SPIRIT OF ELIJAH WANTED.

What we want is the spirit of Elijah, and our God is the same as his God. It was in the power of prayer that he stood before Ahab, and what we want is to get hold of God in prayer, and to have power from heaven—not human power, but power from on high, and God is ready and willing to give us that power. Yes, it is the weak things, it is the despised things that God uses. Those unlearned men from Galilee, Christ called them around Him. The last men that we should have thought of. He called those fishermen out of Galilee, and that little handful of men shook the world. It was these men that went around the world preaching the glorious Gospel and the glad tidings. Why, before He could use Saul He had to change his name, and call him Paul—Little!—little! He had to show him that he was weak before He could use him. And Paul says, "When I am weak then am I strong." It was not enticing words, it was not eloquence that Paul had. Why, he said his speech was contemptible! Yes, contemptible! He did not profess to be an orator, but he preached Christ, the power of God and the wisdom of God, Christ and Him crucified. What London wants the whole world wants, and that is Christ and Him crucified. And the world will perish for want of Christ. Let every man and woman that loves the Lord Jesus begin to publish the tidings of salvation. Talk to your neighbors and your friends. Run and speak to that young man! Talk to him of heaven and of the love of Christ! Tell him that you want to see him saved. And let the Christians of London in this hall to-night rise and take the city. Our God is able. Shout, for the King is in our midst! Let us compass the walls of Jericho, and they will soon come tumbling down. Bear in mind that, that God is far more willing to bless us than we are to have Him. Let us keep close to Christ. That is what London wants. They don't like to have Christ preached faithfully; but it is just what men don't like to have that we must give them. I learnt that long ago. The very medicine we don't like is the medicine that we ought to have, and the very truths that men object to and that make them angry, are the very truths that bring them to the cross of Christ. What we want is to preach Christ in season and out of season,—

"Tell the old, old story,
  Of unseen things above,
Of Jesus and His glory,
  Of Jesus and His love."

Why, the stone that the builders have rejected has become the chief corner-stone. The very stone that they would not have was the very stone that God chose, and upon this stone He is building His Church now—upon the rock of ages. It is Christ, my friends, Christ that they want, and then they will get sure food for eternity.

GOD'S LION A LAMB.

The lion of hell was overcome by a Lion. The Lion of the tribe of Judah is what? A Lamb. God's lion is a Lamb. There was weeping once in heaven. John said he wept much when he got there. There was a book handed to him, and there was not any one worthy to open the book. There was Abel, he was not worthy; there was Enoch, who walked with God, but he was not worthy; there was Noah, but he was not worthy; there was Abraham, the friend of God, but he was not worthy; there was Moses, who went up into the mountain and talked with God, and took the law from God's hands, but he was not worthy. There they were gathered from all climes and all ages, but not one of them was worthy; and John looked down upon the earth, and there was not one worthy, and he wept because there was no one worthy to open the book. At last one touched him, and said, "Weep not, for there is one worthy; the Lion of the tribe of Judah, He hath prevailed, and He is worthy." And he turned round to look at the Lion of the tribe of Judah, and behold, it was a slain Lamb. God's Lion is a lamb slain from the beginning of the world, and what we want is to go out and preach Christ. It is the weakness of God which overcomes the strength of man. It is the weakness of God that we want. The foolishness of God is worth more than the wisdom of the world.

UNION ESSENTIAL TO SUCCESS.

Then there is another thing. If we are going to have success, let us have union. Now there are three classes, I believe, that ought to be in sympathy with this movement. The first class is ministers. Every minister that wants to crown Christ King, that wants to see souls saved, ought to be interested in this work. Why? Because we come here to help you; not to destroy, but to help, to build up, to strengthen every minister's hands, to help him to do his work. Then we ought to have sympathy from every Sabbath-school superintendent. We ought to have the sympathy of every teacher, of every worker, of every missionary, of every colporteur in London, and if we cannot have you here night after night let us have your prayers and your sympathy. Do not be criticising and finding fault; but be much in your closet with God, and God will answer prayer, and there won't be a Sabbath-school, and there won't be a mission district, and there won't be a church in all London, but will be blessed if we are just working together and praying for a great blessing.

APPEAL TO PARENTS.

Then there is another class,—fathers and mothers. You that have children in this city of London ought to have sympathy with a movement of this kind. We ought to have your prayers, we ought to have your council, we ought to have your heartfelt sympathy. We have come here just to try in the name of our Master to win your children to God and to heaven, to win them to a pure life, to save them from the haunts of vice, from going down to a drunkard's grave. When I was in Liverpool the other day, a mother came to me and brought a photograph of a beautiful boy, seventeen years old. He is nineteen now. She said, "That boy has been gone two years, and I do not know where he is. He had trouble, and he fled from home, and my heart is just breaking. I do not know but that he is in London, and I give you his photograph, and if you see him in the audience there, I want you just to try and win him to the Lord, that he may come back to cheer my heart," and the great tears rolled down that mother's cheek. There is many a boy in London like that. We have come here after them, just in hopes that God will win them to Christ, and that they will go back to be a blessing to their parents and to the Church of God. If that young man is here to-night, I bring you good news. Your mother still loves you, and wants you to return. Her heart is just breaking for you. And let me say to every man and woman here to-night that is out of Christ, God wants you; Jesus wants you. There is room in heaven for you; and the Lord has sent us just to invite you to the Gospel feast.

Let me say, before I close, that we want unity among God's people. Where there is union I do not believe any power, earthly or infernal, can stand before the work. When the Church, the pulpit, and the pew get united, and God's people are all of one mind, Christianity is like a red-hot ball rolling over the earth, and all the hosts of death and hell cannot stand before it. I believe that men will then come flocking into the kingdom by hundreds and thousands. "By this," says Christ, "shall all men know that ye are my disciples, if ye love one another." If we only love one another, and pray for one another, there will be success. God will not disappoint us. When General Grant was moving on to Richmond, and his army had been repulsed in the Wilderness, he called together his co-commanders and held a council, and asked them what they thought he had better do. His leading generals and all thought he had better retreat. He heard them through, and then broke up the council of war, and sent them back to their headquarters; but before morning an orderly came round with a dispatch from the General directing an advance in solid column on the enemy at daylight. That was what took Richmond, and broke down the rebellion in our country. Christians of London, let us advance in solid column against the enemy; let us lift high the standard, and in the name of our God let us lift up our voice, and let us work together, shoulder to shoulder, and keep our eye single to the honor and glory of Christ. Let us pray that we may get self out of the way, and that Christ may be all and in all, and then we will have great success. Let our watchword be, "Here am I; send me."

II.

CHRIST SEEKING THE LOST.

"For the Son of man is come to seek and to save that which was lost."—LUKE xix, 10.

To me this is one of the sweetest verses in the whole Word of God. In that little short sentence we are told what the Son of God came into the world for, we are told what His mission to this dark world was. He came not to condemn the world, but that the world through Him might be saved. He did not come to make men wretched, He did not come to make us miserable; but He came to save that which was lost. Now, when a prince, and especially a crown prince, comes to London, what a royal reception you give him; and the question is raised, what is he come for? Now, here is the Prince of Heaven coming down into this dark world. What is He come for? Why, we are told that He came to seek and to save that which was lost. And every man or woman in this audience is either lost or found. You are either saved or lost. And bear in mind that Christ takes the place of the seeker. No sooner did the news reach heaven that Adam had fallen in Eden than God came down to him. Adam ought to have gone up and down the garden of Eden crying out, "My God, where art Thou? I have sinned, I have sinned and fallen." Instead of that he went away and hid himself, and God came down, and His voice was heard in the garden of Eden, "Adam, where art thou?" It was the voice of Christ, the voice of love. But Adam had gone away frightened, and God took the place of the Seeker, and from the time of Adam's fall until the present, God has always taken the place of the Seeker. No man or woman in this audience has been saved but that He sought them first. We do not seek after Him until He first seeks after us. We do not go to Him by nature. Our nature is to go away from Him, as with Adam when he hid away from a loving God. Read what we have in the 15th chapter of the Gospel of St. Luke. It is not the lost sheep from the fold hunting up the Shepherd, but it is the Shepherd seeking after the lost sheep. Whoever heard of a sheep which had strayed from the fold hunting after the shepherd? And so Isaiah has told us, "All we like sheep have gone astray, we have turned every one to his own way," and the Great Shepherd has come down after us.

THE LOST PIECE OF MONEY.

And then, in the portion of Scripture I have read to you, there is that woman who had lost the piece of money. It was not that piece of money seeking its way back into the woman's pocket; but it was the woman lighting a candle and taking a broom, and sweeping diligently until she found it. I can just imagine that some one had paid that woman a bill that day, and had given her ten pieces of silver. When she retired at night, she took the money out of her pocket to count it, and seeing the bulk looked small, she said to herself, "Well, this don't look like ten pieces; I must have lost one piece; where have I lost it?" She begins to think where she has been that day. "I have not been out of the house; it must be somewhere in the house," she says. Then she goes and lights a candle, and gets a broom and sweeps the floor, and raises a great dust. And so it is when the Holy Spirit begins to seek after a soul; there is some great commotion. So she begins to search and grope around; she moves the chairs, the sofa, the table, and all the rest of the furniture, and looks in every corner until she finds the piece. Who was it that rejoiced — the piece of silver or the woman? the sheep that was lost or the shepherd that found it? It was the woman that rejoiced over the lost piece; it was the shepherd that rejoiced over the lost sheep he had found. And so we find it is here. Christ takes the place of the seeker. "For the Son of man is come to seek and to save that which was lost." What Adam lost in Eden, I find in the second Adam. When God put Adam in Eden, He bound him strong to the throne of God with a golden chain. When Satan walked in, he broke the chain; but the second Adam came to seek and to save that which was lost. What the first Adam lost for me I get in the second Adam. He came to seek and to save that which was lost.

THE STORY OF BARTIMEUS.

Now let us go back a little before the text. I am very often blessed in my own soul by taking a text all round. In the 18th chapter of the gospel of St. Luke, you will find Christ is going into Jericho; and as He drew near the gates of Jericho there was a poor blind man who sat by the wayside, begging people to give him a farthing, and crying out, "Have mercy on a blind man!" This poor beggar met a man, who said to him, "Bartimeus, I have good news to tell you." "What is it?" said the blind beggar. "There is a man of Israel who can give you sight." "Oh, no," said the blind beggar, "there is no chance of my ever receiving my sight. I never shall see. In fact, I never saw the mother who gave me birth; I never saw the wife of my bosom; I never saw my own offspring. I never saw in this world, but I expect to see in the world to come." The man said, "Let me tell you, I have just come down from Jerusalem, and I saw that the village carpenter, Jesus of Nazareth, and I saw a man who was born blind, who had received his sight; and I never saw a man with better sight. He does not have to use glasses now, and he was born blind." Then, for the first time in this poor man's heart, hope rises, and he says, "Tell me how the man got his sight?" "Oh," says the other, "Jesus first spat on the ground and made clay, and put it on his eyes"—why that is enough to put a man's sight out, to fill a man's eyes with clay — "and then He told him to go and wash his eyes in the Pool of Siloam, and he would receive his sight. And then, Bartimeus, He does not charge you anything, you have no fee to pay; you just tell Him what you want, and the poor man has as much influence as the rich. It does not need dukes, or lords, or influence; you just call upon Him yourself; and if He ever comes this way, don't let Him go back without your going to see Jesus." And Bartimeus said, "I will indeed do so, and ask Him for my sight." I can imagine him being led by a child to his seat as usual, and that he is crying out, "Please give a blind beggar a farthing." He hears the footsteps of the coming multitude, and inquires, "Who is it passing? What does the multitude mean?" They tell him it is Jesus of Nazareth passing by. The moment he hears that he says, "Why, that is the Man that gave sight to the blind." The moment it reached his ear that it was Jesus of Nazareth, he began to cry out at the top of his voice, "Jesus, Thou Son of David, have mercy upon me." Some of those who went before—perhaps Peter was one of them—rebuked him, thinking the Master was going up to Jerusalem to be crowned King, and did not want to be distracted. They never knew the Son of God when He was here. He would hush every harp in heaven to hear a sinner pray; no music would delight Him so much. But the blind man lifted up his voice and cried still louder, "Thou Son of David, have mercy upon me;" and the prayer reached the ears of the Son of God, as prayer always will, and they led the poor blind man to Him. The Lord grant that each one here in this Agricultural Hall may cry out, "God, have mercy upon me a sinner; God, have mercy upon this soul of mine;" and the Lord Jesus will be good to you. Well, when Jesus heard the blind beggar, He commanded him to be brought unto Him. So they ran to him and said, "Be of good cheer, the Master calleth thee; He hath a blessing for thee." When Jesus saw him He said, "What can I do for you?" "Lord, that I may receive my sight." "You shall have it;" and the Lord gave it to him. "Ask, and ye shall receive." Oh, may sinners cry out to-night, "God, give me my sight; God, be merciful to me a sinner." And now the beggar followed with the crowd, glorifying God. I can imagine he sang as sweetly as Mr. Sankey; no one sang sweeter than he when he shouted, "Hosanna to the Son of David;" no one sang louder than this man who had received his sight. Then he follows on with the crowd, which we see pressing into the gates of the city. I can imagine when he gets into the city he says to himself, "I will go down and see my wife"—having, of course, after those years of blindness, a curiosity to see what his wife looked like, also to see his children.

THE CONVERSION OF ZACCHÆUS.

As he is passing down the street a man meets him and turns round and says, "Bartimeus, is that you?" "Yes, it's me." "Well, I thought it was, and yet I thought my eyes must deceive me. How did you get your sight?" "I just met Jesus of Nazareth outside the walls of the city, and I asked Him to have mercy upon me, and He gave me my sight." "Jesus of Nazareth! is He in this part of the country?" "He is already on His way to Jerusalem. He is now going down to the eastern gate." "I should like to see Him," says the man, and away he runs down the street; but he cannot get a glimpse of Him, being little of stature, on account of the great throng round Him. He runs to a sycamore tree, and says to himself, "If I get up there and hide, without any one seeing me, He cannot get by without my getting a good look at Him." A great many rich men do not like to be seen coming to Jesus. Well, there he is in the sycamore tree, on a branch hanging right over the highway, and he says to himself, "He cannot get by without my having a good look at Him." All at once the crowd burst out. He looks at John—"That's not him;" he looks at Peter—"That's not him." Then he saw One who was fairer than the sons of men. "That's Him." And Zacchæus, just peeping out from amongst the branches, looked down upon that wonderful, yes, that mighty God-Man, in amazement. At last the crowd comes to the tree, and it looks as if Christ was going by; but He stops right under the tree, and all at once He looks up and sees Zacchæus, and He says to him, "Zacchæus, make haste and come down." I can just imagine Zacchæus says to himself,—"I wonder who told Him my name. I was never introduced to Him." But Christ knew all about him. Sinner, Christ knows all about you; He knows your name and your house. Do not think God does not know you. If you would try to hide from Him, bear in mind you cannot hide from Him. He knows where each one of you is; He knows all about your sins. Well, He said to Zacchæus, "Make haste and come down." He may have added, "This is the last time I shall pass this way, Zacchæus." That is the way He speaks to sinners,—"This may be the last time I shall pass this way; this may be your last chance of eternity." He may be passing away from some soul to-night. Oh, sinner, make haste and come down and receive Him. There are some people in this nineteenth century who do not believe in sudden conversions. I should like them to tell me where Zacchæus was converted. He certainly was not converted when he went up into the tree; he certainly was converted when he came down. He must have been converted when he came down. He must have been converted somewhere between the branches and the ground. The Lord converted him just right there. People say they do not believe in sudden conversions, and that if a man is converted suddenly he won't hold out, he won't be genuine. I wish we had a few men converted like Zacchæus in London; it would make no small stir. When a man begins to make restitution, it is a pretty good sign of conversion. Let men give back money dishonestly obtained in London, and see how quick people will believe in conversion. Zacchæus gave half his goods to the poor. What would be said if some of the rich men of London did that? Zacchæus gave half his goods all at once; and he says, "If I have taken anything from any man falsely, I restore him fourfold." I think that is the other half. But to get Christ is worth more than all his wealth. I imagine the next morning one of the servants of Zacchæus going with a check for £100, and saying, "My master a few years ago took from you wrongfully about £25, and this is restitution money." That would give confidence in Zacchæus' conversion. I wish a few cases like that would happen in London, and then people would not go on talking against sudden conversions.

THE COMPLAINTS OF MODERN PHARISEES.

Now Christ becomes the guest of Zacchæus, and while he is in his house the Pharisees begin to murmur and complain. It would have been a good thing if all those Pharisees had died off with that generation; but, unfortunately, they have left a good many grandchildren behind them, living down here now in the afternoon of this 19th century, who are complaining "This man receiveth sinners."

Many men complain because the Lord saves men for nothing, but the Lord deals in sovereign grace. But while they are murmuring and complaining on every occasion, Christ uttered the wonderful words of my text for to-night, "For the Son of Man is come to seek and to save that which was lost." "I did not come to condemn Zacchæus, to make him wretched; he is wretched enough now. I did not come here to torment him, I came to bless and to save him." When Christ commenced His ministry in the wonderful sermon on the mountain, there is blessing, blessing, blessing. He came to bless man, and poor Zacchæus needed blessing, and He first gave it him. If there is some poor Zacchæus here to-night, or if there is some poor blind beggar here to-night, He will bless you. The Son of man is come for that purpose; He left Heaven and a throne for that. He came "to seek and to save that which was lost;" and so the vilest man in London can be saved if he will be. The Lord is able and willing to save. "He is come to seek and to save that which was lost." A man must believe he is lost before he can be saved. One reason why many are not saved is because they do not believe they are lost. They fold their filthy rags of self-righteousness about them, instead of acknowledging that they are miserable sinners.

PREACHING IN THE TOMBS.

While I was occupying the Fulton-street pulpit in New York, the governor of the City Tombs Prison said he would like to have me go down and talk to the prisoners. After the prisoners were all brought in, I found there was no chapel in connection with that prison, and I had to talk to them in their cells. I talked from a little iron railing running right across the narrow passage-way, to some three or four hundred prisoners, and could not see a man. I had never had that experience before. After I had done, I thought I would like to see who I had been talking to, and how they had received the interpretation of the Gospel. I went to the first door and looked in the little window of a cell where the inmates could have best heard me. There were some men playing at cards. No doubt they had been playing all the while I had been preaching. They did not want to hear. Some men come here to-night out of curiosity; they do not want to hear the glad tidings, and they do not believe the Gospel's good news. Well, these men had been playing cards all the while I was preaching. I said, "My friends, what is your trouble?" "Well, stranger, false witnesses appeared against us. We are innocent." I said to myself, "Christ cannot save anybody here; there is nobody guilty." I went to the occupiers of the next cell and asked why they were there. They said, "We got into bad company, and the man who done the deed got clear, and we got caught." I said, "Christ cannot save anybody here." I went to the next cell and asked how it was with them. They said, "False witnesses went into court and swore falsely." I said, "Christ cannot save anybody here." I went to the next cell and said, "How is it with you?" The reply was, "The fact is, the man who done the deed is very much like me. I am perfectly innocent." I never found so many innocent men in a prison in my life. It seemed that the magistrates who sent them there were the only guilty ones.

THE PENITENT PRISONER.

I began to get discouraged, but when I had got almost through I found one man with his elbows on his knees and two streams of tears running down his cheeks. I looked in at the little window, and I said, "My friend, what is the trouble?" He looked up with despair and remorse on his face, and said, "My sins are more than I can bear." I said, "Thank God for that." "Ain't you the man that has been talking to us? I thought you said you was a friend; and you say you are glad my sins are more than I can bear." "Yes." "I don't understand your friendship if you are glad my sins are more than I can bear." "I will explain it to you. If your sins are more than you can bear, you will cast them on One that will bear them for you." "Who is that?" "It is the Lord Jesus;" and I stood there at that prison door and preached Christ, and held up Christ for that poor wounded man, who was believed to be the worst man in the whole prison of the city of New York. After telling him of Christ I got down and prayed. After I prayed I said, "Now you pray." He said he could not pray; it would be blasphemy. But the man put his head on the pavement, and, like the publican, without even lifting his eyes towards heaven, he cried, "God be merciful to me a sinner." After prayer, when he got up, I took his hand, and he gave me a good warm grasp of the hand; a hot tear fell on my hand, which burned down into my soul. I got so interested in the man that before I started for the hotel, I said, "I will pray for you to-night, and I would have you join me in prayer at the same time." That night, while I was praying in my hotel, as I told him I should pray for him at a certain hour, it seemed as if I knew that God was answering my prayer. I could not leave New York and go back to Chicago until I had seen that man. No sooner did I fix my eye on the man's countenance, than I saw that a great change had taken place. Remorse and gloom had fled away, and the face of the man was streaming with celestial light. He seized my hand, and tears of joy trickled over his cheeks. I said, "Tell me all about it;" and he said, "Last night, when in my cell praying—I do not know the exact time, because when I came to prison, they took away my watch, but I think it was about midnight—the Lord Jesus took away the burden, and set me entirely free; and since then I am the happiest man in the whole city of New York." And I believe he was, for he told me of the love, joy, and peace that none but one that had received the Lord Jesus knew anything about. After I had talked and prayed with him some time, I bade him good-bye.

GOOD NEWS FOR THE LOST.

Now, my friends, can you tell me how it was the Lord Jesus came into that prison, and passing one cell after another, went to that one cell and set the captive free? It was because he believed he was lost, that he had sinned and come short of the glory of God. He called to God for mercy, and God dealt him mercies. If there is a man or woman in this audience to-night who believes they are lost, I have good news to tell them,—Christ will come after you. He came to save you, He came to bless you. Now, do not let this night pass, my friends, without just accepting salvation as the gift of mercy from a loving God. He wants to deal out mercy and grace for every soul here. The Son of man is come—what for? To seek and to save. And do you think He is not able to save? And is He not willing to save? There is not a man in this audience but knows deep down in his heart that Christ is able and willing to save. My friends, let Him save you to-night while you are on praying ground; while God is talking to you, and offering you salvation without money and without price. The gift of God is eternal life. That is the gift to-night. Who will have it? Who will take the gift? Who will accept the gift of God? Who will receive it? While I am talking to you, lay hold on eternal life. While I am speaking to you, just receive the gift of God, and go on your way rejoicing. Now, I have no doubt that while I am speaking to you, there are a good many of you that have got friends praying for you; I have received a good many letters, especially from mothers, stating that they have sons in London, and praying that the Holy Ghost might win their souls to Christ. Maybe some of those sons are here now. One of these young men spoke to me last night, and told me his mother was very anxious that he should attend these meetings. I have remarked the great number of young men who come to our services. I never saw more young men at a meeting than I see to-night. I may be speaking to more young men than ever I spoke to in my life. How comes this, that there are so many young men here? Perhaps they have in the country a loving mother praying that the Holy Ghost may convert their hearts; or a loving sister, or a loving father, or brother, pleading hard for their salvation. Well, if you have got these friends that are diligent for your salvation, treat them kindly, for you will not always have them.

A YOUNG MAN'S TALE.

I went to a meeting in Chicago a few years ago, and a young man got up and said, "Will you allow me to speak to these young men?" At first, as he was a stranger, I thought I wouldn't; and then I thought he might have a message from God, so I said, "Say on." And that young man just pleaded with those young men, and said, in closing his speech: "If any of you have fathers, or mothers, or Christian friends, who are diligent for your salvation, treat them kindly, for you will not always have them. I was an only son, and I had a godly father, who went down to his grave praying for me, for I was a wayward boy. After father died, mother began to be more anxious than ever. Sometimes she would weep over me, and say, 'Oh, my boy, if you were only a Christian I should be so happy.' Some nights I heard her in her chamber weeping, and crying to God for her boy. I could not stand it any longer, so I had to leave home. I must become a Christian, or get away from home. So I ran away. It was a long time before I heard of her, and then I was told she was sick, and the thought came stealing over me, 'She may die. I will go back.' And then I thought, 'If I go back home, I will have to become a Christian. I cannot live at home with mother without becoming a Christian; I will not go.' The next time I heard from that mother I heard she was much worse. Then the thought came to me, 'If my mother died, and I should never see her, I should never forgive myself.' So I started off. There was no railway into the village, and I had to take coach. I got to the village about sundown; the moon had commenced to shine. My mother lived about a mile and a half from the little town, and to get home I had to go by the old village churchyard, so I thought I would go and look at father's grave, and see if there was any new-made grave. As I drew near my heart began to quake. I could not tell why. The moon showed me a new-made grave, and then for the first time in my life the question occurred to me, 'Who is going to pray for my lost soul now? Father's gone and mother's dead.' I took up some of the earth and found it was just damp, and I threw myself on my mother's grave, and there I spent the night. I did not move until the break of day; but before I left that grave, my mother's God had become my own. And, young men, I believe God, for Christ's sake, forgave me that night, but I never forgave myself." Young man, if you have a praying mother or a praying father, treat them kindly, for you will not always have them; they will soon all be gone, and that voice which is now pleading day and night for your soul will be hushed in the grave. Therefore, this night, while they are praying, seek the kingdom of God; and it won't take an anxious sinner long to meet an anxious Saviour. Let your hearts be lifted up now, friends, to Christ in united prayer for every unsaved soul in this hall this night; and now let us all have a few moments of silent prayer.

III.

SAVED OR LOST.

THE ONE ALTERNATIVE.

"For the Son of man is come to seek and to save that which was lost."—LUKE xix. 10.

I WANT to call your attention to the same text that we had last night. I had really only just commenced with the text when it was time to close. Now before I begin, I want to ask a question of every man and woman in this room, and I should like every one just to take the question home with him—"Am I saved, or am I lost?" It must be one thing or the other. There is no neutrality about the matter. A man cannot be saved and lost at the same time; it is thoroughly impossible. Every man and woman in this audience must either be saved or lost, if the Bible be true; and if I thought it was not true, I should not be here preaching, and I would not advise you people to come if you think the Bible is not true; but if the Bible is true, every man and every woman in this room must either be in the dark or out of it, either saved or lost.

Last night, a man yonder told me that he was anxious to be saved, but Christ had never sought for him. I said, "What are you waiting for?" "Why," said he, "I am waiting for Christ to call me; as soon as He calls me I am coming." Now, I do not believe there is a man in London that the Spirit of God has not striven with at some period of his life. I do not believe there is a person in this audience but that has been called; I do not believe that there is a person in this audience but that the Son of God has sought for and is seeking for him.

HOW THE SAVIOUR SEEKS.

Now, for a minute or two, let us look and see how He seeks. There are different ways in which the Son of man seeks. He very often seeks though some faithful minister. Many of you have sat under faithful ministers; you have heard heart-searching sermons, and the truth has gone down deep into your hearts; you have been many a time touched, and tears have come down your cheeks, and you have felt "almost persuaded to be a Christian." That is the Son of God seeking for your soul through that minister. You have heard a sermon sometimes that has roused you, so that you could not forget it, and for days you have been under deep conviction. That is the way the Son of God seeks. Some of you have had a tract put in your hand, with a startling title, perhaps, "Eternity! where will you spend it?" and the arrow has gone home. You may have been troubled, and may have pulled out that arrow and tried to forget it. That is the Son of God seeking for your soul through that tract. Perhaps some of you have had a faithful Sabbath-school teacher who has wept over your souls in your earlier days, who prayed for you and plead with you to become Christians. That is the Son of God seeking for your soul through that faithful Sabbath-school teacher. Many of you have had godly, praying mothers, that have prayed all night for your soul. It is through the prayers of such a mother that the Son of God is seeking you. Many of you, perhaps, have been laid away upon a bed of sickness, and have had time to meditate in the silent watches of the night; the Spirit of God has come into your chamber, has come to your bedside, and you have been troubled about eternity and about the grave, and where you would spend eternity, and how it would be beyond the grave! That is the Son of God seeking for your soul. Some of you have lost friends. I doubt whether there is a man or woman in this audience who has not lost some loved one; it may be a praying mother, it may be a loved father, it may be a dear child; and when death came and took that one from you, you were greatly troubled. You might have taken that friend by the hand, and as he or she was dying, you might have said, "I will meet you in heaven." The Spirit of God strove in you for weeks and months and yet the Spirit left you because you strove against and resisted the workings of the Holy Ghost. My friends, that is the way the Son of man seeks. Can you rise in this hall to-night and say that the Son of God never sought for you? Is there a person in this hall that can rise and say, "I have lived twenty, thirty, forty, fifty years, and the Son of man never sought for my soul"? I do not believe that man or woman lives in all London.

PRAYING FOR LONDON.

My friend, He has been calling for you from your earliest childhood, and He has put it into the hearts of God's own people just to call you together in this hall. This hall has been opened at great expense, and prayer is going up all over the Christian world for London. Perhaps there never has been a time in the history of your life when so many were praying for you as at the present time. That is the Son of God seeking for your soul through the prayers of the Church, through the prayers of ministers, through the prayers of the saints, not only in London, but throughout the world. I have received news to-day in a dispatch sent across from America that all the churches nearly, in America, are praying for London. What does it mean? God has laid it upon the heart of the Church throughout the world to pray for London. It must be that God has something good in store for London; the Son of man is coming to London to seek and to save that which was lost, and I pray that the Good Shepherd may enter this hall to-night and may come to many a heart, and that you may hear the still small voice: "Behold, I stand at the door and knock; if any man, will hear My voice and open the door, I will come in unto him and will sup with him, and he with Me." O friends, open the door to-night, and let the heavenly visitor in. Don't turn Him away any longer. Don't say with Felix, "Go thy way this time, and when I have a convenient season I will call for thee." Make this a convenient season; make this the night of your salvation. Receive the gift of God to-night, and open the door of your heart, and say, "Welcome, thrice welcome into this heart of mine;" and He will come. You invite Him, and see how quickly He will come. My friend, He is come. "The grace of God hath appeared, bringing salvation unto all men." Oh, that the loss of a soul may wake us up to-night, that we may know what it means! I believe the world is asleep—and the Church too. I do not believe there would be a dry eye in this audience if we could for five minutes get a glimpse of a lost soul. We mourn with men who have lost health; we pity them, we sympathize with them, and we say, "It is very sad." We mourn with men who have lost wealth, and we think it is very sad. But what is health, what is wealth, compared with the soul?

A TOUCHING CHICAGO STORY.

I was in an eye infirmary at Chicago, on the Sabbath before the great fire. A mother brought her little baby to the doctor—a child only a few months old—and she wanted the doctor to look at the child's eyes. He did so, and he said to the mother, "Your child is blind; it will never see again; you have neglected it; if you had brought it here three days ago I could have saved the sight." The moment the doctor said that, the mother pressed the little child to her bosom, and there was a wail that came from that mother that broke my heart. I wept, the doctor wept; we could not help but weep. She pressed her darling child to her bosom. "My darling," she said, "are you never to see the mother that gave you birth? My child! my child!" It was a sight that would move almost any heart. But what is the loss of sight compared with the loss of a soul? I would rather a thousand times have these eyes dug out of my head, and go through the world blind, than lose my soul. I have a son, and no one but God knows how I love him; but I would see those beautiful eyes dug out of his head to-night rather than see him grow up to manhood and go down to the grave without Christ and without hope. The loss of a soul! Christ knew what it meant. That is what brought Him from the bosom of the Father; that is what brought Him from the Throne; that is what brought Him to Calvary. The Son of God was in earnest. When He died on Calvary it was to save a lost world; it was to save your soul and mine.

THE LOSS OF A CHILD.

A friend of mine in Chicago took his children out one beautiful day in the summer. They were the children of a large Sabbath-school, and they were to have a day in the country. There was a little boy on the platform of the railway-station, and by some mistake he fell down under the wheels, and the whole train passed over him. The train went back, and the body was found so mangled that the superintendent had to take off his coat to tie up the mangled corpse. He left it at the station, and, taking two of the teachers with him, went to the house of the parents. (The little boy was an only one.) When they got to the house one said to the others, "You go in." "No, I can't," was the reply. The superintendent wanted the teachers to go in, because he thought the parents would blame him; but the teachers refused to go. So the superintendent went in. He found the parents in the dining-room at dinner. He called the father out, thinking that he would tell the father first, that he might break the news to the mother. Taking him into another room, he said, "I have sad news to tell you; your little Jemmie has got run over." The father turned deadly pale. "Is he dead?" he asked. "Yes, sir, he is dead." Then the father rushed into the dining-room, and instead of breaking the news gently to his wife, he cried out like a madman, "Dead, dead!" The mother said, "Who?" "Our little Jemmie." Said the young man who told it me the next day, "I cannot tell you what I suffered when that mother came rushing out to me, and said, 'Where is my boy? Where are his remains? Take me to them that I may see him.' I told the mother that the body was so mangled that she could not identify it; and she fainted away at my feet." Said he, "Moody, I would not be the messenger of such tidings as that again if you would give me all Chicago." There is not a mother or a father in this hall but would say it is terrible to lose a beautiful child like that, to have it swept away so suddenly. Well, it is terrible, but, my friend, what is that in comparison with the loss of the soul?

A MORE TERRIBLE LOSS.

Suppose that child had grown up to manhood, and had died a drunkard, and gone down to a drunkard's grave. See the hundreds and thousands in London reeling their way down, not only to the drunkard's grave, but to the drunkard's hell. I tell you, my friend, I would rather have a train a hundred miles long run over my boy, so that I could not find a speck of his body—I would rather have him die in early childhood, than have him grow up to manhood, and die without God and without hope. It is terrible for a man to die outside the Ark. It is a terrible thing for a man to die without hope and without mercy, especially in this Gospel land, where he is exalted to heaven with privilege, where the Gospel is proclaimed faithfully from Sunday to Sunday, yea, from day to day, and one might say, from hour to hour. Through the length and breadth of this great city, the Gospel has been proclaimed as faithfully, and perhaps more faithfully, than in any other city in the world. London, I say, is exalted to heaven with privileges, and it is a sad thing, indeed, that a man should go to hell from London, for then he goes down in the full blaze of the Gospel. He goes down from a Gospel land. He goes down to hell from a land where he has heard the glorious tidings of Christ and Him crucified. Yes; you say it is very sad to see a child like that swept away, or to see a little child lose its sight. You say it is very sad to see a man lose his wealth and become poor. It is very sad to see a man lose his reputation. But, my friends, bear in mind there is hope. A man can come to Christ if he has lost his reputation and his character. Christ will "receive" men who have not got any reputation; Christ will "receive" men who have not got any character; and they may have a seat in the kingdom of God. But, if a man dies without God, then there is no hope. You go to the grave and weep over it, and when the morning of resurrection shall come, that man will rise to everlasting shame and contempt. The star of Bethlehem will not shine over that grave. Oh, my friends, let us wake up, and let us haste to the rescue. Let us, as fathers and mothers, see that our children are brought into the ark, that they are saved, that they are gathered early into the fold of Christ.

THE POOR DRUNKARD.

I was over in this country in 1872. About that time there was a young man who had come from the country to London. He was the only son of a widow. He was her prop and her stay; her hope and her comfort. Oh, how that widow loved that boy! How her prayers went up for him! When he came to this city his employer invited him to the theatre, and invited him to drink. I have met that mother since I have been on this trip, and she told me that the employer discharged that young man after he became a drunkard; that he refused to have him in his employ; that her son came home and died a poor drunkard. That mother is now weeping over that boy, and she mourns as a mother without hope, because it is said that no drunkard shall inherit the kingdom of God. Now, that is terrible. How many mothers have sons in London hastening to ruin! God wants you and me to go and tell them the glad tidings, to invite them to the Gospel feast. And there is not a man in all London so far gone but that Christ will save him. If we will just go and labor for them and pray for them, God will give us the privilege of winning many of them into His kingdom.

"SAVED!"

A few years ago—I think it was only two years this month—a vessel of the White Star line went to pieces on a rock off the coast of Newfoundland, and 500 men went down to a watery grave. There was a young man of great promise, having a large business in Detroit, who was on board that vessel, and soon after she went down there came a dispatch to Detroit to his wife and partner to say that he was lost. The business was suspended, and that young wife was thrown into deep mourning. Her heart was just broken, and the mother's heart was bleeding that her boy had gone down, as they supposed. But in a few hours there came another dispatch over the wires, "Saved!" with his name signed to it. They felt so grateful, that they had the dispatch framed and put up in his office, and there it is. If you go into that man's office now to do business with him, you may see that dispatch, "Saved!" Now, let the news flash over the wires to heaven to-night, sinner, that you want to be saved. You can be saved, if you will. God is able to save. God is willing to save. God is waiting to save. Now, this night, make up your mind that you will be saved. Make up your mind that you will press into the kingdom. God invites you to come. He invites you to come just as you are. "Him that cometh unto Me I will in no wise cast out."

ROWLAND HILL AND LADY ANNE ERSKINE.

I have seen a story in print, I do not know whether it is true or not, but it illustrates a good point. I see our friend Dr. Newman Hall here to-night. The story I speak of is told of a predecessor of his, Rowland Hill. One day he was preaching in the open air to a vast crowd of people, when Lady Anne Erskine was riding by. She said to her footman, "Who is that man who is drawing so vast a crowd?" She was then told that it was the celebrated Rowland Hill. "Drive me," said she, "near the platform, so that I may listen." The man went on preaching, and, seeing by the lady's dress that she belonged to royalty, he turned to some one on the platform and inquired who it was. They told him it was Lady Anne Erskine. He continued his preaching, and then all at once he stopped and said, "My friends, I have got something to-day to put up for sale. I am going to sell it by auction." Everybody of course was startled to think that a man should stop in the middle of a sermon to sell something by auction. Said he, "It is the soul of Lady Anne Erskine. Is any one going to bid for her soul? Hark! Ah! I hear a bid. Who bids? 'Satan.' 'Satan, what will you give for this soul?' I will give riches and honor and pleasure. Yea, I will give the whole world for her soul.' Hark! I hear another bid. Ah! methinks I hear another bid. ·Who bids? 'The Lord Jesus.' 'Jesus, what will you give for this soul?' 'I will give peace and joy and comfort that the world knows not of. Yea, I will give eternal life for her soul.'" He then turned to Lady Anne Erskine, and said, "You have heard the two bidders for your soul. Which bidder shall have it?" It is said that she ordered her footman to open her carriage door, and, rushing in, she began to weep, and said, "The Lord Jesus shall have my soul, if He will accept it." Now that may be true or not; but there is one thing that I *know* to be true—that there are two bidding for your soul and mine. Satan bids, and he offers that which he cannot give. He is a liar and has been from the foundation of the world. I pity the man who is living on the promises of the devil. He will never satisfy. But the Lord Jesus is able to give all that He offers. He offers peace and joy and comfort that the world knows not of. He offers eternal life in the kingdom of God. He offers a seat in His mansions. We are to sit with Him upon His throne. May God help you this night. Make up your minds to-night that you will not leave the Agricultural Hall until the great question of eternity is settled, until you have crossed the borderland, and pressed into the kingdom of God. Make up your mind to this. Make up your mind that this shall be the night of your salvation. "Now is the accepted time; now is the day of salvation."

THE BOY AND THE DARK MOUNTAIN.

A few years ago there was a little story going through the American religious press that touched my heart as a father. It was about the death of a little boy. The mother thought her little boy was safe in the arms of Jesus. She thought he was trusting sweetly in Christ; but one day as he drew along towards the chambers of death, she came into his room, and he said, as he was looking out of the window, "Mother, what are those mountains that I see yonder?" The mother said, "Eddie, there is no mountain in sight of the house." "Don't you see them, mother?" said he; "they're so high and so dark. Eddie has got to cross those mountains. Won't you take him in your arms and carry him over those mountains?" The mother said, "Eddie, I would if I could, but I cannot." Now, I want to say to you that there is a time coming when your mother cannot help you. There is a time coming when your friends cannot help you. When you come to the mountain, if you have not Christ, you must take that journey alone, for there will be no one to help you then. What will you do in the swelling Jordan, without a Savior, without Christ? but, if you have Him, He won't leave you. What does He say? "Yea, though I walk through the valley of the shadow of death, I fear no evil. Thy rod and Thy staff they comfort me." Now, this little boy, instead of being troubled by a valley, was troubled by a mountain. The mother prayed with that little boy. Then she said, "Eddie, you must take your eyes off your mother. You must have your eyes upon Jesus. He will help you." The mother again prayed with him, and tried to get his little mind off from the dark mountain. All at once he said, "Mother, hark! don't you hear them call?" "Hear who, Eddie?" "Don't you see the angels just on the other side of the mountain? They are calling for me. Take me, mother, and carry me over the mountain." The mother said again, "Why, my boy, I cannot go with you; but Christ will be with you. He will take you safe over the mountains if you trust Him." Again the mother prayed for her little boy, for she could not bear to have him die in that state of mind, so troubled about the mountain. At length he closed his eyes and he prayed, "Lord Jesus, be with me, and take me over the mountains." Then he opened his little eyes, and said, "Good-bye, mamma; Jesus is coming to carry me over the mountains;" and the little sufferer was gone. Sinner, Christ has come to-night to carry you over the mountains. He will carry you safe. He will carry you over the mountains of unbelief, if you will only let Him. Oh! may God help you this night to press into His kingdom.

IV.

MAN SEEKING FOR GOD.

"Seek ye the Lord while He may be found, call ye upon Him while He is near."—ISAIAH lv. 6.

FOR the last two nights some of you may remember I have been speaking about the Son of man seeking that which was lost. Now to-night I want to bring out the other side—man seeking for God. "Seek ye the Lord while He may be found, call ye upon Him while He is near." Now in the 29th chapter of the prophecy of Jeremiah we are told how we may find God. God says, "Ye shall find Me when ye shall search for Me with all your heart." Now it won't take a great while for an anxious sinner to meet an anxious God. It won't take a great while for a man who is really in earnest about the salvation of his soul to find peace in Him. I never yet found a man really with his heart set upon this one thing—to find God,—but that he soon found his way into the Kingdom of Heaven. The great trouble with men is, that they are not really in earnest. Men don't seek for God as they seek for wealth and position down here in this world. Suppose I should say to-night that I lost last night in this hall a diamond worth £20,000—which I didn't do; but suppose I should say I did, and that I would give any one £10,000 that found it. I would not give much for the sermon. You would be thinking about the diamond all the evening; you would be thinking, "I wish I could find that diamond. I should like that £10,000." And I can imagine as soon as the meeting was over—and some of you would not wait for that—you would look about and search this hall. How earnestly you would seek for that diamond! Well, is there a man or woman in this audience that will say that salvation is not worth more than all the diamonds in the world, and that it is not worth all the goods of this world? Now, is it not a fact that every man and woman here to-night can find God if they will? "Seek ye the Lord while He may be found." Why, the way that text reads implies that there is a time coming when man cannot find God, when men shall seek and not be able to find. We read of such a thing. We read of their knocking at the door when it is shut. We read that they sought with tears, but sought too late. Not but what there was a time. I believe there is a time in the history of every man when he can accept the gift of God, if he will; when he can press into the kingdom of God, if he will.

A LESSON FROM THE GOLD FIELDS.

When a man becomes really in earnest about the salvation of his soul, when a man seeks for that pearl of great price, the kingdom of God, as men seek for wealth and honor down here, and as men seek for position, then it is we will see hundreds and thousands pressing into the kingdom of God. I was out on the Pacific Coast three or four years ago, and the news would come to a town that there was gold found sometimes three or four hundred miles away. Then that town would be full of excitement; the fever would rise very high, and in course of twenty-four hours you would see men pack up, and away they would go, leaving their wives, their children, their homes, comfort and everything, and go off these hundreds of miles to seek for wealth. There was a report that a silver mine had been found in the Rocky Mountains, and men would go off thousands of miles, and go down into the bowels of the earth, working hard, day and night, in order to find a little of the silver. I could not help but admire their earnestness. I am quite sure you like to see a man in earnest, and going into business with all his heart. You like to see a man, whatever he does, do it with a whole heart. You do not like a half-hearted man. Why should not we be just as much in earnest about the salvation of our souls as men are to accumulate wealth? Let a war break out, and there is a chance of men achieving honor on the battle-field, and men come forward and volunteer. They leave their homes, their wives, and their children, and go thousands of miles away, to India or China, or all round the world, if there is a chance of getting honor, just for a few short fleeting years, down here in this world. If men are willing to give up everything for wealth or honor in this life, how much more ought we to be willing to give up everything for the life to come! The gift of God is life without end, eternal life. May every anxious man and woman get it to-night. Of course you admit that this is worth more than everything else. If you do, why don't you ask for it? God says you shall find Him if you seek Him with all your heart.

A LIFEBOAT STORY.

I remember reading some time ago of a vessel being wrecked at sea. The lifeboats were lowered, and there were not enough boats to take all on board. There was one man in the water who, anxious to save his life, swam up to the lifeboat, and reached out his hand and wanted to get in, and one of the men in the boat drew a sword and cut off that hand. The man swam up again and laid hold of the boat with the other hand, and the man in the boat drew the sword again and cut off the other hand. But the man was so anxious and in earnest about saving his life, that he swam up again and seized the boat with his teeth. And then the men in the boat relented; they would not do any more to him; they could not cut off his head, so they took him into the boat and saved his life, because he was so much in earnest. See what men will do for their lives. Let a man be in a wrecked vessel who is worth a million sterling, and if that vessel can be saved from going down only by his giving away that million sterling, he will give it in a minute to save his life for a few short fleeting years. But here is everlasting life, eternal life; that is the gift of God, and the Lord God says you shall have it when you call on Him. "Seek ye the Lord while He may be found, and call ye upon Him while He is near."

TWO DUBLIN BROTHERS.

When we were in Sheffield a few weeks ago, there were two brothers that came from Dublin to Sheffield, and they came to the inquiry-room, and some one said to them as they were talking, "What brought you here? how is it you have come from Dublin?" Then the men said that they had had a brother converted in a meeting while we were there, and such a great change had come over him. They found that he possessed something that they had not got, and they had come to Sheffield in hopes of finding what their brother had found. Do you think these men did not find what they had come for? I never knew a man that tried but what he found peace.

AN ANXIOUS MOTHER.

I see right before me to-night a mother with her boy, a young man sitting by her side. A little while ago the mother left London and went to Dundee, because her heart was set upon the conversion of her two boys. I remember how she used to plead for these two boys. I remember talking with one of them one night, and I plead with him to accept Christ. The next morning we left for Glasgow, and they left for London; and when I came up to London a few days ago, that mother came to tell me that boy had been taken away by death. Now, you know, that mother took the boy off to Dundee that he might hear and be saved. It was a good thing to see a mother burdened for the salvation of her boy. It was God that put it in her heart to take that boy off to that distant city, that he might find peace. Mother, if you have a son out of Christ, let your prayers be going up to Christ day and night. Father, let us have your prayers that your son may be converted. Let us pray together and be in earnest about this great question of salvation. God was in earnest when He sent us the prophets, and we killed them. Then He sent us His only begotten Son, and we murdered Him. And after Christ went back to heaven He sent down the Holy Ghost. Oh, my friends, God is in earnest about the salvation of man's soul; and I would to God that London might wake up, and that men throughout this city would be in earnest about the salvation of their souls. Is the Bible a myth, is the Bible a sham? If it is, why, then, away with it! But if it is true that man is lost, and that Christ is come to save him, then let us earnestly seek the kingdom of God. "Seek ye the Lord while He may be found." That is the text, and it applies to everyone here. These little children—I am glad to see them here night after night—may they seek the kingdom of God to-night. Little girl, seek the kingdom of God to-night. Young man, young lady, seek the kingdom of God to-night. And you who are advancing in the evening of life—your natural force is becoming abated; your eye is growing dim—my friend, make haste and get into the kingdom of God. Do not put off the salvation of your soul for another night; but let this be the night, let this be the hour that you accept of the gift of God. Make up your mind you will not leave this Hall to-night until you have sought the kingdom of God; until the great question of eternity is settled.

A PREPARATION FOR JUDGMENT.

Eighteen years ago, in America, there was a wave of blessing passed over the land. A great many people thought it was all excitement. You could get into the train at New York and go clear into the western prairie, and all along you would see the churches all open and lit up, and crowds of worshipers. The whole nation was moved, and half a million were brought into the Church of God. Little did we know what was taking place. God was preparing that nation for a baptism of blood. After a few years the terrible civil war broke out. It was God calling His children in. It was God preparing the nation for the great affliction brought upon us for our sin. I have noticed very often grace precedes judgment. I do not know what is going to take place. My friends, I am one of those men who believe God is working in a wonderful way. The tidings from every city is this—the people are praying. It is a question in my mind if there was ever so much prayer going up to God as at the present. Not only here, but all round the world, we have God's people making their hearts burdened for the salvation of souls. And is it not God working? Will there ever be a better time for you to seek the kingdom of God than the present, when there is such a great awakening, and when there is such a spirit of expectation; when the Church of God is coming up as one man, and the spirit of unity prevails? Think of the praying ones here. Do you believe there were ever so many men and women praying for your soul as there are here to-night? Look over this audience—what are these Christians doing now? They are silently praying God. I can see they are praying. There is a young man with his mother sitting by his side. That mother is pleading, "God save my boy to-night!" May it go down deep into his soul! "Seek ye the Lord while He may be found." I call on you to come to God to-night. Can He be found to-night? Let me ask you that question. Do you believe it?

AN APPEAL TO THE MINISTERS.

I appeal to these ministers here — Do you believe that God can be found here to-night? I ask you ministers of religion whether you believe that God can be found here to-night? (The ministers on the platform loudly answered, "Yes.") Do you believe God can be found here to-night? (Replies of "Yes," from the audience.) Then, if it is true, do not trifle, do not make light. Call upon Him while He is here. Is He here? We have very good reason to believe that last night there were some that found Christ. If it is true He came into this hall last night and saved some, will He not save you to-night, if your heart is set upon that one thing, "I must be saved?" Is He not able and willing, and anxious to save you? And now, would you just ask Him? Let there be a cry from the very depths of your heart to-night, "Lord, God, save my soul. Lord, God, save me just now." Call upon Him while He is near. If He is near won't you call upon Him? Don't let that scoffing man next to you keep you out of the kingdom of God. There is a scornful look upon the face of that man next to you; perhaps he is making light of what I am saying. Don't mind him, don't look to him, but just look right up to God, and ask Him to save you. Now, every true friend—and you all have friends—every true friend, if you could get their advice to-night, would tell you to be saved now. Ask that minister sitting next to you, "Had I better seek the kingdom of God to-night?" What would he tell you? "By all means, don't put it off another minute." Ask that godly, praying mother by your side, "Mother, is it best to seek the kingdom of God to-night?" Would she say put it off one week, or put it off one month? Do you think that mother would say that? There is not a Christian mother in this hall who would say it. I doubt if there is any unconverted mother either here whose advice would be to put off becoming a Christian. Ask that praying sister of yours, ask that praying brother, ask any friend you have here—if you are sitting near one—if it is not the very best thing you can do. And then shout up to heaven and ask Him who is sitting at the right hand of God, and who loves you more than your father or your mother, or anyone on earth—who loves you so much that He gave Himself for you; ask Him what He will have you do, and hear His voice rolling down from the throne of God, "Seek ye first the kingdom of God." That is what He will say. And then shout down to the infernal regions, and ask those down there—you may have some acquaintances gone there—and what will they say? What did that man say we hear of there? "Send someone to my father's house, for I have five brethren, that he may testify unto them, lest they also come into this place." Heaven, earth, and hell unite in this one thing—"Seek first the kingdom of God." Don't put it off. Call upon Him while He is near. And if you call upon Him in real earnest He will hear that call.

THE PENITENT THIEF.

Look at the thief on the cross. It may have been the thief had a praying mother, and that mother may have taught him this 55th chapter of Isaiah. That mother may have taught him this very verse; and while on the cross it flashed across him, and he remembered the teaching of his mother, and said, "This is truly the God man." And he heard Christ make that wonderful prayer—"Father, forgive them, for they know not what they do;" and then the thief called upon Him, "Lord, remember me when Thou comest into Thy kingdom." And no sooner did that prayer fall upon the ears of the Son of God, than He answered it, "This day shalt thou be with Me in paradise." He called, and the Son of God answered him. Blind Bartimeus when the Son of God went by called out, and they tried to hush him, but the Lord heard his call and prayer, and He answered his prayer and gave him his sight. Oh, my friends, to-night call upon Him, and He will have mercy upon you, and open your eyes and give you sight. Look at that poor leper. He cried, "Lord, if Thou wilt Thou canst make me clean." And the Lord answered him, "I will; be thou clean." Don't forget, my friends, that there is a time coming when your calling and your prayer will not be answered; when you will call to the mountains and hills to cover you from the wrath of God. We are hasting on to the time when London and other places will pray; when scoffers will pray and call on the mountains and rocks to cover them from the wrath of the Lamb. Their prayer has already been put in print for that terrible day of the Son of man which is coming. What does God say? "Seek the Lord while He may be found." My friends, seek Him to-night. He can be found here to-night.

TOO LATE.

I have no doubt that those who would not pray when the ark was building, prayed when the flood came, but their prayer was not answered. I have no doubt that when Lot went out of Sodom, Sodom cried to God, but it was too late, and God's judgment swept them from the earth. My friends, it is not too late now, but it may be at twelve o'clock to-night. I cannot find any place in this Bible where I can say you can call to-morrow. I am not justified in saying that. There it is said, "Behold, now is the accepted time, now is the day of salvation." Those men of Jerusalem, what a golden opportunity they had, with Christ in their midst. We see the Son of God weeping over Jerusalem, His heart bursting with grief for the city as He cried, "Oh! Jerusalem, Jerusalem! thou that stonest the prophets, how often would I have gathered thee as a hen gathereth her brood, but ye would not." He could look down forty years, and see Titus coming with his army, and besieging that city. Eleven hundred thousand people perished in that city. They called upon God then, but it was too late. To-night it is a day of mercy. It may be I am talking to someone to-night whose days of grace may be short, to someone who may be snatched away very soon. There may be someone here to-night who will never hear another Gospel sermon; someone who may be hearing the last call. My friends, be wise to-night. Make up your mind that this shall be the night, and this the very hour that you will seek the kingdom of God, and seek it with all your heart.

AN INCIDENT AT NIAGARA.

A few years ago, two young men were seen on the Niagara river in a little canoe. They were drinking champagne, and having what they called a jolly time. Some of the men on the shore saw their danger, as they were hastening on towards the cataract, and they lifted up their voices and warned them; but the young men only mocked them, as the current swept them along. They were not rowing, they were doing nothing. Some people think if they are doing nothing they will be saved. My friends, if you just neglect salvation you will be lost. You are in the current of time which is sweeping you on towards eternity. Well, these young men were not rowing, but the current was taking them on. It wasn't long before someone else saw them and lifted up their voices; but they only made light of them. And the same is happening here to-night. The current bore them on towards death. At last a third party saw them further on, and lifted up their voices, and told them the cataract was not far away. But the men made a mock of them, and made light of what they said; and the current still swept them on. At last one of them said, "Hark! I hear the rapid." They seized the oars and pulled against the current; but it was too late. They had got too far down, and in a little time two men were seen going down the fall, and they leaped over into the jaws of death. How many in London are doing the same thing? You are warned to-night. My warning to-night is that you seek the kingdom of God. Make up your mind not to leave this hall until you have accepted the gift of God, which is eternal life. God wants to bestow it upon every one. Do not neglect the appeal of to-night; but "turn ye, for why will ye die, oh, house of Israel?" Oh! that all of you may turn unto the Lord and live.

A STORY FROM THE FARM.

A few years ago, before I had left the farm, I was talking one day to a man who was working there, and who was weeping. I said to him, "What is the trouble?" And he told me a very strange story. When he started out in life, he left his native village, and went to another town to find something to do, and he said he was unsuccessful. The first Sabbath he went to a little church, and there the minister preached from this text: "Seek ye first the kingdom of God;" and he said that he thought the text and the sermon were for himself. The sermon made a deep impression upon him, and he could not forget it for some days. But he said he did not want to become a Christian then. He wanted to get rich, and when he was settled in life he would seek the kingdom of God. He went on, and the next Sabbath he was in another village, and he went to church again, and he made a point of going to church every Sunday morning. It was not long before he heard another minister preach from the same text, "Seek ye first the kingdom of God." He thought surely some one must have been speaking to the minister about him. For the minister just pictured him out. But he said he would not seek the kingdom of God then; but when he got settled in life, and had control of his time, and was his own master, he would then seek the kingdom of God. Some time after he was at another village, and he went to church again, and he had not been going a great while when he heard the third minister preach from the same text: "Seek ye first the kingdom of God and His righteousness, and all things else shall be added." He said it went right down into his soul; but he calmly and deliberately made up his mind that he would not become a Christian, that he would not seek the kingdom of God, until he had got settled in life, and owned his farm, and then he would attend to the salvation of his soul. Many a man thinks he can't make money if he becomes a Christian. How the devil deceives you! This man said, "Now I am what the world calls rich. I go to church every Sunday, but I have never heard a sermon from that day to this which has ever made any impression upon my heart. My heart is as hard as a stone." As he said that tears trickled down his cheeks. I was a young man at that time and did not know what it meant.

IN THE ASYLUM.

When I was converted the first man that came to my mind was this man, and I thought when I should go back home I would see him, and tell him, and preach Christ to him. When I went back home I said to my widowed mother, naming this man, "Is he still living in the same place?"

My mother said, "Didn't I write to you about him?" I said, "No, you didn't; what about him?" "He is gone mad, and has been taken away to the insane asylum, and everyone that goes up to see him he points his finger at and says, 'Seek ye first the kingdom of God.' I thought I should like to see him, but I found he was so far gone that it would do no good, and therefore I didn't go. The next time I returned home my mother told me he was at home idiotic. I went to the house to see him, and there was that vacant look in his eye when I went in. I said, "Do you know me?" He pointed his finger at me, and said, "Young man, seek ye first the kingdom of God." God had driven that text into his mind, but his reason was gone. The next time I returned home he was gone; and three years ago last autumn, when I visited my father's grave, I noticed a new grave-stone had been put up. I stopped to read it, and found it was my friend's. The autumn wind was making a mournful noise, and I seemed to hear the wind whispering the text, "Seek ye first the kingdom of God." Now, my friend, will you seek the face of God? Will you seek the kingdom of God to-night, with all your heart? Seek the Lord while He may be found. You have heard these witnesses say that He may be found here to-night. Is not it the very worst thing you can do to go out of this hall without obtaining salvation, without being saved? Young man, make up your mind this night that you will seek the kingdom of God now. Behold, now is the accepted time; behold, now is the day of salvation. Christ is inviting you to come — " Come unto Me all ye that labor and are heavy laden, and I will give you rest." Oh, may we all find rest in Christ to-night.

V.

THE CALL TO SELF-EXAMINATION.

"Where art thou?"—GENESIS iii. 9.

THIS was the first question put to man after his fall. As I said the other night, Adam would have gone up and down Eden crying out, "My God, my God, where art Thou?" but God took the place of the seeker. God came down, and indeed you hear His voice ringing and echoing through the Garden of Eden, "Adam, where art thou?" It was the voice of grace, it was the voice of love, it was a loving God seeking after His lost one. "Adam, where art thou?" Six thousand years have rolled away, and yet this text has come rolling on down the ages. I doubt whether there has been any one of Adam's sons who has not heard this text at some period or other of his life—sometimes in the midnight hour stealing over—" Where am I? Who am I? Where am I going? and what is going to be the end of this?" I think it is well for a man to pause and ask himself that question; and will you do it to-night, my friends? I do not ask you where you are in the sight of your neighbors; I do not ask you where you are in the sight of your friends; I do not ask you where you are in the sight of the community in which you live. It is of very little account where we are in the sight of one another, it is of very little account what men think of us; but it is of vast importance what God thinks of us—it is of vast importance to know where men are in the sight of God; and that is the question to-night. It was the first question put to man after his fall. It was a very small audience that God had, Adam and his wife, and God was the preacher. Adam, like a great many of us, in fact, like the whole of the human race, thought he could hide from God. "After he had fallen he went away and hid." Many and many a one here to-night in this hall undoubtedly has the same thought. He thinks that his life is hid; he thinks that God does not know anything about him, that God's eye is not upon him. But, my friends, God knows all about him; God knows our lives a great deal better than we do; God's eye is bent upon us from our earliest childhood up. By day and by night God's eye is bent upon every one in this hall. And now, my friends, I would just ask you to ask yourselves, "Where am I?" Ah! "Where art thou?" I wish I could get this audience just to stop five minutes and think. That is just what the devil does not want you to do. Think, my friends, think.

I want to divide this audience into three classes, and I think it is a proper division. There are just three classes here to-night. The first class are those that are Christians or professing Christians; and the second class those that have wandered from God, and are what the Bible calls backsliders; and the last class are those that never have been saved.

THE CURSE OF CHRISTIANITY.

1. Now, would the Christians here to-night in this hall, each one, just take that question home. I do not mean to stand here and say that I am on a higher platform than the rest of you, or that I do not mean to apply this question to myself; but it is good for Christians to stop sometimes and ask, "Where am I in the sight of God? What am I doing? Is my life here as it should be in the community I live in? Am I a light in this dark world?" Now Christ says, "Ye are My witnesses." Christ was the Light of the world, and the world would not have the true Light; the world rose up and put out the Light, and now Christ says, "I leave you down here to testify here of Me; I leave you down here as My witnesses." That is what the apostle meant when he said that Christians are to be living epistles, known and read of all men. I will venture to say that if I just ask those who profess to be Christians—those that really are Christians—to rise in this hall, and by that act say, "I am on the Lord's side," there would not be many of you who would be with me; there would be many of you who would say, "I do not think it is right to have it put in that way." But Paul tells us to be ready to give a reason for the hope that is within us. I do not have much hope of there being a reformation until we get a division between the Church and the world. If a man is for God, let him say so. If a man is for God, let him come out and be on God's side; and if he is for the world, let him be in the world. This serving God and the world at the same time—this being on both sides at the same time—is just the curse of Christianity at the present time. It retards the progress of Christianity more than any other thing. "If any man will be My disciple, let him take up his cross and come after Me." And in another place He says, "Let him take up his cross daily and follow Me." Now, what does London want?

THE QUESTION OF A DUBLIN MAN.

I see a man on this platform—I do not know if he remembers it—but when I was here in 1867, there was a merchant who came over from Dublin, and was talking with this business man in London; and as I happened to look in, this business man in London introduced me to the man from Dublin. The Dublin man said to the London man, alluding to me, "Is this young man all O O?" Said the London man, "What do you mean by O O?" Said the Dublin man, "Is he Out-and-Out for Christ?" I tell you it burned down into my soul. This friend said I was a little ashamed, but I thought I was not, though I was a young man then. It means a good deal to be O O for Christ, but that is what all Christians ought to be, and their influence would be felt in this city very soon, if men who are on the Lord's side would come out and take their stand, and lift up their voices in season and out of season.

Professed child of God, where art thou? Now take it home with you; take it to heart to-night; ask, Where am I? There are a great many in the Church who make one profession, and that is about all you hear of them; and when they come to die you have to go and hunt up some musty old church records to know whether they were Christians or not. God won't do that. I have an idea that when Daniel died, all the men in Babylon knew whom he served. There was no need for them to hunt up old books. His life told his story. What we want is men with a little courage to stand up for Christ. When Christianity wakes up, and every child that belongs to the Lord is willing to speak for Him, is willing to work for Him, and, if need be, willing to die for Him, then Christianity will advance, and we shall see the work of the Lord prosper. There is one thing which I fear more than anything else, and that is the dead, cold formalism of the Church of God. Talk about the *isms!* Put them all together, and I do not fear them so much as dead, cold formalism. Talk about the false *isms!* There is none so dangerous as this dead, cold formalism which has come right into the heart of the Church. There are so many of us just sleeping and slumbering while souls all around are perishing.

AN AMERICAN ANECDOTE.

There was a little story going the round of the American press that made a great impression upon me as a father. A father took his little child out into the field one Sabbath, and he lay down under a beautiful shady tree, it being a hot day. The little child ran about gathering wild flowers and little blades of grass, and coming to its father and saying, "Pretty! pretty!" At last the father fell asleep, and while he was sleeping, the little child wandered away. When he awoke, his first thought was, "Where is my child?" He looked all around, but he could not see him. He shouted at the top of his voice, and all he heard was the echo of his own voice. Running to a little hill, he looked around and shouted again, but all he heard was the echo of his own voice. No response! Then going to a precipice at some distance, he looked down, and there upon the rocks and briers, he saw the mangled form of his loved child. He rushed to the spot, took up the lifeless corpse and hugged it to his bosom, and accused himself of being the murderer of his own child. While he was sleeping his child had wandered over the precipice. I thought as I heard that, what a picture of the Church of God! How many fathers and mothers, how many Christian men are sleeping now while their children wander over the terrible precipice a thousand times worse than that precipice, right into the bottomless pit of hell. Father, where is your boy to-night? It may be just out here in some public-house; it may be reeling through the streets of London, drunk; it may be pressing on down to a drunkard's grave. How many fathers and mothers are there in London—yes, praying Christians too—whose children are wandering away while they are slumbering and sleeping? Is it not time that the Church of God should wake up and come to the help of the Lord as one man, and strive to beat back those dark waves of death that roll through our streets, bearing upon their bosom the noblest young men we have? Oh, my God, wake up the Church! And let us trim our lights and go forth and work for the kingdom of God.

THE GOSPEL TAKING A FRESH START.

My friends, I believe there has never been a time, in our day at least, when work for Christ was more needed than at present. I do not believe there ever was in your day and mine a time when the Spirit of God was more poured out upon the world. There is not a part of Christendom where the work is not being carried on, and it looks very much as if the glad tidings were just going to take, as it were, a fresh start and go round the globe. It is time for you, Christians here in London, to rise as one man. You live at the very centre of the world, and if London is moved, the world is moved. May the London Christians come up as one man. Thank God you are here to-night, and may God fire up every heart. It is not only brains that are wanted, but the heart on fire, and when the heart is on fire and filled with the Holy Spirit, and with the love of God, then God can use us and work through us.

A SCOTTISH BACKSLIDER.

2. But the other class—backsliders—where are you? I can just imagine over there a young man who came up to London five years ago. Perhaps he came from Scotland. He was a member of the Church there; he was a teacher in a Sabbath-school; but when he came to London he found society a little different from what it was in Scotland. He found himself among strangers, and he thought he would not just take a class at once in the Sabbath-school. So he gave up teaching in the Sabbath-school; he gave up all work for Christ. It may be a few months ago he was invited to go to a theatre; and although your conscience said you ought not to go, you went. And then you were invited into a public-house. It may be you got to drinking; it may be you are under the influence of liquor here to-night. Young man, "where art thou?" Come now, backslider, tell me, are you happy? Have you had a happy hour since you left Christ? Does the world satisfy you? Do those husks that you have got far off in a foreign country satisfy you? I have traveled much for a young man, but I never found a happy backslider in my life. I never saw a man that was really born of God, and born again, and born of the Spirit, that ever could find the world satisfy him afterward. I pity the backslider, but I want to tell you that the Lord Jesus pities you a good deal more than any one else can pity you. He knows how bitter your life is, He knows how dark your life is, and He wants you to come home. Oh, backslider, come home to-night. I have come with a loving message from your Father. He will receive you with joy and gladness, and He will say as of him mentioned in Luke xv., "Bring out the best robe, and put it upon him, kill the fatted calf, put a ring on his hands and shoes on his feet, and let us rejoice and be glad, for the wanderer is come home, the dead is alive again." Oh, prodigal, come home to-night. Backslider, while I am speaking, say down in the depths of your

heart, "I will come back to-night." Say as the prodigal of old did, "I will arise and go to my Father," and He will receive you. I never heard of a backslider coming home but God received him. I never heard of a prodigal with his face toward home but God was ready to receive him. Did you ever read of such? Never. I defy any man to say he ever knew a really honest backslider want to get home but God was willing to take him in. And He takes you back just as you are. He will restore His love unto your heart to-night if you will only come.

A CHICAGO BOY AND THE GAMBLERS.

A good many years ago, before Chicago had become a large city, it was a grain market. There were no railways running there then, and the grain used to be shipped on the lake. There was a man living out in the Western prairies, a good many miles from Chicago, a farmer and a minister (that was a very common thing in those early days out in the West), and he sent his only son into Chicago with a load of grain. He waited and waited for his boy to return, but he did not come home. At last the father could wait no longer, so he saddled his horse and went into Chicago. He went round to the places where he had sent his boy to sell grain, and he found that he had sold it. Then he feared that some one had murdered him, and he got detectives on his track. They tracked him into a gambling den, where he had gambled away the whole of his money. After he had done that the men said, "Sell your horses and machine and then you can get all the money back again and go home to your father, and no one will know anything at all about it." That is the way the devil leads men on. He sold his horses and machine, and gambled that money away too. Like the man who was going to Jericho, they stripped him, and then they cared no more about him. What could he do? He was ashamed to go home to meet his father, and he fled. The father knew what it all meant. He knew the boy was ashamed to come home. He was grieved to think that the boy should have such feelings towards him. That is just exactly like the sinner. He thinks because he has sinned God will have nothing to do with him. My friend, if you have sinned, come and ask God to forgive you, and He will forgive you. What did that father do? Did he say, "Let the boy go?" No; he went after him. He arranged his business and started after the boy. And I want to say to you that from the time when Adam fell to the present time God has been seeking after His children. That man went from town to town. When he got into the pulpit to preach, when he had finished his sermon he told the story of how he had lost his boy, and described him, and he asked any of the audience who might ever meet with him to write and let him know. At last he found that he had gone to California, thousands of miles away. Did that father say, "Let him go"? No, off he went to the Pacific coast, seeking the boy. He went to San Francisco, and he advertised in the paper that he would preach at such a church on such a day. When he had preached he told his story, in hopes that the boy might have seen the advertisement and come to the church. When he had done, away under the gallery there was a young man who waited until the audience had gone out; then he came towards the pulpit. The father looked, and saw it was that boy, and he ran to him, and pressed him to his bosom. The boy wanted to confess what he had done, but not a word would the father hear. He forgave him freely, and took him to his home. My friends, you have been enticed away by the devil; now, God is inviting you to come home to-night. Don't go out of this hall until you have returned to your Father's house. Come home, oh, backslider. Oh, wanderer, return to-night.

ALL SEEKERS FINDERS.

3. The last class I want to speak to for a few minutes are those that have no God, no hope, no Christ, no peace, no joy. I want to tell you to-night how you can be saved if you will. If you really want to pass from death to life, if you want to become an heir of eternal life, if you want to become a child of God, make up your mind this night that you will seek the Kingdom of God; I tell you upon the authority of this Word that if you seek the Kingdom of God you will find it. No man ever sought Christ with a heart to find Him who did not find Him. Now stop a moment. Let us be still just for a moment; for if there is any time in a man's life when he wants to think, it is on an occasion like this. Now, friends, you that are not Christians, just ask yourselves where you are. Ask, "Where am I?" Here you are, surrounded by a praying circle. Young man, right by your side, it

may be, is your father, and at this very minute he is lifting his prayers to God for you. I have received numbers of letters from mothers, stating that their young men would be here to-night, and they are praying for you. Young man, will you not yield to that praying mother? Will you not go home to-night and make her heart glad by telling her that you have given yourself to Jesus, that her God is your God? While the minister is offering salvation, there are men praying for your salvation. Just lay hold on eternal life. Make up your mind that you will not go away until the great question is settled. I never knew a man make up his mind to have the question settled, but it was settled soon. This last year there has been a solemn feeling stealing over me. I am what they call in the middle of life, in the prime of life. I look upon life as a man going up a hill, and then down again. I have got to the top of the hill, if I should live the full term of life—three-score years and ten, and am just on the other side. I am speaking to many here who are also on the top of the hill, and I ask you, if you are not Christians, just to pause a few minutes, and ask yourselves where you are. Let us look back on the hill that we have been climbing. What do you see? Yonder a gravestone; it marks the grave of a praying mother. Did you not promise her when she was dying that you would meet her in heaven? Am I not speaking to some here to-night who made that solemn promise? Young man, have you kept it? Look a little further up the hill. There is a gravestone that marks the grave of a little child —it may have been a little lovely girl— perhaps her name was Mary; or it may have been a boy, Charlie; and when that child was taken from you, did you not promise God, and did not you promise the child, that you would meet it in heaven? Is the promise kept? Think! Are you still fighting against God? Are you still hardening your hearts? I would to God that you would to-night settle this question. Now, look down the hill. What do you see? Yonder there is a grave: we cannot tell how many days, or years, or weeks it is away; we are hastening towards that grave. It may be the coffin is already made that this body shall be laid in; it may be that the shroud is already waiting. My friend, is it not the height of madness to put off salvation so long? Undoubtedly I am speaking to some who will be in eternity a week from now. In a large audience like this, during the next week death will surely come and snatch some away; it may be the speaker, or it may be some one who is listening. Why put off the question another day? Why say to the Lord Jesus, again to-night, "Go Thy way this time, and when I have a more convenient season I will call for Thee?" Why not let Him come in to-night? Why not open your heart, and say, "King of Glory, come in?" He will receive you.

THREE STEPS TO PERDITION.

You know there are three steps to the lost world; let me give you their names. The first is Neglect. All a man has to do is to neglect salvation, and that will take him to the lost world. Some people say, "What have I done?" Why, if you merely neglect salvation you will be lost. I am on a swift river and lying in the bottom of my little boat; all I have to do is to fold my arms, and the current will carry me out to sea. So all that a man has to do is to fold his arms in the current of life, and he will drift on and be lost. The second step is Refusal. There are many who have got on the first step, neglect. If I met you at the door and pressed this question on you, you would say, "Not to-night, Mr. Moody, not to-night." But there are others of you who, if I said, "I want you to press into the kingdom of God," would politely refuse:—"I will not become a Christian to-night; I know I ought, but I won't to-night." Then the last step is to despise it. Some of you have already got on the lower round of the ladder. You despise Christ. I see some of you looking at me with scorn and contempt. You hate Christ, you hate Christianity; you hate the best people on earth and the best friends you have got; and if I were to offer you the Bible you would tear it up and put your foot upon it. Oh, despisers! you will soon be in another world. Make haste and repent and turn to God. Now, on which step are you, my friend—neglecting, or refusing, or despising? Bear in mind that a great many are taken off from the first step: they die in neglect. And a great many are taken away refusing. And a great many are on the last step, despising salvation. I wish I could settle this question for you. I wish I could bleed for you. Won't you come? Everything that is pure and holy and lovely is beckoning us to a world of love and peace; everything that is polluted and vile and hellish and carnal is beckoning us down. I set before you life and death; which will you choose? When Pilate had Christ on his hands, he said, "What shall I do with Him?" and the multitude cried out, "Away with Him! crucify Him!" Young men, is that your language to-night? Do you say, "Away with this Gospel! Away with Christianity! Away with your prayers, your sermons, your Gospel sounds! I do not want Christ?" Or will you be wise and say, "Lord Jesus, I want Thee, I need Thee, I will have Thee?" May God bring you to that decision!

THE CHILD ANGEL.

I will tell you an anecdote now, because the man of whom the story is told may represent many in this audience to-night. A few years ago I was attending a Sabbath-school convention in a little town, where a man to whom I was a stranger took me into his house. It was a warm day and the curtains were down, so that the room was dark. His wife was in bed, and he excused himself because he had some matters to attend to. I was left alone. It was so dark that I could not read, and I walked up and down the room till I felt lonely. Presently he came in, and I said, "Have you no children?" I am very fond of children, and I thought if he had any I could play with them. He said no; he had one, but God had taken her from him; she was in heaven, and he said he was glad of it. I said, "Glad that your only child is dead?" "Yes," he said. "How is that?" I asked. "Was she deformed, or was anything wrong with her?" "No," said he, "she was as perfect as could be;" and he got up and brought me one of those old fashioned daguerreotypes—a portrait of a beautiful girl, with golden curls falling down her neck, more like an angel than a child. I asked how old she was. "Seven." "What do you mean by saying you are glad she is in heaven?" "Well," said he, "I worshipped that child, that child was in all my plans, I was making money for my child, and every Sunday I spent hours with her; she was the idol of my heart, but I did not know it. One day I found my child sick. I did not think it was dangerous, but in a few days she died, and I accused God of being unjust in sparing the families of others and taking away my child, and I refused to be reconciled. I would have torn God from His throne if I could. For three days and nights I neither ate, nor drank, nor slept. I was almost mad. On the third day I buried her, and when I came home, as I walked up and down the room, I thought I heard the voice of my little one; but then I thought, 'No, that voice is hushed forever.' Then I thought I heard her little feet coming towards me, but then I said, 'No, I shall never hear those little feet again.' At last I threw myself on my bed, and began to weep. Nature gave way, and I fell asleep. I had a dream. I suppose it was a dream; but it has always seemed to me more like a vision. I thought I was crossing a waste, barren field, and I came to a river that looked so cold and dark and dreary that I drew back from it; but, looking across, I saw the most beautiful land my eyes had ever rested upon; and as I gazed I thought that death and sickness and disease could never enter there. Then I saw a company on the other side, and among them my own darling child. She came to the bank of the river, and waving her little angel hand, said, 'Father, come right this way; it is so beautiful here;' and she beckoned me to the world of light. I then went to the water's edge, and thought I would plunge in, but it was too deep for me—I could not swim. I thought I would give anything to cross. I tried to find a boat, but there was no ferryman. I looked for a bridge, but there was none; and while I was wandering up and down the little angel voice came across the stream, 'Come right this way, father; it is beautiful here!' All at once I heard a voice as if it came from heaven, saying, 'I am the Way, the Truth, and the Life. No man cometh unto the Father but by Me.' The voice awoke me from sleep. I thought it was my God calling me, and that if I would ever see my child again I must come to God through Jesus Christ. That night I knelt beside my bed and gave myself to God. Now I no longer look upon my child as sleeping in her grave, but I see her with the eye of faith in that beautiful land, and every night when I lie down I hear her sweet voice saying, 'Come right this way, father,' and every morning I hear her repeating the same words. Now my wife is converted. I am superintendent of the Sabbath-school, and eight children have been converted, and I am trying to get as many converted as I can to go with me to that beautiful land." Undoubtedly I am speaking to some father to-night with a lost one in that world. If that child could speak to you, would it not say, "Come right this way, father?" And many a young man is here who has a sainted mother or sister in heaven. If she could now speak from the battlements of heaven, would not the words be, "Come right this way, my brother," "Come right this way, my son"? Oh, thank God that we have all got an elder Brother across the stream. The Son of God stands on the banks to-night, calling to every one, "Come this way, my child." Young man, won't you rise and go to your Father to-night? May God call you home, wanderer! May every backslider return and press into the kingdom. I beg of you as a friend, do not leave this hall to-night until you have sought the kingdom of God. Make up your mind this night and this hour that you will press into the kingdom.

VI.

THE NEW BIRTH.

"Except a man be born again, he cannot see the kingdom of God."—JOHN iii. 3.

MUCH less inherit it. He can't even get a glimpse of the kingdom of God except he be born again. I believe that we have the most important subject before us to-night that will ever come before us in this world. I don't believe there is any truth in the whole Bible so important as the truth brought out in the third chapter of the Gospel of John.

It is the A B C of God's alphabet. If a man is unsound on regeneration, he is unsound on everything. That is really the foundation stone; and he must get the foundation right. If he don't, what is the good of trying to build a house? Now, He says plainly, "Except a man be born again." And although regeneration, or the new birth, is taught so plain in the third chapter of John, I don't believe there is any truth in the whole Bible that the church and the world are so mixed up on, and in such great darkness about, as this great truth in the third chapter of John. There are a great many that are, as it were, like the man that saw men as trees walking. Many Christians do not seem to be just in a mind about this new birth.

BORN A CHRISTIAN.

Only this afternoon, as I was in the inquiry-room, a person came in, and I said, "Are you a Christian?" "Why," she says, "of course I am." "Well," I said, "how long have you been one?" "Oh, sir, I was born one." "Oh! indeed, then I am very glad to take you by the hand; I congratulate you; you are the first woman I ever met who was born a Christian; you are more fortunate than others; they are born children of Adam." She hesitated a little, and then tried to make out that, because she was born in England, she was a Christian. There are a great many who have the idea that, because they are born in England or a Christian country, they have been born of the Spirit. Now, in this third chapter of John, the new birth is brought out so plain, that if any one will read it carefully and prayerfully, I think their eyes will soon be opened. That which is born of the flesh is flesh; it remains flesh; and that which is born of the Spirit is spirit, and that remains spirit. So, when a man is born of God, he has God's nature. When a man is born of his parents, he receives their nature, and they received the nature of their parents, and you can trace it back to Eden. We have received the nature of the first Adam, but when a man is born of God, or born from above, or born of the Spirit — that is the way the Holy Ghost puts it in that third verse—he receives God's nature, and then it is he leaves the life of the flesh for the life of the Spirit.

SATAN GOING TO CHURCH.

Before I go on I want to say one thing: and that is, what this new birth, or being born of the Spirit, is not. A great many think they have been born again because they go to church. A great many say, "Oh, yes, I am a Christian; I go to church every Sabbath." Let me say here that there is no one that goes to church so regularly in all London as Satan. He is always there before the minister, and he is the last one out of the church. There is not a church in London, or a chapel, but that he is a regular attendant of it. The idea that he is only down in the slums and lanes and alleys of London is a false idea. The idea that he is only in public-houses — I will confess I think he is there, and that he is doing his work very well — but to think that he is only there is a false idea. He is wherever the Word is preached; it is his business to be there, and catch away the seed. He is here to-night. Some of you may go to sleep, but he won't. Some of you may not listen to the sermon, but he will. He will be watching, and when the seed is just entering into some heart he will go and catch it away. May God rebuke Satan to-night, and may the Word of God fall deep into the hearts of many. May many be called to-night.

A CHRISTIAN BECAUSE BAPTIZED.

Another class say, "Oh, yes, I am a Christian, because I was baptized." Now, I want to say here that baptism is one thing, and being born again is another. Because a person is baptized, you would not say that that is new birth. Would you call that being born from above? You cannot baptize a man into the kingdom of God. Now, bear that in mind. If I could save men by baptizing them, you would not catch me preaching. I would get water and baptize them; that would be the quickest way. It would be no use to be praying and pleading for men to flee from the wrath of God. But, you can never get them into the kingdom of God by baptism. Baptism is all right in its place. I am not here crying down church ordinances; I am talking about the new birth, and there are a great many, I believe, being deceived on this one point, that because they have been baptized at some time of their life they have become Christians. But that is not new birth; that is not being born from above and of the Spirit. Do not let Satan deceive you, my friends, on that point, for it is a very important truth; and we want to have every one here to-night to understand, and I hope the Spirit of God will make plain the difference between baptism and conversion, or regeneration, or being born of the Spirit, or being born again.

JOINING THE CHURCH.

There is another class that say, "Oh, yes, I became a Christian when I joined the church—the day when I united with the church." That ain't being born again. What is that to do with the new birth, being united with the church on earth? There are a great many united with the church who are on their way to death and ruin. A great many have no hope of eternal life who are members of the church. One of the twelve Christ chose to follow Him turned out a hypocrite and a traitor; he was not loyal to Christ at heart. My friends, don't just build your hope of heaven upon some profession of your faith; but bear in mind it is the being born of God. Now just let me stop a minute, and you just think, and ask yourselves that question, "Have I been born again?" It is the most solemn question that will ever come before you down here—"Have I been born from above? Have I been born of the Spirit?" It ain't making some new resolutions. You have made enough of them. That ain't the new birth. I never met any one who had not made some good resolutions in their life. It ain't trying to do good. A great many say, "I try to do the best I can and I think it will come out all right." What is that to do with the new birth and the new creation? It don't say to him that tries to do the best he can, but to him that believeth or that is born of the Spirit; and "Except a man be born again he cannot see the kingdom of God."

INSTANTANEOUS CONVERSION.

Now, I believe this birth is instantaneous. I have met a great many people who cannot tell the day or the hour of their conversion; but there must have been a time when they passed from death unto life—when they were born of the Spirit.

There must have been a time when their names were written in the Book of Life. They may not be conscious of the day, or the hour, or the week, or the month, or the year; but, my friends, I beg of you to be sure that they have been born of the Spirit. Don't be deceived upon this one truth, because Christ Himself says, "Except a man be born again he cannot see the kingdom of God."

As I said before, when I was born of my parents I received their nature, I received the nature of the flesh; and I cannot serve God in the flesh. "God is a Spirit, and they that worship Him must worship Him in spirit and in truth." And before a man can worship God he must be born of God; he must be born of the Spirit. Then with this new birth, with this new life he can serve God. Then the yoke is easy; then the burden is light. A man may as well try to fly to the moon as to serve God before he has been born of the Spirit. It is utterly impossible. The natural man is at enmity against God; his natural heart is at war with God; it always has been, and it always will be. And not only that, but you cannot make it better. Somebody said that God never mends. God creates anew; therefore don't be trying to patch up that old Adam nature. God says, "It shall never come into My presence." Therefore God has just set it aside. But He tells us how we are to come into His presence, and how we are to get into His kingdom. This is worthy to be borne in mind. You cannot educate men into it. That is what the world is trying to do. But he that climbeth up by some other way than the Lord's way the same is a thief and a robber. You had better be born into it in God's way.

FOREIGNERS HAVE NO RIGHT TO COMPLAIN.

We have a law in America that no man shall be President of the United States that has not been born on American soil. We have a great many Englishmen come to America; but I have never heard one complain about that law. We have a great many Germans, Scotchmen, Irishmen, and Welshmen, in fact men from all parts of the world, who come to America, and yet I have never heard one complain about that law. They say America has the right to say who shall be President. I come here to your country, and I do not complain because you have a Queen to reign over you. What right have I to complain? Has not England a right to say who shall rule it, and who shall be its Queen? Foreigners have no right to interfere. And I would like to ask you this question, Has not God a right to say who shall come into His kingdom, and how we shall come? Now, my friend, God tells us here we are to come into His kingdom by the new birth. We must be born from above, born of the Spirit, and then we get a nature that goes out towards God. If you take a drunken man, and put him on the very pavement of heaven, he will not be happy there. The drunkard doesn't want heaven. What is he to do there? He has no whiskey to drink there, and he has none of his old companions. What is he to do? He would say, "This is hell to me. I don't want to sit here." A man that cannot spend one Sabbath on earth among God's people, what is he to do with that eternal Sabbath, with those that have washed their robes and made them white in the blood of the Lamb? A man must have a spiritual nature before he wants to go to heaven. Heaven cannot have any attractions to a man until he is born out of heaven of the Spirit.

A WORD FOR THE MORALISTS.

Now let us go back to this man that Christ said these words here to. I often rejoice He didn't say this to that woman at the well, nor to that woman Mr. Sankey has been singing about to-night. If He had said to them, people would have said, "Oh, that poor woman needs to be converted; but I am a moralist; I don't need to be converted. Regeneration will do for harlots, thieves, and drunkards, but we moralists do not need it." But who did Christ say it to? He said it to Nicodemus. Who was he? He belonged to the house of bishops. He would have been a bishop if he had been here. Nicodemus stood very high; he was one of the church dignitaries; he stood as high as any man in Jerusalem, except the high priest himself. He belonged to the seventy rulers of the Jews; he was a doctor of divinity, and taught the law. There is not one word of Scripture against him; he was a man that stood out before the whole nation as of pure and spotless character. What does Christ say to him? "Except a man be born again, he cannot see the kingdom of God." I can see a scowl on his forehead. He says, "What do you mean by being born again—born from above, born of the Spirit? Now I am old, can I a second time enter my mother's womb, and be born again?" Jesus saith, "Verily, verily, I say unto thee, Except a man be born of water, and of the Spirit, he cannot see the kingdom of God." He didn't take back what He had said, but he just repeated it—"Except a man be born again, he cannot see the kingdom of God." I can just imagine Nicodemus was like tens of thousands of men in London to-day. The moment you talk to them about regeneration or conversion, there is a scowl on their forehead. They say, "I don't understand it." Of course, the natural man don't understand spiritual things. It is a matter of revelation. I hope God will reveal Himself to many a soul here to-night. A great many men try to investigate and find out God. Suppose you spend a little of your time in asking God to reveal Himself to you.

A TALK IN THE SMOKING-ROOM.

I heard, some time ago, of some commercial travelers who went to hear a man preach. They came back to the hotel, and were sitting in the smoking-room, talking, and they said the minister did not appeal to their reason, and they would not believe anything they could not reason out. There was an old man sitting there listening, and he said to them, "You say you won't believe anything you can't reason out?" "No, we won't." The old man said, "As I was coming on the train, yesterday, I noticed some sheep and cattle and swine and geese, all eating grass. Now, can you tell me by what process that same grass was turned into feathers, hair, bristles, and wool?" "Well, no, we can't just tell you that." "Do you believe it is a fact?" "Oh, yes, it is a fact." "I thought you said you would not believe anything you could not reason out?" "Well, we can't help believing that; that is a fact we see before our eyes." "Well," said the old man, "I can't help but believe in regeneration, and a man being converted, although I cannot explain how God converted him." I have no doubt, if a man spoke about this to me 21 years ago, I should have said it was all Greek, and that I did not understand what the man was talking about. There may be a good many in this hall to-night wondering what that American is talking about. Born again; born of the Spirit! I do not understand it. But I understand it now (and I can call hundreds of witnesses here)—why? Because I have been born of the Spirit.

THE GREAT TEACHER'S ILLUSTRATION.

Now, the illustration which Christ used to Nicodemus was the wind. "The wind bloweth where it listeth, and no man knoweth whence it cometh nor whither it goeth." Now, you cannot see the spirit of God work in this audience; but I hope and pray He may be working now in the hearts of many, convincing them of sin! Do you believe more than ever that you are a sinner? Well, that is the work of the Holy Ghost. The devil never told you you are a sinner; he tries to make you believe you are good enough. If you believe to-night that you have sinned against God, that is the work of the Holy Ghost. He is here to-night at work. We cannot see Him, but there are a great many who know He is here. Suppose I should say I don't believe in the wind, and that it must be all imagination. I have lived thirty-seven years, and have never seen the wind. No one ever saw the wind. It is all imagination; it is folly for men to talk about the wind. I can just imagine that boy there saying, "Why, I know more than that man; I know there is wind, for it blew my hat off this very day into the mud; and I have often felt it blowing in my face." My friends, you have never felt the wind more than I have felt the Spirit of God. You have never seen the effects of the wind more than I have seen the effects of the Spirit of God, and of the workings of the Holy Ghost, and there are hundreds of witnesses here to-night who would testify the same thing.

AN APPEAL TO THE DRUNKARD.

It may be that I am talking now to some poor drunkard here. When he comes into his house, his children listen, and hear by the footfall that their father is coming home drunk, and the little things run away and hide from him as if he was some horrid demon. His wife begins to tremble. Many a time has that great, strong arm been brought down on her weak, defenceless body. Many a day has she carried about marks from that man's violence. He ought to be her protector, support, and stay; but he has become her tormentor. His home is like hell upon earth; there is no joy there. There may be one such here to-night who hears the good news that he can be born again, and receive a nature from heaven, and receive the Spirit of God. God can give him power to hurl the infernal cup from him. God will give him grace to trample Satan under his feet, and the drunkard will then become a sober man. Go to that house three months hence, and you will find it neat and clean. As you draw near that home you hear singing; not the song of the drunkard; that is gone; all things have become new. He has been born of God, and is singing one of the songs of Zion:

"Rock of ages, cleft for me,
Let me hide myself in Thee."

Or perhaps he is singing that good old hymn that his mother taught him when he was a little boy:

"There is a fountain filled with blood,
Drawn from Immanuel's veins;
And sinners plunged beneath that flood
Lose all their guilty stains."

He has become a child of God, an heir of heaven. His children are climbing up his knee, and he has his arms round their neck. That dark home is now changed into a little Bethel on earth. God dwells there now. Yes, God has done all that, and that is regeneration. May God convert the drunkard! I hope many a drunkard will be converted to-night.

Christians, lift up your hearts for the poor drunkards of London. If they try to lead a better life, One mighty to save, Christ the Lord, will give them the victory; for, strong as driuk may be, His grace is stronger. May the Christians make haste and tell the glad news to the drunkards of London!

THE WORTH OF GOOD RESOLVES.

Then some of you may have been saying, "I wish Mr. Moody would tell us how we are to become Christians; for he says that we cannot be Christians by trying to do good and by making new resolutions." Many a time you have been at a meeting like this, and have resolved to turn over a new leaf, and you may now form another good resolution. If you do, you will break it. I would not give that for all your resolutions. What are you going to do? If it is a new birth you are to have, you cannot create life. Can you bring life to a dead fly? All the wise men in London cannot do it. God alone is the author of life; and if you have new birth, it must be God's work. When the Jubilee Singers were in the north of England my family went to see them, and my little boy asked why they didn't wash the black off their faces. I told him it was because they were born black.. The Ethiopian cannot change his skin, nor the leopard his spots. You cannot save yourself. There is a man dying—can you put new life into him? Or can you raise up a dead body by saying, "Young man, arise"? That is the work of God. Your souls are dead in trespasses and sin. May the Lord Jesus Christ speak life. God said, "Let there be light;" and there was light. And if He says, "Let there be life,' there will be life.

THE BEGGAR AND THE PRINCE OF WALES.

I imagine some of you will say, "I haven't anything to do." Well, you haven't. Salvation has been worked out for you by another. Many go all round the world in search of honor or possessions. Salvation is worth thousands of times more; but you don't get it that way. God has but one price for salvation. Do you want to know what it is? It is without money and without price. Rowland Hill said that most auctioneers found they had hard work to get the people up to their price, but that he had hard work to get people down to his. "The wages of sin is death, but the gift of God is eternal life." Who will have it to-night? I say to you, young man, will you have that gift to-night? Suppose I was going over London Bridge, and saw a poor, miserable beggar, bare-footed, coatless, hatless, with no rags hardly to cover his nakedness, and right behind him, only a few yards, was the Prince of Wales with a bag of gold, and the poor beggar was running away from him as if he was running away from a demon, and the Prince of Wales was hallooing, "Oh, beggar, here is a bag of gold!" Why, we should say the beggar had gone mad, to be running away from the Prince of Wales with the bag of gold. Sinner, that is your condition. The Prince of Heaven wants to give you eternal life, and you are running away from Him. "The wages of sin is death, but the gift of God is eternal life." Then you say, "If I have nothing to do, what is going to become of me? If it is not by working in earnest, how am I to be saved?"

THE CHEAP AND SIMPLE REMEDY.

It is God's work entirely how you are to be saved. I will tell you; Scripture will tell you—that is better. Take the illustration Christ used to Nicodemus; you could not have a better. He took him to the remedy:—"As Moses lifted up the serpent in the wilderness, even so must the Son of man be lifted up, that whosoever believeth in Him should not perish, but have eternal life." Now there is the remedy. How am I to be saved? By looking for life, eternal life; just by looking. It's very cheap, isn't it? Very simple, isn't it? Little girl, just look away to the Lamb of God to-night and be saved. What says the great wilderness preacher? "Behold the Lamb of God, which taketh away the sins of the world." You might say the whole plan of salvation is in two words— Giving; Receiving: God gives; I receive.

MR. MOODY AS AN ARMY CHAPLAIN.

I remember, after one of our terrible battles—I was in the army, tending soldiers—and I had just laid down one night, past midnight, to get a little rest, when a man came and told me that a wounded soldier wanted to see me. I went to the dying man; he called me chaplain, but I was not. He said, "Chaplain, I wish you to help me to die." I said, "I would help you to die if I could. I would take you on my shoulders and carry you into the kingdom of God if I could; but I cannot. I can tell you of One that can." And I told him of Christ being willing to save him; and how Christ left heaven and came into the world to seek and to save that which was lost. I just quoted promise after promise, but all was dark, and it almost seemed as if the shades of eternal death were gathering around his soul. I could not leave him, and at last I thought of this third chapter of John, and I said to him, "Look here, I am going to read to you now a conversation that Christ had with a man that went to Him when he was in your state of mind, and inquired what he was to do to be saved." I just read that conversation to the dying man, and he laid there with his eyes rivetted upon me, and every word seemed to be going home to his heart, which was open to receive the truth. When I came along down to the verse where it says, "As Moses lifted up the serpent in the wilderness, even so must the Son of man be lifted up, that whosoever believeth in Him should not perish, but have eternal life," the dying man cried, "Stop, sir. Is that there?" "Yes, it is all here." Then he said, "Won't you please read it to me again?" I read it the second time. "As Moses lifted up the serpent in the wilderness, even so must the Son of man be lifted up, that whosoever believeth in Him should not perish, but have eternal life." The dying man brought his hands together, and he said, "Bless God for that. Won't you please read it to me again?" I hope you will pardon me for reading it the third time, but I want the Spirit of God to impress it on your hearts to-night. "As Moses lifted up the serpent in the wilderness, even so must the Son of man be lifted up, that whosoever believeth in Him should not perish, but have eternal life." I read the next verses: "For God so loved the world that He gave His only begotten Son, that whosoever believeth in Him should not perish, but have everlasting life. For God sent not His Son into the world to condemn the world, but that the world through Him might be saved."

THE DYING SOLDIER.

I read through the whole chapter, but long before the end of it he had closed his eyes. He seemed to lose all interest in the rest of the chapter, and when I got through it his arms were folded on his breast, he had a sweet smile on his face; remorse and despair had fled away. His lips were quivering, and I leaned over him, and heard him faintly whisper from his dying lips, "As Moses lifted up the serpent in the wilderness, even so must the Son of man be lifted up, that whosoever believeth on Him should not perish, but have eternal life." He opened his eyes, and fixed his calm, deathly look on me, and he said, "Oh, chaplain, that is enough; that is all I want." And in a few hours he pillowed his dying head upon the truth of those two verses, and rode away on one of the Saviour's chariots, and took his seat in the kingdom of God. Oh, sinner, you can be saved to-night if you will. Look and live. May God help every lost soul here to-night to look on the Lamb of God, which taketh away the sins of the world.

VII.

A SERMON ON ONE WORD.

I SHALL take for my text to-night the one word, "Gospel." I do not think there is a word in the English language that is so little understood in this Christian land of England as this very word "Gospel." We have heard it from our earliest childhood up. There is not a day, and with many of us not an hour during the day, but that we hear the word "Gospel." And yet I say a partaker of the Gospel is a long time before he really knows the meaning of the word. It means "good tidings." I think it would do us good sometimes to get a dictionary and hunt up the meaning of some of the words we use so often; some of those Bible words, such as "Gospel" and "Christ." I think it would change our ideas. I think this would be a very joyful meeting to-night if every one really believed that the Gospel is good news. Why, you let a man or a boy bring a dispatch into this audience and hand it to any one here, and if that brings good news you can see it immediately in the man's face; his face lights up when he opens the dispatch. You can see he really believes it. And if it is really good news, if it brings him the tidings of a long-lost boy got or coming home, why, if his wife is sitting next to him, he passes the dispatch to her; he wants her to have knowledge of it. He does not wait for her to ask for it; he does not wait till they get home. So when I preach, those who really believe the Gospel, if I am near enough to look into their eye, I see their face lights up and looks remarkably sharp; but those who do not believe it put on a long face, and look as if you had brought them a death warrant, or invited them to attend a funeral. If you go to hear some dull and stupid sermon or lecture, that is not the Gospel.

THE ANGELIC REVIVALISTS.

The Gospel is good tidings of great joy. No better news ever came out of heaven than the Gospel. No better news ever fell upon the ears of the family of man than that Gospel. Hark! hear those shepherds talking to one another after the angels had gone away. They believed the message, and they were full of joy. You can see them on the way now to Bethlehem. They said, "Let us go and see what has taken place." And what was the message that the angels brought to those shepherds? "Behold, I bring you good tidings of great joy, which shall be to all people. For unto you is born this day in the city of David a Saviour." Now, if those shepherds had been like a good many people at the present time, they would have said, "We do not believe it is good news. Do not believe it. It is all excitement. Those angels want to get up a revival. Those angels are trying to excite us. Don't you believe them." That is what Satan is saying now. "Don't you believe the Gospel is good news." Because he knows the moment a man believes good news, he just receives it. I never saw a man in all my life that did not like good news. I never saw a man in all my travels that did not like good news. There is not any one here to-night but what likes good news. And every man and woman that is under the power of the devil does not believe the Gospel is good news. The moment you are out from under his power and influence then you believe it. May God bring you out to-night, that the Gospel may sink deep into your heart.

"GOD'S SPELL."

It is the best news that ever came to this sin-cursed earth. It means "Good spell," or in other words, "God's spell." We are dead in trespasses and sin, and God wants us to be reconciled. It is a Gospel of reconciliation, and God is shouting from the heights of glory, "Oh, ye men, I am reconciled, now be ye reconciled." We have glorious news to tell you—God is reconciled and beseeches His subjects to be reconciled. The great apostle says, "We beseech you in Christ's stead, be ye reconciled." The moment a man believes down go his arms of rebellion, and he just believes the Gospel. The unequal controversy is over. A light from Calvary crosses his path, and he can walk in unclouded sun, if he will. It is the privilege of every man and woman in this vast assembly from this hour to walk in unclouded sun if they will. What has brought darkness into the world? Darkness came because man would not believe the Gospel that Christ is the light of the world. Now I want to tell you why I like the Gospel. It is because it has been the very best news I have ever heard. That is just the reason I like to preach it. Because it has done me so much good. I do not think a man can preach the Gospel until he believes it himself. A man must know it down deep in his own heart before he can tell it out; and then he tells it out very poorly.

POOR AMBASSADORS.

We are very poor ambassadors and messengers; but never mind the messenger, take hold of the message—that is what you want. If a boy brought me good news tonight, I would not care about the look of the boy; I would not care whether he was black or white, learned or unlearned. The message is what would do me good. A great many look at the messenger instead of the message. Never mind the messenger. My friends, get hold of the message to-night. The Gospel is what saves, and what I want now is that you just believe the Gospel.

Paul says in this 15th chapter of the 1st Corinthians what the Gospel is. He says, "I declare unto you the Gospel." And the first thing he states in the declaration to these Corinthians is this:— "Christ died for our sins according to the Scriptures." That was the old-fashioned Gospel. I hope we never will get away from it. I don't want anything but that old, old story. Some people have itching ears for something new. Bear in mind there is no new Gospel. Christ died for our sins. If He did not, how are we going to get rid of them? Would you insult the Almighty by offering the fruits of this frail body to atone for sin? If Christ did not die for our sins, what is going to become of our souls? And then he goes on to tell that Christ was buried, and that Christ rose again.

DEATH AND THE REDEEMER.

That is what he is trying to bring out in the 15th chapter of the 1st book of Corinthians. He burst asunder the bands of death. Death could not hold Him. I can imagine when they laid Him there in Joseph's sepulchre, if our eyes could have been there, we should have seen Death sitting over that sepulchre, saying, "I have Him; He is my victim. He said He was the resurrection and the life. Now I have hold of Him in my cold embrace. Look at Him. There He is; He has had to pay tribute to me. Some thought He was never going to die. Some thought I would not get Him. But He is mine." But look again! The glorious morning comes, and the Son of man bursts asunder the bands of death, and came out of the sepulchre. We do not worship a dead God, but a Saviour who still lives. Yes, He rose from the grave; and then they saw Him ascend. That is what Paul calls Gospel. Not only Christ's death and burial, but they saw Him ascend into heaven. He went up and took His seat at the right hand of God, and He will come back again. The Gospel consists of five things—Christ's death, burial, resurrection, ascension; and "I will come again," says He. Thanks be to God, He is coming back by-and-by. He will come and take the kingdom; He will sway His sceptre from the rivers to the ends of the earth by-and-by. A little while and He shall rule and reign. Let us lift up our heads and rejoice that the time of our redemption draweth near.

CHRIST'S DEATH THE GOSPEL.

Let us get back to the simple Gospel—Christ died for our sins. We must know Christ at Calvary first, as our Substitute, as our Redeemer; and the moment we accept of Him as our Saviour and our Redeemer, then it is that we become partakers of the Gospel. The moment I believe on the Lord Jesus Christ as my Substitute, as my Saviour, that moment I get light and peace. To-night I know some people say, "Oh, it is not Christ's death; it is Christ's life. Do not be preaching so much about the death of Christ; preach about His life." My friends, that never will save any one. Paul says, "I declare unto you the Gospel. Christ died"—not Christ lived—"Christ died for our sins, who His own self bare our sins in His own body, on the tree." Now, when I accept of Christ as my Saviour, as my Substitute, then I am justified from all things which I could not be by the law of Moses.

PERSONAL REMINISCENCES.

As I was going to say a few minutes ago, the reason I like the Gospel is that it has taken out of my path the worst enemies I ever had. My mind rolls back to twenty years ago, before I was converted, and I think very often how dark it used to seem at times as I thought of the future. There was death — what a terrible enemy it seemed! I was brought up in a little village in New England. It was the custom there when a person was buried to toll out the age of the man at his funeral. I used to count the strokes of the bell. Death never entered that village and tore away one of the inhabitants, but I always used to count the tolling of the bell. Sometimes it would be away up to seventy, or between seventy and eighty—beyond the life allotted to man, when man seemed living on borrowed time when cut off. Sometimes it would be clear down in the teens, and childhood, and death would take away one of my own age. It used to make a solemn impression on me. I used to be a great coward. When it comes to death some men say, "I do not fear it." I feared it, and felt terribly afraid, when I thought of the cold hand of death feeling for the cords of life; and being launched out to eternity, to go to an unknown world. I used to have terrible thoughts of God; but they are all gone now. Death has lost its sting. And as I go on through the world I can shout now, when the bell is tolling, "Oh, death, where is thy sting?" And I hear a voice come rolling down from Calvary, "Buried in the bosom of the Son of God." He just robbed death of its sting. He just took the sting of death into His own bosom. And if you take a wasp, and just take the sting out of that wasp, you will not be afraid of it any more than you would of a little fly. The sting has been taken out. And you need not be afraid if you are in Christ. Christ died for your sin. The penalty, the wages of sin is death. Christ received the wages on Calvary, and therefore there is no condemnation. All that death can get now is this old Adam. I do not care how quickly I get rid of it. I will get a better body, a resurrected body, a glorified body, a body much better than this. Yes, my friends, "To die," says the apostle, "is gain."

THE FEAR OF DEATH.

If a man is in Christ, let death come. Suppose death should come stealing up into this pulpit, and should lay his cold, icy hand upon my heart, and it should cease to throb; I should rise to another world, and should be present with the King. I should be absent from the body, but present with the Lord. That is not bad news. There is no use in trying to conceal it, death is an enemy to a man's rest. What a glorious thought to think that when you die you will sink into the arms of Jesus, and that He will carry you away to yon world of light! A little while longer here, a few more tears, and then you can gain an unbroken rest in yon world of light. The Gospel turns that enemy into a friend, and you even shout for death. Well, then, I used to go and look into the cold, silent grave, and I used to think of that terrible hour when I would have to be laid down in the grave, and this body would be eaten up with the worm. But now the grave has lost its terror and gloom; I can go and look down into the grave and shout over it, and cry out, "Oh, grave, where is thy victory?" And I hear a shout coming up from the grave; it is the shout of the Conqueror, of Him who has been down and measured the depth of it, of my Lord and Saviour : "Because I live, you shall live also." Yes, the grave has lost its victory. The grave has no terror to the man in Christ Jesus. The Gospel takes that enemy out of the way.

SIN PUT AWAY.

And then there was the terrible name of sin. I thought all my sins would be blazed out before the great white throne; that every sin committed in childhood and in secret, and every secret thought, and every evil desire would be just blazed out before the assembled universe; that everything done in the dark would be brought to light. But thanks be to God, the Gospel tells me my sins are all put away in Christ. Out of love to my soul, He has taken all my sins, and cast them behind His back. That is a safe place to have sin, behind God's back. God never turns back; He always marches onward. He will never see your sins if they are behind His back. That is one of His own illustrations. Out of love to my soul, He has taken all my sins upon Him. Not a part. He takes them all out of the way. There is no condemnation to him that is in Christ Jesus. You may just pile up your sins till they rise up like a dark mountain, and then multiply them by ten thousand for those you cannot think of; and after you have tried to enumerate all the sins you have ever committed, just let me bring one verse in, and then that mountain will melt away : "The blood of Jesus Christ, His Son, cleanseth us from all sin." The blood covers the sin.

WHAT GOD CANNOT DO.

In Ireland, some time ago, a teacher asked a little boy if there was anything that God could not do; and the little fellow said, "Yes; He cannot see my sins through the blood of Christ." That is just what He cannot do. The blood covers them. Is it not good news to get rid of your sin? You come here a sinner, and if you believe the Gospel your sins are taken away. "Believe on the Lord Jesus Christ, and thou shalt be saved." You shall be justified from all things; which you could not by the law of Moses. By believing, or by receiving the Gospel, Christ becomes yours. Only think, young man; say, think of it. You just are invited to accept of the Gospel. You are invited to make an exchange; to get rid of all your sins, and to take Christ in the place of them. Is not that wonderful? What a foolish young man you will be not to make the bargain. The Lord says, "I will take your sins, and give you Myself in the place of them." But a great many say, "No"; and just hug the sin to their bosom. May God help you to come up, sinner, to-night, and receive the Lord Jesus Christ as your way, your truth, and your life!

THE FEAR OF JUDGMENT.

There is another name which used to haunt me a great deal—the great Judgment Day. I used to think that was a terrible day when I should be summoned before God, and could not tell till then whether I should have a seat on His right hand or on His left. Until I stood before the great white throne of judgment I could not tell whether I should hear the voice of God saying, "Depart from Me, ye cursed," or whether God would say, "Enter thou into the joy of the Lord." But the Gospel tells me that question is already settled: "There is now no condemnation to him that is in Christ Jesus." Listen to this verse: "Verily, verily;" and when you see that word "Verily, verily" in Scripture, you may know there is something very important coming. It means, "Mind what I tell you," or "truly, truly." "Truly, truly I say unto you, he that heareth My Word, and believeth on Him that sent Me, hath"—h-a-t-h, hath; lay hold of that little word hath to-night—"hath eternal life, and shall not come into condemnation;" that means into judgment—"but is passed from death unto life." Well, now, I am not coming into judgment for sin. The question has been settled, because Christ was judged for me, and died in my stead, and I go free. Is not that good news?

A PRAYER FOR MR. MOODY.

Why, I heard of a man praying the other day that I might lay hold of eternal life. I could not have said Amen to that. I laid hold of eternal life nineteen years ago, when I was converted. What is the gift of God, if it is not eternal life? And that is what God wants to give to everyone in this hall to-night, and it is the greatest gift that can be bestowed on anyone down here in this dark world. If an angel just came straight from the throne of God onto this platform, and proclaimed to this vast assembly that God had sent him here to offer to this audience any one thing they might ask, that each one should have his own petition granted—what would be the cry in this audience? There would be but one cry coming up from you, and the shout would make heaven ring: "Eternal life! eternal life!" Everything would float away into the dim past. There is not anything a man values more than his life. Let a man worth a million sterling be on a wrecked vessel, and if he could just save his life for six months by giving that million, he would give it in an instant. There is life without end. The gift of God is eternal life; and is it not one of the greatest marvels that men have to stand and plead, and pray men to take this gift? May God help you to take it now. Do not listen to Satan any longer. Reach out the hand of faith and take it now. Young man, "Believe on the Lord Jesus Christ, and thou shalt be saved." Trust Him to save you now, and then there will be no condemnation. Death will have lost his sting, the grave and its victory will be safe out of the way, and the judgment will be passed for you.

A JUDGMENT OF MERCY.

"Oh," but do you say, "what do you make out of that passage in Corinthians which says, 'Every man must give an account of the deeds done in his body'?". But that is a judgment of mercy, it is not a judgment of sin; that period is past to the believer. Oh, my friends, to-night I beg of you, do not go out of this hall unsaved. Believe the Gospel to-night. Lay hold of eternal life while God is offering it to you. Be reconciled to-night. Take your stand hard by the cross, and you are saved for time and eternity. I am told that at Rome, if you go up a few steps on your hands and knees, that is nine years out of purgatory. If you take one step now you are out of purgatory for time and eternity. You used to have two steps into glory—out of self into Christ, out of Christ into glory. But there is a shorter way now with only one step—out of self into glory, and you are saved. May God help you to take the step now! Flee, my friends, to-night to Calvary, and get under the shadow of the cross.

THE FIRE ON THE PRAIRIE.

Out in our western country in the autumn, when men go hunting, and there has not been for months any rain, sometimes the prairie grass catches fire, and there comes up a very strong wind, and the flames just roll along twenty feet high over that western desert, and go at the rate of thirty or forty miles an hour, consuming man and beast. When the frontiersmen see it coming, what do they do? They know they cannot run as fast as the fire can run. Not the fleetest horse can escape from that fire. They just take a match and light the grass around them and let the flames sweep, and then they get into the burnt district and stand safe. They hear the flames roar as they come along; they see death coming towards them; but they do not fear, they do not tremble, because the fire has peaced over the place where they are, and there is no danger. There is nothing for the fire to burn. There is one mountain peak that the wrath of God has swept over; that is Mount Calvary, and that fire spent its fury upon the bosom of the Son of God. Take your stand here by the cross, and you will be safe for time and eternity. Escape for your life, young man and young lady; flee to yon mountain, and you are saved this very minute. Oh, may God bring you to Calvary to-night, under the shadow of the cross to-night! Then let death and the grave come! You will shout, "Glory to God in the highest!" We will laugh at death and glory in the grave, and just know this: that we are safe, sheltered by the precious blood of the Lamb. There is no condemnation to him that is in Christ Jesus. God wants to pardon every one here to-night. God is coming down and beseeching you to take the pardon. Every man and woman here has broken the law, and he that has broken the least of the laws is guilty of all. I am sure I am not talking to one man or woman in this audience to-night who can say they have not broken the law.

A WORD THE DEVIL FEARS.

You have all sinned and come short of the glory of God, but God comes and says, "I will pardon you. Come, now, and let us reason together." "Now" is one of the words of the Bible the devil is afraid of. He says, "Do not be in a hurry; there is plenty of time; do not be good now." He knows the influence of that word "now." "To-morrow" is the devil's word. The Lord's word is "now." God says, "Come, now, and let us reason together. Though your sins are as scarlet, they shall be white as snow. Though they be red as crimson, I will make them as wool." Scarlet and crimson are two fast colors; you would not get the color out without destroying the garment. God says, "Though your sins are as scarlet and crimson, I will make them as wool and snow. I will do it." That is the way God reasons. He puts the pardon in the face of the sinner the first thing. That is a queer way of reasoning, but God's thoughts are not our thoughts; and so, my friends, to-night, if you want to be saved, the Lord says He will pardon you.

THE GOVERNOR IN THE CONDEMNED CELL.

A few years ago, when Pennsylvania had a Christian Governor, there was a young man down in one of the counties who was arrested for murder. He was brought before the Court, tried, found guilty, and sentenced to death. His friends thought there would be no trouble in getting a reprieve or pardon. Because the Governor was a Christian man they thought he would not sign the death warrant. But he signed it. They called on the Governor and begged of him to pardon the young man. But the Governor said, "No; the law must take its course, and the man must die." I think the mother of the young man called on the Governor and plead with him; but the Governor stood firm and said, "No; the man must die." A few days before the man was executed, the Governor took the train to the county where the man was imprisoned. He went to the sheriff of the county and said to him, "I wish you to take me to that man's cell, and leave me alone with him a little while; and do not tell him who I am till I am gone." The Governor went to the prison and talked to the young man about his soul, and told him that although he was condemned by man to be executed, God would have mercy upon him and save him, if he would accept pardon from God. He preached Christ, and told him how Christ came to seek and to save sinners; and, having explained as he best knew how the plan of salvation, he got down and prayed, and after praying he shook hands with him and bade him farewell. Some time after the sheriff passed by the condemned man's cell, and he called him to the door of the cell, and said, "Who was that man that talked and prayed with me so kindly?" The sheriff said, "That was Governor Pollock." The man turned deadly pale, and he threw up both his hands and said, "Was that Governor Pollock? was that kind-hearted man the Governor? Oh, sheriff, why did you not tell me? If I had known that was the Governor I would have fell at his feet and asked for pardon; I would have plead for pardon and for my life. Oh, sir, the Governor has been here, and I did not know it." Sinner, I have got good news to tell you. There is One greater than the Governor here to-night, and He wants to pardon every one. He does not want you to go out from here to-night condemned. He wants to bring you from under condemnation; to pardon every soul here. Will you have the pardon, or will you despise the gift of God? Will you despise the mercy of God and His offer of mercy? Oh, this night, while God is beseeching you to be reconciled, let me join with your praying mother, with your praying father, with your godly minister, with your Sabbath-school teacher, and all your praying friends; let me join my voice with theirs to plead with you to-night to be reconciled. Make up your mind now, while I am speaking, that you will not cross your threshold until you are reconciled, and there will be joy in heaven to-night over your decision. Oh, may God bring hundreds to a decision to-night. May Christians keep praying for this one thing. Let there be a united prayer to God now, that thousands may be reconciled to God to-night, and spend eternity in yon world of light.

An Englishman told me some time ago

A LITTLE STORY OF RECONCILIATION,

which illustrates this truth. We want to preach the Gospel of reconciliation; the good news that God is reconciled. God does not say He can do, but He has done it. You must accept what He has done. The story is this:—There was an Englishman who had an only son; and only sons are often petted, and humored, and ruined. This boy became very headstrong, and very often he and his father had trouble. One day they had a quarrel, and the father was very angry, and so was the son; and the father said he wished the boy would leave home and never come back. The boy said he would go, and would not come into his father's house again till he sent for him. The father said he would never send for him. Well, away went the boy. But when a father gives up a boy, a mother does not. You mothers will understand that, but the fathers may not. You know there is no love on earth so strong as a mother's love. A great many things may separate a man and his wife; a great many things may separate a father from a son; but there is nothing in the wide world that can ever separate a true mother from her child. To be sure, there are some mothers that have drunk so much liquor, that they have drunk up all their affection. But I am talking about a true mother; and she would not cast off her boy.

THE MOTHER AND THE MURDERER.

We had a case in our country of a young man who had committed murder. His father would have nothing to do with him, but his mother went down into his cell every day. When the trial came on, the papers tried to write him down, and seemed determined that the boy should be put to death. Because he was the son of a wealthy man, they thought the judges and the courts would have mercy upon him; and there was a hissing, as it were, going up from all America against that young man.

But that mother was not ashamed to be seen in the courts with him. She took her seat as near him as she could. She would have taken the boy's place, and laid down her life to have saved her boy. Look at a mother watching her sick child; she would take the disease out of the child into her own bosom if she could. When the boy was found guilty, no one seemed to feel the blow as that mother. A mother would, perhaps, not go to see that worthless boy executed; but if she could get that body she would cover it with kisses; she would go to the grave and cover it with flowers; she would cherish the memory of that boy as long as she lives. Why, a mother's love is stronger than death; death cannot tear down a mother's love. But, my friends, a mother's love is not anything to be compared with God's love. You never saw a mother that loved her child as God loves you sinners. God loves you thousands of times more than your mother. God loves you more than you love yourselves. He has His heart set upon you, and wants to save and bless you.

Well, the mother of the boy who had quarrelled with his father began to write and plead to the boy to write to his father first, and his father would forgive him; but the boy said, "I will never go home till father asks me." She plead to the father, but the father said, "No, I will never ask him."

THE MOTHER'S DYING WISH.

At last the mother came down to her sick bed, broken-hearted, and when she was given up by the physicians to die, the husband, anxious to gratify her last wish, wanted to know if there was not anything he could do for her before she died. The mother gave him a look; he well knew what it meant. Then she said, "Yes, there is one thing you can do. You can send for my boy. That is the only wish on earth you can gratify. If you do not pity him and love him when I am dead and gone, who will?" "Well," said the father, "I will send word to him that you want to see him." "No," she says, "you know he will not come for me. If ever I see him you must send for him." At last the father went to his office and wrote a dispatch in his own name, asking the boy to come home. As soon as he got the invitation from his father he started off to see his dying mother. When he opened the door to go in he found his mother dying and his father by the bedside. The father heard the door open, and saw the boy, but instead of going to meet him he went to another part of the room, and refused to speak to him. His mother seized his hand—how she had longed to press it! She kissed him, and then said, "Now, my son, just speak to your father. You speak first, and it will all be over." But the boy said, "No, mother, I will not speak to him until he speaks to me." She took her husband's hand in one hand and the boy's in the other, and spent her dying moments and strength in trying to bring about a reconciliation. Just as she was expiring, she could not speak, so she put the hand of the wayward boy into the hand of the father, and passed away. The boy looked at the mother, and the father at the wife, and at last the father's heart broke, and he opened his arms, and took that boy to his bosom, and by that body they were reconciled. Sinner, that is only a faint type, a poor illustration, because God is not angry with you. God gives you Christ, and I bring you to-night to the dead body of Christ. I ask you to look at the wounds in His hands and feet, and the wound in His side. My friends, gaze upon His five wounds. And I ask you, "Will you not be reconciled?" When He left heaven, He went clear down to the manger that He might get hold of the vilest sinner, and put the hand of the wayward prodigal into that of the Father, and He died that you and I might be reconciled. If you take my advice you will not go out of this hall to-night until you are reconciled. "Be ye reconciled." Oh, this Gospel of reconciliation! My friends, come home to-night. Your father wants you to come home to-night. Say as the prodigal did of old, "I will arise and go to my father," and there will be joy in heaven.

VIII.

THE MASTER'S PARTING COMMISSION.

"And He said unto them, Go ye into all the world, and preach the Gospel to every creature. He that believeth and is baptized shall be saved, but he that believeth not shall be damned,"—MARK xvi. 15, 16.

Go ye into all the world, and preach the Gospel to every creature. I wish you just to mark that text. It does not say, "Go ye into all the world, and preach the Gospel to the elect;" it does not say, "Go ye into all the world, and preach the Gospel to the rich," or to the learned, or to the unlearned; but "Go ye into all the world, and preach the Gospel to every creature." And I am one of those men that believe that God means what He says; that when God says, "Go and preach to every creature," He means that every man shall be invited to the Gospel feast, and that none need to be excluded, or that none need to stay away. And if a man does not come it will be because he is not willing to accept of the invitation. As Christ says, "Ye will not come unto Me that ye might have Me." It is not because men cannot come; it is because they will not come.

SATAN HINDERING.

There are a few boys who want to go out, disturbing the meeting, but if the friends will just be kind enough to give me their attention we will go back to the text. The devil does not want you to hear the text. That is just what gives life, the Word of God. The text is worth more than the sermon. Hear the proclamation, "Go ye into all the world, and preach the Gospel to every creature. He that believeth and is baptized shall be saved; but he that believeth not shall be damned." That is plain language, is it not? It is so plain, that there is not any one here need misunderstand it; and, as I said before, Christ means what He says. He sends out His messengers to proclaim the glad tidings. Gethsemane is behind, the empty grave is behind; Calvary, in all its horrors, is now past; He is on His way back home to take His seat at the right hand of the Father. His little church is gathered round Him—a little handful of men; and He breathes upon them the Holy Ghost; and now this is His parting commission, "Go ye into all the world, and preach the Gospel to every creature." I thank God for that text; I thank God that the commission is for us to proclaim it to every creature, and that every person in this wide, wide world is invited to the Gospel feast.

THE PROMISE FOR ALL.

Every one of God's proclamations are connected with that word "whosoever." I think it was Richard Baxter said he thanked God for that "whosoever." He would a good deal rather have that word "whosoever" than Richard

Baxter; for if it was Richard Baxter, he should have thought it was some other Richard Baxter who had lived and died before him; but "whosoever" he knew meant him. I heard of a woman once that thought there was no promise in the Bible for her; she thought the promises were for some one else, not for her. There are a good many of these people in the world. They think it is too good to be true that they can be saved for nothing. This woman one time got a letter, and when she opened it she found it was not for her at all; it was sent to another woman, or it was meant for another woman that had her name; and she had her eyes opened to the fact that if she should find some promise in the Bible directed to her, she would not know whether it meant her or some one else that bore her name. But you know the word "whosoever" means every one in this house; that boy down there, that grey-haired man, and that young man right in the blush of youth. "Go ye into all the world, and preach the Gospel to every creature." It does not leave out one. Go and proclaim the glad tidings to every man.

PARDON FOR THE PRISONERS: AN OHIO STORY.

I was in Ohio a few years ago, and was invited to preach in the State prison. Eleven hundred convicts were brought into the chapel, and all sat in front of me. After I had got through the preaching, the chaplain said to me: "Moody, I want to tell you of a scene which occurred in this room. A few years ago, our commissioners went to the Governor of the State, and got him to promise that he would pardon five men for good behavior. The Governor consented, with this understanding—that the record was to be kept in secret, and that at the end of six months the five men highest on the roll should receive a pardon, regardless of who or what they were; if they were there for life they should receive a pardon. At the end of six months the prisoners were all brought into the same chapel where I had been preaching; and the commissioners came up, and the president of the commissioners stood upon the platform, and put his hand in his pocket, and brought out some papers, and said, 'I hold in my hand pardons for five men.'" And the chaplain told me he never witnessed anything on earth like it. Every man was as still as death; many were deadly pale, and the suspense was something awful. The commissioner went on to tell them how they had got the pardon; but the chaplain said to the commissioner, "Before you make your speech, read out the names. This suspense is awful." So he read out the first name, "Reuben Johnson will come and get his pardon;" and he held it out, but no one came forward. He said to the Governor, "Are all the prisoners here?" The Governor told him they were all there. Then he said again, "Reuben Johnson will come and get his pardon. It is signed and sealed by the Governor. He is a free man." The chaplain told me he looked right down where Reuben was, and he was looking all round to see the fortunate man who had got his pardon. Finally the chaplain caught his eye, and he said, "Reuben, you are the man." Reuben turned round and looked behind him to see where Reuben was. The chaplain said the second time, "Reuben, you are the man," and the second time he looked round, thinking it must be some other Reuben. Now, men do not believe the Gospel is for them. They think it is too good, and pass it over their shoulders to the next man. But *you* are the man to-night. This boy, this grey-haired man, this reporter, and every creature are all invited. Well, the chaplain could see where Reuben was, and he had to say three times, "Reuben, come and get your pardon." At last the old man got up and came along down the hall, trembling from head to foot, and when he got the pardon he looked at it and went back to his seat and buried his face in his hands, and the prisoners heard him weep to think he was a free man. When the prisoners got into the ranks to go back to the cells Beuben got into the ranks too, and the chaplain had to call to him, "Reuben, get out of the ranks; you are a free man, you are no longer a prisoner." And Reuben stepped out of the ranks. That is the way men make out pardons. They make out pardons for good character or good behavior. But God makes out pardons for men that have not got any character, and who have been very, very bad. He has got a pardon for every sinner in London if he will take it. I do not care who he is or what he is like. He may be the greatest libertine that ever walked the streets of London, or the greatest blackguard that ever lived, or the greatest drunkard, or thief, or vagabond; but I come to-night with glad tidings, and preach the Gospel to every creature, "and whosoever will, let him take the water of life freely." Every man is invited.

A GLASGOW LADY ANSWERED.

A lady came to me in Glasgow, and said, "Mr. Moody, you are always saying, 'Take, take, take.' Is there any place in the Bible where it says, 'Take,' or is it only a word you use? I have been looking for it in the Bible, but cannot find it." I said, "It is almost the last word in the Bible. 'And the Spirit and the Bride say, Come. And let him that heareth say, Come. And let him that is athirst come. And whosoever will, let him *take* the water of life freely.'" God says, "Let him take." Who can stop him if God says "Take?" All the devils in hell cannot stop a poor soul from taking if God says "Take." All the powers on earth cannot hinder him. That little boy can come, and all the powers infernal and all the powers in the world cannot hinder him. God says to-night you may take the water of life freely. It is offered free to every one. Every one can be saved if they will.

ANECDOTE OF DR. WILLIAM ARNOT.

When the Rev. Mr. Arnot, that is now in Edinburgh, was pastor of a church in Glasgow, he heard of a woman that he knew being in trouble. She could not pay her debts and she could not pay her rent; so he went round to her house, thinking he would help her. He knocked at the door, and listened, and thought he heard some one inside; so he knocked again, but no one came. He knocked the third time very loud and listened, but did not hear any one; all was still. After waiting some time, he made a great noise, and at last left the house. Some few days after, he met the woman in the street, and he said to her, "I was round at your house the other day. I heard you were in trouble, and could not pay your rent, and I went to help you." The woman said, "Was that you? I was in the house all the time, but I thought it was the landlord come for the rent, and as I had not got the money, I kept the door locked." That woman represents a sinner. A sinner thinks God is coming to demand something. Instead of that, God comes to give and to bless. Christ comes to pay the debt. Christ comes to pay the rent. You all owe God a debt you cannot pay; and the Gospel is that Christ comes and offers to pay it for you. You had better pull back the bolt and let Him in to-night.

A DUBLIN DOOR AND THE SINNER'S HEART.

When we were in Dublin, I went out one morning to an early meeting, and I found the servants had not opened the front door. So I pulled back a bolt, but I could not get the door open. Then I turned a key, but the door would not open. Then I found there was another bolt at the top, then I found there was another bolt at the bottom. Still the door would not open. Then I found there was a bar, and then I found a night-lock. I found there were five or six different fastenings. I am afraid that door represents every sinner's heart. The door of his heart is double-locked, double-bolted, and double-barred. Oh, my friends, pull back the bolts to-night, and let the King of glory in! He wants to bless you; He wants to pay the debt; He wants to cancel the debt; He wants you to be reconciled; He wants you to be saved. He does not wish the death of any, but that all may turn unto Him and live. What said the angels to those shepherds on the plains of Bethlehem? "Behold, I bring unto you good tidings of great joy, which shall be to all people. For unto you is born this day, in the city of David, a Saviour." Now, I contend that men can hear no better news than that—that a Saviour has been given, and that God wants to save men; not that men shall be lost, not that men shall perish, but that a Saviour has been given to save us from our sins. Christ did not come into the world to condemn the world, but that the world through Him might be saved. Look at Him going back to Nazareth; what did He do when He turned into the synagogue one Sabbath? He opened the book at the place where it is written, "The Spirit of the Lord is upon Me; because He hath anointed Me to preach the Gospel to the poor; He hath sent Me to heal the broken-hearted." My friends, think of the broken hearts in London; and Christ says He is come to heal the broken-hearted. "He hath sent Me to heal the broken-hearted, to proclaim liberty to the captive." Think, you poor drunkards in London, slaves to the infernal cup, slaves to strong drink. I bring you good news to-night. The Son of God can set your soul free, and can make you free men. He says: "He hath sent Me to proclaim liberty to the captive, sight to the blind, liberty to them that are bruised, and to proclaim the acceptable year of the Lord." Is not that good news? Christ was anointed for that purpose. God sent Him to proclaim the glad tidings. I would to God that every man in this vast assembly would believe the Gospel and be saved! Oh, that you would just receive the Lord Jesus as your way, your truth, and your life. All you have to do is just to take Him.

ALL THE SINNER HAS TO DO.

This afternoon in the inquiry-room there were a great many that came up to inquire what they must do to be saved. A young lady among the number said to me, "Mr. Moody, I want to be saved. I wish you would tell me how." The tears trickled down her cheeks, and she said, "You do not know how I want to be saved!" I said, "My friend, you would know how to take a gift, would not you? If I offered you my Bible, you would know how to take it, would not you?" "Yes, sir," she said, "I should." "Salvation is a gift, and just as you would take a present, you take God's present. And God's present to you is His Son from heaven. You just receive Him." She said, "Mr. Moody, is that all I have got to do?" I said, "Yes, that is all you can do. You receive Him first." "But," said she, "have not I to ask for Him?" I said, "You need not do it. What is the use of asking for what God is offering?" Suppose I say to this boy here, "Look here, I want to give you my Bible," and the boy says, "I wish you would make me a present of the Bible. Will you give it me?" And I say, "Take it, take it," and he keeps asking for it. Now, God is behind every sinner offering salvation. You have nothing to do but to take it. Who will take salvation as a gift to-night?

HELPING HIMSELF IN THE ORCHARD.

I was out on the Pacific coast in California, two or three years ago, and I was the guest of a man that had a large vineyard and a large orchard. One day he said to me, "Moody, while you are my guest, I want you to make yourself perfectly happy, and if there is anything in the orchard or in the vineyard you would like, help yourself." Well, when I wanted an orange, I did not go to an orange-tree and pray the oranges to fall into my pocket, but I walked up to a tree, reached out my hand, and took the oranges. He said, "Take," and I took. God says "Take," and you do it. God says, "There is my Son." "The wages of sin is death; the gift of God is eternal life." Who will take it now?

A DEVICE OF SATAN.

Satan is down in the audience working while I am preaching. Satan says, "If you take it, you will have to give up too much. Do not you let that man get a power over you to-night. Do not believe that man. If you become a Christian, you have got to give up so much." Let me say—mark the words—God does not come here and ask any man to give up anything. The first thing God wants you to do is to take; and after you have taken the new life, and got a new nature, old things pass away, and all things become new. I tried to stop swearing before I was converted, and the more I tried the worse I became. But one night, when Jesus met me, I just received Him, and I have had no desire to swear since. It stopped itself—I got something better. The things I once loved I now hate; and the things I once hated I now love. There was a perfect change, a revolution in my life, when God revealed Himself to me; and since then His yoke is easy and His burden is light. God does not come down and say, "Young man, give up this and that;" but he says, "There is my Son; take Him." There is the gift, and I tell you that there is not anything that God can give us that is worth more than the gift of eternal life. It you were allowed to choose yourself, you would ask for eternal life. You would rather have that gift than any other; and that is the gift that God wants to bestow upon you. God says, "Here it is all in my Son. If you receive Him here, he will 'receive' you yonder. If you reject Him here He will reject you yonder." He came unto His own, the Jews, and they would not have Him. "His own received Him not; but as many as received Him, to them gave He power to become the sons of God, even to them that believe on His name." Now, the moment you receive Christ, you get power to serve Him; the moment you receive the Lord Jesus, you get power to live for Him.

DOMESTIC ANECDOTE.

My wife had a schoolmate that had a little boy about four years old, and this beautiful little boy was one day cutting a piece of string with a penknife, and the knife went into his eye and put it out. My wife was therefore very careful about the children not using a knife. But if you tell a child he shall not have a thing, that is the very thing he wants. A good many people say they would like to have had Adam's chance. If they had they would have gone down like Adam. If you put a thousand children into this building with a great number of toys, and put one little thing in a room and shut it up, and if you said to the children, "I shall be gone a few hours; do not go near that room," that is the very first place they would go to. They would want to see what was in there. If you tell a child he shall not have a thing, that is the very thing he wants. My wife went out one day, and my little boy, two years old, got hold of a pair of scissors. My little girl knew he ought not to have them, and she went to him and tried to get them away; but the little fellow held on to the scissors, and would not give them up. She was afraid of sticking them into his eyes, so she ran off to another room, and got an orange, and came running in, and held it up, and said, "Willie, do not you want the orange?" and the little fellow dropped the scissors, and went for the orange. If you will allow me the illustration, God comes here, and says, "Here is my Son, take Him." He saves the sinner; and the moment we get Him, these things we love so much are gone; they float away into the dim past. Christ is worth more than all the world; and God comes and says, "Here is my Son, take Him, and believe on Him." And the moment you receive Him, you get power over the flesh, the world, and the devil; and you do not get the power until you receive life from Christ, until you believe on the Lord Jesus Christ. May God help you to believe now, and to receive the Gospel to-night! "Go ye into all the world and preach the Gospel to every creature." May every man and every woman in this room to-night believe the Gospel and be saved!

THE RICH EVANGELIST AND THE PEOPLE'S DEBTS.

I will give you another illustration, for illustrations are better than dry sermons. I heard of an Englishman that was converted some time ago, and when the Lord converted him, he had a great desire to see every man converted; and I would not give much for that man's conversion who did not have that desire. This man Christ had such a hold upon, that he wanted to go out and publish the good tidings. So he went into one town, and gave notice that he would preach in such a place. It got noised round that the man was rich, so a great many went to see him out of curiosity. He had a great audience the first night, but, as he was not a very eloquent man, the people did not get interested. Men looked at the messenger instead of the message; but never mind the messenger. The next night hardly any one was there. Then he got out great placards, and placarded the town, and he stated that if any man in that town owed any debt, if they would come round to his office between nine and twelve o'clock on a certain day, he would pay the debt. Of course that went through the town like wild-fire. One said to the other, "John, do you believe that?" "No, I am not going to believe that any stranger is going to pay our debts." Not any one believed it, although there were a good many, no doubt, that would have liked to get their debts paid. Well, the day came, and at nine o'clock the man was there. At ten o'clock none had come. At eleven o'clock a man was seen walking up and down, looking over his shoulder, and finally he stuck his head in the door and said, "Is it true that you will pay any man's debt?" "Yes; do you owe any debt?" "Yes." "Have you brought the necessary papers?" The placard had told them what to do. "Yes." So the man drew a cheque and paid the other's debt, and he kept him and talked with him till twelve o'clock; and before twelve o'clock two other men came and got their debts paid. At twelve o'clock that man let them out, and the people outside said to them, "He paid your debts, did not he?" "Yes, he did," they answered. But the people laughed and made fun of them, and would not believe it till they pulled out the cheque, and said, "There it is. He has paid all the debt." And then the people said, "What fools we were we did not go in and get our debts paid!" But they could not; it was too late; the door was closed; the time was up. And then the man as before preached the Gospel, and great crowds went to hear him; and he said, "Now, my friends, that is what God wants to do, but you will not let Him do it. Christ came to pay our debts, and that is the Gospel." I could not have a better illustration of the Gospel than that. Every man owes God a debt he cannot pay. Would you insult the Almighty by offering the fruits of this frail body to atone for sin? Isaiah says, "He was wounded for our transgressions; He was bruised for our iniquity; the chastisement of our peace was upon Him; and with His stripes we are healed." Paul says, "I declare unto you the Gospel; Christ died for our sins, according to the Scriptures." My friends, will you believe the Gospel to-night, and be saved?

CHRIST'S COMMISSION TO PETER.

I can imagine when Christ said to the little band around Him, "Go ye into all the world and preach the Gospel," Peter said, "Lord, do you really mean that we are to go back to Jerusalem and preach the Gospel to those men that murdered you?" "Yes," said Christ to Peter; "go, hunt up that man that spit in My face, and tell him he shall have a seat in My kingdom if he will accept of salvation as a gift. Yes, Peter, go, hunt up that man that made that cruel crown of thorns and placed it on My brow, and tell him I will have a crown ready for him when he comes into My kingdom, and no thorns in it. I will give him a crown of life. Hunt up that man that took a reed and brought it down over the cruel thorns, driving them into My brow, and tell him I will put a sceptre in his hand, and he shall rule over the nations of the earth if he will accept salvation. Hunt up that man that spit in my face, and tell him I forgive him freely, and will have a crown ready for him if he will accept of salvation. Peter, go hunt up that man that drove the spear into my side, and tell him there is a nearer way to my heart than that. Tell him I forgive him freely, and that he can be saved if he will accept of salvation as a gift. Hunt up the men that drove the nails into My hands and feet, and tell them I forgive them freely, and tell them they shall have a seat in My kingdom if they will accept of it. Go ye into all the world and preach the Gospel to every creature." Oh, may God help you to hear the Gospel to-night and to be saved! Christ died for our sins. Think of the sins represented by this vast body of men. But, thanks be to God, they can all be laid on His Son to-night if you will lay them on Him. He came to take away the sin of the world. Look yonder! see what it says! "The blood of Jesus Christ, His Son, cleanseth us from all sin." Look yonder! "Behold the Lamb of God which taketh away the sins of the world." May God help you to lift your eye to the Lamb of God to-night! Look, sinner, now! "Behold the Lamb of God which taketh away the sin of the world;" and if you go out of this world unsaved, it will be your fault. If you go out from here to-night without Christ as your Saviour, it will be your fault: you will do it at the peril of your soul. May God help you to look now and live.

IX.

POPULAR PRESENT-DAY EXCUSES.

TO-NIGHT I am going to call your attention to the same subject as last night, when we took up some of the popular excuses of the present day. We had time only to speak of a few, and to-night we want to follow up the same subject. Our friend has been singing about heaven, the home of the soul, and I read to you a few verses in Revelations about that upper and better world. And now to-night I want every one in this audience to believe that they really have an invitation to that world of light. It is God that is inviting every soul within this assembly to that feast. It is not an invitation of mine, it is not a text that I have manufactured, it is not an invitation that is got up by man, but it comes from the living God Himself to every soul here. Every person here is invited to the feast, and now the question comes, "What are you going to do with the invitation?"

THE YOUNG MAN FROM BRADFORD.

I was made glad to-day to hear of a young man that came to this meeting last night. He came up from Bradford, and as he came into this hall, he said, "If Christ can be found here, I am determined to have Him;" and the moment there was an opportunity given to go into the inquiry-room, that young man went in, and after a friend had talked with him some time, to all human appearance he accepted Christ, and went on his way rejoicing. I hope there will be many such here to-night, who have said to themselves, "If Christ can be found here to-night, by the grace of God, I'll find Him." If there are any such, let me say to you, "My friends, I bring an invitation to each one of you to be present at the marriage supper of God's beloved Son."

And now, are you going to join with the three men that we were speaking of last night, and say, "I pray thee, have me excused?" Are you going to make excuse?

THE SCEPTICS AND INFIDELS OF LONDON.

I want to come to some of the excuses that we meet with every night in the inquiry-room, and the excuse I have met in London, more perhaps than any other—for I have found more sceptics and more infidels the few days I have been in London than in any place I was ever in; young men coming into the inquiry-room full of infidelity, darkness, and doubt, and one of the greatest objections they have; one of the excuses that they are hiding behind, is the Bible. They are giving that as the reason why they do not accept the invitation to be at the marriage supper of the Lamb. Now, I want to say I never met a sceptic or infidel who had read the Bible through. I heard a man say the other day to another man, "Have you read such a book?" "Yes." "Well, what is your opinion of it?" "Well, I only read it through once; I would not like to give my opinion without reading it more carefully." But men can give their opinion about God's Book without reading it. They read a chapter here and there, and say, "Oh, the Book is so dark and mysterious;" and because they cannot understand it by reading a few chapters, they condemn the whole of it. The Word of God tells us plainly that the natural man cannot understand spiritual things. It is a spiritual book, and speaks of spiritual things, and a man must be born of the Spirit before he can understand the Bible. What seems very dark and mysterious to you now will all be light and clear when ye are born of the Spirit.

THE MYSTERIES OF THE BIBLE.

I can remember some portions of Scripture that were very dark and mysterious to me when I was converted, but now they are very clear. I can remember things that ten years ago were very dark and mysterious, but as I have gone on I understand them better, and the more we know of God, and the more we study the Word, the plainer it will become; but the idea of an unconverted man is to take up the Bible and condemn it before he has been born of the Spirit. Why, when a man is born of the Spirit then he will understand the Word of God, and not before. You say, "If that is so, how am I to understand how to be saved?" I will tell you. When God puts salvation before a sinner He puts it so plain that a man that runs can read, and a wayfaring man, though a fool, need not err therein. There are a great many things in the Book which are dark and mysterious, but when it comes to the plan of salvation God has put it so plain that that little girl ten years old can understand it if she will. You understand what it is to come. "Come unto Me, all ye that labor." You know what it is to take a gift. "He came unto His own, and His own received Him not. But as many as received Him, to them gave He power to become the sons of God." "The wages of sin is death, but the gift of God is eternal life." That is taking a gift. You know what it is to believe in a man; well, "believe in the Lord Jesus Christ, and thou shalt be saved." You know what it is to put trust and confidence in a man; now, put your trust and confidence in the living God, and you are saved. You are saved by casting yourself unreservedly upon the Lord Jesus Christ. When God puts salvation before a man He puts it so plain and simple that if he is willing to come as a little child he can come.

THE CHILD AT SCHOOL.

Supposing I should send my little boy, five years old, to school to-morrow morning, and when he came home I should say, "Can you read, write, spell? Do you understand all about arithmetic, geometry, algebra?" The little fellow would look at me, and say, "Why, what do you talk in that way for? I have been trying all day to learn the A B C." Supposing I replied, "If you have not finished your education you need not go to the school any more;" why, what would you say? You would say, "Moody has gone mad." Well, there is about as much sense in that as in the way that infidels talk about the Bible. They take it up, read a chapter, and say, "Oh, it is so dark and mysterious we cannot understand it." This blessed Book is given to be a lamp to our feet and a light to our path, to guide the way to those eternal mansions. It never was given to keep men out of the kingdom of God. That is the devil's work, trying to make you believe the Word of God is not true. I tell you the only way we can overcome the enemy of our soul is by the written Word of God, and the devil knows that, and so he comes up and says, "It is full of lies, it is dark and mysterious, it contradicts itself; don't you believe it." He knows the moment a man goes to the Word of God and believes it, he gets liberty to his soul, and he gets beyond Satan's reach; he gets a weapon in his hand with which to conquer the devil; he overcomes the enemy of his salvation. The devil does not want you to find that out, and whispers this lie, and you believe it rather than the Word of God. Young man, your mother is right, the Bible is true, and you had better take that.

WHAT ENGLAND OWES TO THE BIBLE.

Why, these infidels that want to take away the Bible from us, what are they going to give us in its place? What has made England but the Word of God? I heard a most eloquent man in America a few years ago say, "You look back in history a few years and you see England and France moving along abreast in the march of nations. France closed the Bible and would not give it to its people. England opened the Bible, and what is the result? Why, the English language is spoken round the world, and the sun never sets upon the Queen's dominions." And look and see how the English language has gone round the world. See what the Bible has done for England, and look and see what has become of France. Poor France closed its Bible, and it has gone down, and every nation that puts down the Bible has to go down, and every nation that exalts the Bible and lifts it up, God lifts it up and blesses them. Oh, my friends, let us cling close to the Bible. What are you going to do without it? What are you going to give us in the place of it? Do not give that for an excuse. Keep this in mind: you will never stand up before the bar of God and say the Bible kept you out of the kingdom. It may sound very well here now; you may be satisfied to give that for an excuse down here in the Agricultural Hall to-night; you will not be satisfied to give that in the Courts of Heaven—in fact, you never will get there; you will not stand up in the great judgment day and say the Bible kept you out of the kingdom.

HYPOCRITES IN THE CHURCH.

Then there is another class. Some people say, "I have not any doubt about the Word of God, but the fact is that there are some men in the church that ought not to be there; therefore, I do not purpose to go into the church." I am not asking you to come into the church—not but what I believe in churches; but I am asking you to the marriage supper of the Lamb, and am inviting you to this feast, and we will talk about the church by-and-by. We want you to come to Christ first, then we will talk to you about the church. But you say, here are some hypocrites. So there are, and I can imagine you saying, "Oh, yes, there is a man up here in one of the churches that cheated me out of £5 a few years ago, and you are not going to catch me in the company of hypocrites." Well, my friend, if you want to get out of the company of hypocrites, you had better get out of the world as quick as you can. One of the twelve apostles turned out to be a hypocrite, and there is no doubt there will be hypocrites in the church to the end of time. But "what is that to thee?" says Christ to Peter; "follow thou Me." We do not ask you to follow hypocrites, we ask you to follow Christ; we do not ask you to believe in hypocrites, we ask you to believe in Christ. Another thing, if you want to get out of the company of hypocrites you had better make haste and come to Christ. There will be no hypocrites at the marriage supper of the Lamb; they will all be in hell, and you will be there with them if you do not make haste and come to Christ. That excuse would sound strange, would it not? We very often hear men give it down here, but it would sound very strange before Jehovah, a man saying, "I know you invited me to be at the marriage supper of your Son, but I did not accept it because I knew there were some hypocrites that professed the Gospel." Man will have no excuse when he comes to stand before God; his mouth will be sealed.

THE PRESSURE OF BUSINESS.

There is another class who say, "I know there are hypocrites, but they don't have any influence over me," and if I could go to the door as you go out to-night, and take you by the hand and say, "My friend, why not accept of the invitation to-night?" you would say, "I pray to be excused to-night,

I have not time. I have got some very pressing business to-morrow morning to attend to, and I have to go home to bed as quick as possible, to get my night's rest. You will have to excuse me;" and the mothers here would say, "I have to go home and put the children to bed, you will have to excuse me.;" "very pressing business;" "have no time." Thousands of men in London say they have not time. Thanks be to God, it don't take time, it takes decision. But what have you done with all the time God has given you? Your locks are turning grey, your eye is growing dim, and that temple of your body is coming down — what have you done with all those years? Is it true you have not time? What did you do with the 365 days last year? No time during those 365 days — what have you done with all those hours?· Have not you had. time to accept of this invitation? Why, men spend 15 or 20 years to get an education that they may go out to earn a living for this frail body, that is soon to be eaten up with worms, or 5 years to learn a trade that they may earn a living; and yet they have not five minutes to seek their souls' salvation! You "have no time." Is it true? You know it is a lie, and if you go out to-night unsaved it will not be because you have not time, but because you won't accept the invitation. God says, "Seek first the kingdom." That is the first thing to do. Let the children sit up a little late to-night, let your business be suspended to-morrow. Supposing you do not get so much money to-morrow and get Christ, is not that worth more than money? Better for a man to be sure of salvation than to have the wealth of the world rolled to his feet. If you take my advice you will

JUST TAKE TIME TO-NIGHT,

and just make up your mind — this night the question of eternity must be settled.

But there is another excuse coming up from some one in the gallery. A man says, "My heart is so hard." Well, that is just the very reason you ought to come. If you had not a hard heart you would not need a Saviour. Can you soften your heart? Can you break your heart? Did not God invite the hard-hearted? Did not Christ come to seek and to save that which was lost? It is just because men's hearts are hard that they need a Saviour, and that is no excuse at all. God invites you, and you won't stand up and say to the Great King you did not accept the invitation because you had a hard heart. He invites "whosoever," and you can come along with your hard heart.

CHRIST BREAKING THE CHAIN.

In the north there was a minister talking to a man in the inquiry-room. He says, "My heart is so hard, it seems as if it was chained, and I cannot come." "Ah," says the minister, "come along, chain and all;" and he just came to Christ hard-hearted, chain and all, and Christ snapped the fetters, and set him free right there. So come along. If you are bound hand and feet by Satan, that is the work of God to break the fetters; you cannot break them. Thanks be to God, He can snap the fetters and set the captive souls free to-night. I do not care how hard the heart is: the Lord can save to the uttermost, and He bids you come just as you are. Oh, this old excuse—

"I AM SO BAD!"

Paul said he was the "chief" of sinners, and if the chief has gone up on high there is hope for everybody else. The devil makes us believe that we are good enough without salvation if he can; and if he cannot make us believe that, he says, "You are so bad the Lord won't have you;" and so he tries to make people believe because they are so bad Christ won't have anything to do with them. God invites you to come just as you are. I know a great many people want to come, but they are trying to get better and to get ready to come. Now mark you, my friend, the Lord invites you to come just as you are, and if you could make yourself better you would not be any more acceptable to God. Do not put these filthy rags of self-righteousness about you. God will strip every rag from you when you come to Him, and He will clothe you with glorious garments. When our war was going on we would sometimes go to the recruiting office and see a man come in with a silk hat, broadcloth coat, calfskin boots—his suit might be worth £100; and another man would come in whose clothes were not worth a pound; but they both had to strip and put on the uniform of the country. And so when we go into Christ's vineyard we must put on the livery of heaven and be stripped of every rag.

However bad you are, come along just as you are and the Lord will receive you.

THE ARTIST AND THE BEGGAR.

I read some time ago of an artist who wanted to find a man that would represent the prodigal. One day, walking up the streets, he met a poor beggar, and the thought occurred to him, "That man would represent the prodigal." He told him what he wanted, and found the beggar was ready to come to his place of business and sit for his painting if he would pay him for his time. The man appeared on the day appointed, but the artist did not recognize him. He said, "You made an appointment with me." "No," says the artist, "I never saw you before." "You are mistaken; you did see me, and made an appointment with me." "No, it must be some other artist. I have an appointment to meet a beggar here at this hour." "Well," says the beggar, "I am the man." "You the man?" "Yes." "What have you been doing?" "Well, I thought I would get a new suit of clothes before I got painted." "Well," says the artist, "I don't want you;" he would not have him then. And so if you are coming to God, come just as you are. Do not go and put on some garments of yours, and think the Lord will accept you because you have some good thoughts and desires. Come along just as you are. I do not care how bad you are; this Man receiveth sinners and eateth with them, and all you have to do is to prove that you are a sinner, and I wil prove to you that you have a Saviour, and the greater the sinner the more need of a Saviour.

PREJUDICE AGAINST SPECIAL SERVICES AND LAY PREACHERS.

Some say, "I would like to become a Christian, but I have a prejudice against these special meetings, and against Americans, and against a layman too. If it was a regular minister, and it was our regular minister, I would accept the invitation." If that is your difficulty, I can help you out of that. You can just get right up, and go out of the hall, and run right over to your minister, and have a talk with him. And if you say you do not want to be converted in a special meeting, there are regular meetings in all the churches throughout London, and your minister would be most glad to see and talk and pray with you. But if you say, "There is a great awakening here in London," and you do not want to be converted in that way, you can jump into a train, and go to some town where there is no revival. We can find you some place where there is no revival, and some church where there is not much of the revival spirit. If you really want to go, don't give that for an excuse. How wise the devil is! When the church is cold, and everything is dead, men say, "Oh, well, if there was only some life in the church I might become a Christian, if we could only just have a wave from heaven." Then when the wave does come they say, "Oh, no, we are afraid of excitement, and afraid of these special meetings. We are afraid there will be something done that won't be just in accordance with our ideas of propriety." My friend, it is God who is working. Come along just as you are. Do not wait another minute, but accept the invitation and accept it right here to-night.

A WORD THAT SHOULD BE ABOLISHED.

There is another class here who say, "I would like to come, but then I do not feel." That is, I think, the very worst excuse, and the most common excuse we have. I wish sometimes the word could be abolished—feel, feel. You go into the inquiry-room. "Well, Mr. Moody, I do not feel this and that." Why, supposing my friend Mr. Stone should invite me to go to his house to-morrow to dinner, and I say to Mr. Stone, "I would like to go very much, but I don't know as I feel right." "Well," he says, "what do you mean? Do you mean you don't want to go to my house?" "Oh, no, I want to go." That is what men say —"Oh, yes, we want to be saved." "What do you mean, Mr. Moody? Do you mean that you do not know as you will be well to-morrow? Do you think you will be sick?" "Oh, no, I expect to be well to-morrow if I live." "Well, what do you mean by feeling?" "Well, I do not know just how I'll feel. I would like very much to go to your house to dinner to-morrow, but I don't know as I will feel just right." "I don't understand you, Mr. Moody; I am not talking about feeling; I invite you to come to my house to dinner." "Well, I would like to come very much, but the fact is I do not know how I will feel to-morrow." I can imagine my friend Mr. Stone saying, "What has come over Moody? I think the fellow has gone mad. I asked him to my house to dinner, and he says he would like to come, but he does not know as he will feel right, and he talked about feeling all the time." Of course you would say he has gone mad. That is the way people talk now. You talk to them about coming to the kingdom of God, and they say, "I do not know as I feel just right."

AWAY WITH YOUR FEELINGS.

God is above feeling. Why, can you control your feelings? If I could I would feel good all the time—never catch me feeling bad at anything. I am sure if I could control my feelings I never would have any bad feelings; I would always have good feelings. Bear in mind Satan may change our feelings fifty times a day, but he cannot change the Word of God; and what we want is to build our hopes of heaven upon the Word of God. When a poor sinner is coming up out of the pit, and just ready to get his feet upon the Rock of Ages, the devil sticks out a plank of feeling, and says, "Get on that," and when he puts his feet on that, down he goes again. Take one of these texts—"Verily, I say unto you, he that heareth My word and believeth on Him that sent Me hath everlasting life, and shall not come into condemnation, but is passed from death unto life." My friend, that is worth more than all the feelings that you can have in a lifetime. I would a thousand times rather stand on that verse than on the best frame and feeling. I took my stand there twenty years ago. The dark waves of hell have come dashing up against me; the waves of persecution have dashed up around me; doubts, fears, and unbelief have assailed me; but I have been able to stand right there. It is a sure footing for eternity. It was true 1,800 years ago, and it is true to-night. That Rock is higher than my feeling. What we want is to get our feet upon the Rock, and then the Lord will put a new song into our mouths.

NOT A MISFORTUNE, BUT A SIN.

There is another class who say they cannot believe. Not long ago, a man said to me, "I cannot believe." I said "Who?" "Well, I cannot believe." I said, "Who?" He stammered and stuttered, and I said, "Who cannot you believe—God?" "Oh, yes, I believe God, I cannot believe myself." "Well, you do not want to believe yourself. Your heart is deceitful above all things, and desperately wicked. Put no confidence in the flesh. Don't believe yourself, make yourself a liar, and God to be true. Believe in God, and say as Job said, "Though He slay me I will trust Him." Some men seem to talk as if it was a great misfortune that they do not believe. Bear in mind it is the damning sin of the world. "When He, the Holy Ghost, is come, He will reprove the world of sin, and of righteousness, and of judgment; of sin, because they believe not on Me." That is the sin of the world — "because they believe not on Me." Why, that is the very root of sin, the very tree, and all the fruit. This is the tree that brings forth this bad fruit — it is the tree of unbelief. May God open your eyes to-night to see that God is true, and that you may be led to put your trust in Him now.

A PACK OF LIES.

I wish I had time to go on with these excuses, for they are as numerous as the hairs on our heads. But if I could go on and exhaust them all, the devil would help to make more. You can just take them, tie them up in one bundle, and mark them a pack of lies, the whole of them. Not one of them is true. And let me say, if your excuse is a good one, if you have an excuse that will stand the light of eternity, do not give it up for anything I have said. Hold it firm, take it to the bar of God, and tell it out to Him. But if you have an excuse that won't stand the piercing eye of God, I beg of you as a friend, give it up—let your excuses go. Let them go to the four winds of heaven, and accept of the invitation now. It is a very easy thing for a man to excuse himself into hell, but you cannot excuse yourself out.

And another very solemn thought is, God will excuse you if you want to be excused. He does not want to do it. "As I live, saith the Lord, I have no pleasure in the death of the wicked; but that the wicked turn from his way and live. Turn ye, turn ye from your evil ways; for why will ye die, oh house of Israel?" God wants you to come to His feast. Come just as you are; accept the invitation. Let the shop be closed till you accept this invitation. Let business be suspended till you accept this invitation. Let the oxen stand in the stall till you accept of this invitation. Let everything else be laid aside until the great question of eternity is asked, until you can look up and say, "God is my Father, Jesus Christ is my Saviour, and heaven shall be my future home."

I wish I had time to call your attention to who will be at the marriage supper of the Lamb.

LIFT YOUR EYES HEAVENWARD

to-night, mothers; you have got loved children that have gone on before you, and they will be at the marriage supper of the Lamb, they will sit down with Abraham, Isaac and Jacob, in the kingdom of God — will you be missing? Fathers and mothers that have loved ones that have gone on before you, if you could hear them—they are shouting from the battlements of heaven, "Come this way." Young man, you have a sainted mother there, a loved father there: they are beckoning you heavenward to-night. They have been gathering from the time the holy Abel went up — for 6,000 years they have been gathering out of the four corners of the earth. The purest and best of earth are not down here, they are in heaven, and God wants you and I to be there. Blessed is he that shall be at the marriage supper of the Lamb. Oh, by the grace of God I mean to be there. My friends, let us to-night every one accept of the invitation. God invites rich and poor, high and low, learned and unlearned, all alike to come to the feast. Do not make light of the invitation.

THE REPLY TO THE ROYAL INVITATION.

Suppose we should just write out the excuse to the King of Heaven: "While sitting in the Agricultural Hall, March 24, 1875, I received a very pressing invitation from one of your messengers to be present at the marriage supper of your only-begotten Son. I pray Thee, have me excused." Would you come up and sign that? Would you take your pen and put your name down to that excuse? I can imagine you saying you would let your right hand forget its cunning, and your tongue cleave to the roof of your mouth first. I doubt whether there is a man in this room that could be made to sign this excuse; but what will you do? Many of you will get up and go out of this hall, making light of the preacher, laughing at everything you have heard, paying no attention to the invitation. I beg of you, do not make light of this invitation. It is a loving God that invites you to a loving feast, and God is not to be mocked. Go play with the forked lightning, trifle with any pestilence, any disease, rather than with God. God is not to be trifled with. It is God that invites you. Young lady, what will you do with the invitation to-night? Young man, what will you do with the invitation to-night? Will you accept of it? Oh, may God help you now to say from the very depths of your heart, "By the grace of God I will accept."

Just let me write out another. "To the King of Heaven: While sitting in the Agricultural Hall, March 24, 1875, I received a pressing invitation from one of your servants to be present at the marriage supper of your only-begotten Son. I hasten to reply, By the grace of God I will be present." Who will sign that? Who will set to their seal to-night that God is true? Be wise to-night, and accept of the invitation. Make up your mind you will not go away till the question of eternity is settled. May God bring hundreds to a decision to-night is the prayer of my heart.

Few of Mr. Moody's addresses have excited more interest than the two which follow. They were first reported in full in the "Christian World" in April last. There is no better evidence of the power of these themes over the people than is found in the fact that wherever Mr. Moody goes, these addresses always command the deepest attention of his great audiences.

X.

A SERMON ABOUT HEAVEN.

I HAVE for my subject to-night, heaven. I was going to a meeting some time ago, and a friend said to me on my way, "What is your subject?" I told him I thought I should talk about heaven. I noticed a scowl on his forehead, and said, "What makes you look in that way?" He said he was in hopes I was going to give them something practical, that there would be time enough to talk about heaven when we got there. But there is a passage in Timothy which says that "all Scripture is given by inspiration of God, and is profitable for doctrine," and if God did not want us to talk about and think about heaven He would not have so much written about it. And I think if people talked more about heaven they would have more of a desire to go there. When we were compiling this little hymn-book I wanted to put in two or three more hymns about heaven. My friend said, "I think you have too many about heaven." I don't know, I may be wrong, but I cannot help but like those hymns wonderfully. "That beautiful land on high,"—I have heard it the last ten or twelve years very often, and I have not got tired of it yet. I love to hear those sweet hymns about heaven, for it seems to me we cannot hear too much about heaven. If you were going to America to live and spend the rest of your days, and it was given out I was going to talk about America here to-night, I can imagine how anxious you would be to listen to all I said about that country, about its climate, and about its inhabitants. You could not hear too much about a country which you were going to, to live a few years even, because our life here is but a vapor compared with that life beyond this. Well now, if we are going to spend eternity in heaven, can we hear too much about it? I think not.

THE INFIDEL'S QUERY.

I remember soon after I was converted an infidel got hold of me and wanted to know why it was I always addressed my prayer upwards. He said God was everywhere, He was no more above me in heaven, as I called it, than He was here; He was the God of nature. And so I find infidels and sceptics, they reason away hell, they reason away heaven, and they would even reason away God. Now I will admit that God is here, the same as we say the sun has been shining in London to-day, but it is 95,000,000 of miles away, and so God may be here to-night, but at the same time God is a Person. God has a dwelling-place, and it is right that we should address our prayers upward. I think it is in the 26th of Deuteronomy we read, "Look down from Thy holy habitation from heaven, and bless Thy people Israel and the land which Thou has given us, as thou swarest unto our fathers, a land that floweth with milk and honey." And in Genesis we read that God "went up" from talking with Abraham. In the 3d of John we read Christ said He "came down from heaven." And then we find that when He was here on earth, in one place it is said He looked up towards heaven; in that wonderful prayer in the 17th of John He "lifted up His eyes to heaven," it is said. So we find we have some authority for addressing our prayers upwards; heaven is located above.

THE HOME OF GOD.

Then we find that it is the dwelling-place of God. Would you turn to 1 Kings viii. 30: "And hearken Thou to the supplication of Thy servant and of Thy people Israel when they shall pray toward this place; and hear Thou in heaven, Thy dwelling-place, and when Thou hearest, forgive." Heaven is the "dwelling-place" of God. God has a home, God has a throne, God has a dwelling-place — "hear Thou in heaven, Thy dwelling-place." Now, how far away heaven is I do not know; I have not been able to find out. There is one thing that I do know, it is not so far away but God can hear us when we pray. God can hear every prayer that goes up from this sin-cursed earth. We are not so far from Him but that He can see our tears and hear the faintest whisper when we lift our heart to Him in prayer. In Daniel we read that Gabriel was caused to fly swiftly and come to Daniel. I do not know how long it took him to come, but as near as I can find out it took him about four minutes. If we could find out how fast he flew we might find out how far heaven is. It does not take long for these angels of light to come to our rescue and help if we need them. In 2 Chronicles vii. 14, we read, "If My people, which are called by My name, shall humble themselves, and pray, and seek My face, and turn from their wicked ways, then will I hear from heaven, and will forgive their sin, and will heal their land." That is God's own word, "I will hear from heaven," and then when Christ's disciples came to Him and said, "Lord, teach us how to pray as John taught His disciples," He taught them to pray thus: "Our Father who art in heaven,"— not down here. That is His dwelling-place. God has a throne, and God has a dwelling-place, and let us make heaven real. I believe heaven is a city quite as real as London is. What we want is to make heaven real, and hell real, and God real, and Christ real, and then live as if we believed these things to be real.

THE CURTAIN LIFTED.

Now, we have it established that God is in heaven, that that is His dwelling-place, that He has a throne there. Then would you just turn to the 7th of Acts, for we want to find out who is there and what company we are going to be in when we get there—the 55th verse: "But he being full of the Holy Ghost, looked up steadfastly into heaven, and saw the glory of God, and Jesus standing on the right hand of God." When a man is full of the Holy Ghost heaven does not seem far away; he can see by the eye of faith clearly into the city, and can see Christ standing at the right hand of God. Stephen was full of faith and of the Holy Ghost, and the curtain was lifted and he looked in, and there he saw his blessed Lord and Saviour, whom he loved, standing at the right hand of God. Heaven was real to Stephen, Christ was real, He was a real living person, and he saw Him there. And I think that is what is going to make heaven so attractive to us—Christ will be there.

THE VISION OF THE KING.

One Christian asked another what he expected to do when he got to heaven, and he said he expected to take one good long look of about 500 years at Christ, and then he would want to see Paul and Peter and John and the rest of the disciples. Well, it seems to me one glimpse of Christ will pay us for all that we are called upon to endure here — to see the King in His beauty, to be in the presence of the King. And then the sweet thought is we shall be like Him when we see Him, and we shall see Him in His beauty, we shall see Him high and exalted. When He was down here it was the time of His humiliation, cast out from the world, spit upon and rejected; but God hath exalted Him and put Him at the right hand of power, and there He is, and there, my friends, we shall see Him by-and-by. A few more tears, a few more shadows, and then God shall say, "Come up hither, and into the presence of the King we shall come. It may be I am talking to some one to-night that will see the King before the sun shall rise tomorrow morning — some one in this audience may be summoned away and be there with the Lord Jesus. Yes, it won't be the pearly gates that will be so attractive, it won't be the jasper walls, it won't be the streets paved with transparent gold,—that is not what is going to make heaven so attractive; but it is the thought that Jesus, who loved us, and gave Himself for us, will be there, and we shall see Him, we shall look upon Him. Oh, that will make heaven glorious, to think that we shall see Him ourselves, that we shall behold Him and gaze upon Him, and hear that loved voice. Ah, methinks I would rather hear that voice, and look into those lovely eyes, and gaze upon that face than to see all the world. Yes, that is what God calls us to, that we may be in the presence of His beloved Son.

STORY OF A MOTHERLESS CHILD.

I was reading, some time ago, of a little child whose mother was sick, and the child was not old enough to understand about the sickness of the mother. It was taken away, and when the mother died, they thought they would rather have the child remember its mother as she was when she was well, and so they did not take her back till after the mother was buried. They then brought the child home and she ran into the drawing-room to meet her mother, and her mother was not there. The little thing was disappointed, and ran into all the rooms, but could not find her mother. She began to cry, and asked them to send her back; she did not want to stay; home had lost its attraction because mother was not there. What is going to make heaven so delightful? It won't be the pearly gates; it won't be the jasper walls; but it will be that we shall see the King in His beauty, and shall behold Him, and not only Him, but those that have gone on before us.

THE ANGELS OUR COMPANIONS.

Then look to the 10th verse of the 18th of Matthew. We have God the Father and Christ the Son; they will be with us and we shall be with them. Then we read in this verse, "Take heed that ye despise not one of these little ones; for I say unto you, that in heaven their angels do always behold the face of My Father which is in heaven." So we will have the angels for our companions, we will have the society of angels when we get in that world of light. You may say, "Oh, that is visionary to talk about guardian angels." But you know when Peter was out of prison the damsel who went to the door came back and said it could not be him, it was his angel. Why, I believe the early Christians believed it, and then the Scriptures teach that the angels encamp round about them that love God. I would not be surprised to find that there are more angels in this hall than there are human beings. God has given His angels charge over us to keep us. Look at that servant of Elisha on the mountain; when his servant was alarmed and Elisha prayed God to open his eyes, he found the mountain was filled with angels and chariots and horsemen. They were down from the Eternal City just to shield that one servant of the living God. Oh, my friends, let us cheer up and remember God thinks so much of us that He sends angels down to guard us, but in that world we will be companions of theirs, we will see them face to face, we will talk with them then. We cannot be brought into fellowship with them now, but then we shall be taken into the presence of these very angels.

When Gabriel came down to tell Zacharias what was going to take place he said, "I am Gabriel, who stands in the presence of God." Yes, there are angels in the presence of God, and we will have them for our society.

THE REUNION OF THE REDEEMED.

Just turn to John xii. 26:—"If any man serve Me let him follow Me, and where I am there shall also My servant be; if any man serve Me, him will My Father honor." The servant and the master shall be together. "If any man serve Me, that servant shall be with Me," He says. A great many people come to me and want to know if I really think their friends that have died in the Lord are with the Lord. Some have an idea that they are separated from the Lord. Now, there are a few passages of Scripture that I think give us strong reason to believe that our departed friends that have died with Christ are safe with Him, and so we have not only God the Father, Christ the Son, and angels, but the redeemed saints are there. Would you just look to the 2d Corinthians v. 1, where Paul says, "For we know that if our earthly house of this tabernacle were dissolved, we have a building of God, an house not made with hands, eternal in the heavens." Then in the 8th verse:—"We are confident, I say, and willing rather to be absent from the body, and to be present with the Lord." Yes, if this earthly house were dissolved, we have a building not made with hands, eternal in the heavens. Then he says, "Absent from the body, present with the Lord." I believe Paul thought when he left the body he should see the King in His beauty, that he would behold the Lord Himself. Then turn to Philippians i. 23, "For I am in a strait betwixt two, having a desire to depart and to be with Christ, which is far better." I think these 1,800 years that Paul has been gone from the earth he has been with Christ. Christ would not be separated from him. Then we find other passages—we have not time to dwell upon them, but it seems to me we have strong reason to believe that those friends that have died safe in Christ are with Him to-night. Then would you turn to Revelations vii, 9, "After this I beheld, and lo, a great multitude, which no man could number, of all nations, and kindreds, and people, and tongues, stood before the throne and before the Lamb, clothed with white robes and palms in their hands, and cried with a loud voice, saying, Salvation to our God, which sitteth upon the throne, and unto the Lamb." There they are, redeemed saints, redeemed out of every kindred, every nation under heaven, around the throne, singing the song of Moses and the Lamb. Yes, they sing much sweeter than you can sing on earth. And if we are redeemed, and our garments are washed in the blood of the Lamb, we shall join in that chorus, by-and-by, and sing much sweeter than we can here upon earth; we shall shout, Glory to the Lamb that redeemed us with His precious blood! So now we have redeemed saints there.

THE SEVENTY REVIVALISTS.

There is another thought I want to bring out, and that is, it is the privilege of every child of God in this vast assembly to know that their names are written in the Book of Life, and believe we can have that assurance that our names have gone on before us, and are registered in heaven. Christ sent out His disciples, seventy of them, and told them to go into the towns and villages, and preach the kingdom of God, and tell the glad tidings to the inhabitants; and when these men came back they had had wonderful success. Why, they said that the very devils were subject to them. All they had to do was to command the devils to leave men, and the devils fled before them. They were all elated with their wonderful success; revivals had followed everywhere they had been; they were revival preachers; they were evangelists going into the towns and preaching. I have not any doubt but that there was a good deal of prejudice against them, but they went on preaching the glad tidings, and when they came back, Christ says, "Well, now, do not rejoice at that; I will tell you what to rejoice over. Rejoice that your names are written in heaven." And I would like to ask every one in this audience to-night this question, Is your name there? Can you rejoice to-night that your name is written in heaven, that your name is in the Book of Life? Says Christ to His disciples, 'Rejoice that your names are written in ieaven."

NAMES IN THE BOOK OF LIFE.

Not long ago there was a man complaining about my talking about names written in the Book of Life, he did not believe in it. It took some time to look the subject up, and I was amazed to find so much in Scripture about names being written in the Book of Life. In the 12th of Daniel we read, "And at that time shall Michael stand up, the great prince which standeth for the children of thy people: and there shall be a time of trouble such as never was since there was a nation, even to the same time, and at that time Thy people shall be delivered, every one that shall be found written in the Book; and if our names are written in the Book of Life God will care for us, God will protect us." Not one whose name is written in the Lamb's Book of Life shall perish. If Christ did not want us to know that our names were written there, do you think He would have told His disciples to rejoice that their names were already there? My friend, I believe it is the most important question that can come before us in this world. It is a thousand times better that we have our names written in God's Book than in all the books in the world—a thousand times better that our name shines out upon God's Book of Life, and is written there, than it is to be written in any church record in London. It is a great deal better that we make sure that our name is written in the Book of Life than that it is written in your ledgers with great sums attached to your names. It is a thousand times better to be sure that our name has been written in heaven than to have the wealth of the world.

TELEGRAPH FOR A ROOM.

Two years ago a friend of mine that was in London was going back to America. She went to Liverpool with a party of American friends, and they were talking about what hotel they would stop at, and decided to go to the North-Western. The hotel was full, and as they were starting to find another, they said to my friend, "Are not you going with us?" My friend said, "No, I am going to stay here." "Oh, no," they said, "you cannot stay here." But my friend said, "I am going to stay." "How is it?" "I have got a room." "Where did you get it?" "Why, I sent my name on ahead." She had telegraphed a few days before and secured a room. And that is just what the children of God are doing now; they are sending their names on ahead and getting them down in the Book of Life. They are not waiting for the dying minute. My friend,

SEND YOUR NAME ON AHEAD

to-night, and if you really want it there God will put it there. Yes, every one whose names are written in the Book of Life shall not perish, but shall be saved.

Turn to Philippians iv. 3 : "And I entreat thee also, true yokefellow, help those women which labored with me in the Gospel, with Clement also, and with other my fellow-laborers, whose names are in the Book of Life." There is Paul writing to those "whose names are in the Book of Life." Now, suppose I should ask every one in this audience to rise that have reason to believe that their names are in the Book of Life, would you rise? Supposing a letter should come to you addressed in the way Paul addressed this letter to those women whose names are in the Book of Life, could you say that was for you? Oh, it is the privilege of every child of God to have his name there, and to know that it is there. I find so many people

LIVING IN DOUBTING CASTLE.

Why, it is salvation by doubts nowadays instead of by faith; there are so few that dare to say, " I know that my Redeemer liveth, I know in whom I have believed." We find most Christians nowadays shivering and trembling from head to foot; they do not know whether they are saved or not. Yes, Christ never would have told His disciples to have rejoiced unless they had known that their names were there. Turn to Hebrews xii. 23 for a minute: "To the general assembly and church of the first-born, which are written in heaven, and to God the Judge of all and to the spirits of just men made perfect." A man sometimes asks another man what church he belongs to. Why, I belong to the general assembly and the church of the first-born, which are written in heaven. It is a good thing to belong to that church, because your name will be written in the Book of Life. You will be sure to get into heaven if you belong to that church. You may belong to a great many churches on earth, and not get in. Be sure that you belong to the general assembly of the first-born, and that your names are written in heaven. Make sure of this one thing if you are not sure of anything else. It is better that you fail in health or in business, it is better that you go to some asylum, it is better for you to go to heaven from some poor-house or from some mad-house than to go to hell in a gilded chariot. Make sure that your name is written in heaven; then you have something worth rejoicing over.

THE DREAD ALTERNATIVE.

There is something said in Revelations about the names being written in the book, the 20th chapter and 15th verse: "And whosoever was not found written in the Book of Life was cast into the lake of fire." Young man, is your name in the Book of Life? If it is not, and you should be cut down by death to-night, where would your soul be to-morrow? Only think of it. Say, mother, is your name written in the Book of Life? Are you sure it is there? Just listen to these words again, "And whosoever was not found written in the Book of Life was cast into the lake of fire." May God send home the truth to-night, and may every one in this audience be sure that your name is written in the Book of Life. Let business be suspended, let everything wait till you have made sure of your soul's salvation. Do not let a scoffing, laughing, mocking world cheat you out of heaven. Do not let anything stand between you and this one great question. Look to the 21st chapter and the 27th verse : " And there shall in no wise enter into it anything that defileth, neither whatsoever worketh abomination or maketh a lie; but they which are written in the Lamb's Book of Life." Almost the last words in Scripture are about this Book of Life—they whose names are written in the Lamb's Book of Life Now, my friend, would you just ask your self the question, and may God press it home upon you, and may it sink deep into every heart here, " Is my name written in the Book of Life?" It seems to me the great work is to be sure that our names are there; then we are ready to go and work for others; we are not ready until we know that our names are in the Book of Life. And then these mothers and these fathers, why, what a work we have to do to get our children's names in that Book ! It seems to me every parent ought to be more anxious to have the names of their children written in that Book than to have them written high in some school, than that they should stand highest in their class, —a thousand times better that they shou'd stand well in heaven, and that their names should be written in the Book of Life. And not only that, but I believe these little children can have their names written there, and we as parents can know that our children have their names there, if we work for it, if we pray for it, and that is our aim. Let us be

FAITHFUL WITH OUR CHILDREN

while they are young. I see some children here to-night; I do not know why they should not become Christians now. I do not know why their parents should not labor for their salvation. I believe there is a good deal of infidelity got into the Church of God at the present time. I do not believe we, as parents, realize how young these children can become true disciples of Christ; if we did we would labor more for the salvation of little children.

A MISSIONARY'S TOUCHING STORY.

I was urging this one time in a meeting in America, and an old man got up at the close and said, "I want to endorse every word that has been said. I believe in the conversion of little children. Sixteen years ago I was in a heathen country laboring as a missionary, and my wife died and left me with three little motherless children. On the Sabbath after her death my eldest girl, ten years old, came to me and said, 'Papa, shall I take the children into the bedroom and pray with them as mother used to?'" That is the power of example; the mother was dead, and gone, and little Nellie, ten years old, wanted to follow in her footsteps. The father said yes, she might if she liked, and she led them off to the chamber to pray. He said when they came out he noticed that they had all been weeping, and asked what they had been weeping about. "Well, father," said the little girl, "I prayed just as mother taught me to pray, and then "—naming her little brother — "he prayed the prayer that mother taught him to pray; but little Susie, she was too young, mother had not taught her a prayer, and so she made a prayer of her own, and I could not help but weep to hear her pray." "Why," said the father, "what did she say?" "Why, she put up her little hands, and closed her eyes, and said, 'O God, you have come and taken away my dear mamma, and I have no mamma to pray for me now—won't you please make me good just as my dear mamma was, for Jesus' sake, Amen;'" and, said the old missionary, God heard that prayer. That little child before she was four years old gave evidence of being a child of God, and for sixteen years she was in that heathen country leading little children to the Lamb of God that taketh away the sin of the world. Mother, do you believe your child can come too early? Do you believe your child can have his name written in the Book of Life too early? Oh, may God help us to labor for it, to call our children into the ark! May God give us our children, and may their names be written in the Book of Life!

THE DYING SOLDIER AND THE ROLL-CALL.

A soldier lay on his dying couch during our last war, and they heard him say "Here!" They asked him what he wanted, and he put up his hand and said, "Hush! they are calling the roll of heaven, and I am answering to my name," and presently he whispered, "Here!" and he was gone. That great roll is being called. My friends, your name may come to-night—mine may come. Is your name in the Book of Life? If it is we will go up from earth with a shout of victory upon our lips; it will be no sad summons. But to die without God, without hope, without our names written in the Book of Life, oh, how sad, how dark, how terrible! May God help you to-night, each and every one that are without God and without hope, to press into the kingdom is the prayer of my heart.

XI.

THE BLOOD.

THE subject I wish to call your attention to this afternoon is "The Blood." In the first place would you turn to Genesis iii. 21?—"Unto Adam also, and to his wife, did the Lord God make coats of skins, and clothed them." In this verse we get the first glimpse of blood. Certainly the Lord could not have clothed Adam and Eve with the skins of beasts unless He had shed blood. There we have the innocent suffering for the guilty—the doctrine of substitution in the Garden of Eden. God dealt with Adam in government before He dealt in judgment. Death came by sin. Adam had sinned, and now the Lord comes down to make the way of escape. God came to him as a loving friend, and not to hurl him from the earth. Adam could have said to Eve, "If the Lord has driven us out of the Garden of Eden, He loves us." God put a lamp of promise into his hand before He drove him out, for He said, "The seed of the woman shall bruise the serpent's head." Did you ever think what a terrible state of things it would be if man was allowed to live for ever in his lost, ruined state? It was out of love to Adam that God drove him out of Eden, that he should not live for ever. God put the cherubim there; and now Christ has taken the sword out of his hand, and opened wide the gate, so that we can come in and eat. Adam might have been in Eden ten thousand years, and then be led astray by Satan; but now our lives are hid with Christ. Man is safer with the second Adam out of Eden than with the first Adam in Eden. Would you turn to Genesis iv. 4?—"And Abel, he also brought of the fatlings of his flock, and of the fat thereof. And the Lord had respect unto Abel and to his offering." These two boys were brought up outside of Eden, and had the same parents. Undoubtedly on the morning of creation God marked out the way a man might come to Him; and Abel walked in God's way, and Cain in his own. Perhaps Cain said he could not bear the sight of blood, and he took that which God had cursed and laid it upon the altar. And there are a good many

CAINITES IN THE CHURCH

to-day; and some have got into the pulpit, and they preach that it is not the doctrine of the blood, and that we can get to heaven without the blood. From the time Adam went out of Eden there have been Abelites and Cainites. The Abelites came by the way of the blood—the way God has marked out for them. The Cainites came of their own way. They want to get out of the doctrine of the blood. Some preach they don't believe in the blood, and they say it does not atone for sin. It is better to take God's word than man's opinion; therefore, turn to Genesis viii. 20—"And Noah builded an altar unto the Lord; and took of every clean beast, and of every clean fowl, and offered burnt offerings on the altar." We have thus passed over the first two thousand years, and have come to the second dispensation. The thought I want to call your attention to is this. The first things Noah did when he got out of the ark was to build an altar and slay the animals, thus putting blood between him and his sin. The second dispensation is founded upon blood, and it is most important that these animals were taken through the flood expressly for this purpose. We find Noah walking by that highway, and all the men of God have been walking that way, for it is the blood that atones for sin.

ABRAHAM SAW CHRIST.

Would you turn to Genesis xii. 13?—"And Abraham lifted up his eyes and looked; and, behold! behind him a ram caught in a thicket by his horns; and Abraham went and took the ram and offered him up for a burnt offering in the stead of his son." We find here another type. The ram was typical; he was offered up in the place of Abraham's son. God loved Abraham so much that He spared his son; but God loved us so much that He did not spare His Son, but freely gave Him up for us all. Here we find that mountain-peak sprinkled with blood. Abraham was willing to do all the Lord had told him, for he took the knife, and was ready to give all to God. Then it was that God gave him the secret of heaven, and told him what he was to do. He saw Christ and was glad. Jehovah opened the curtain of time, and Abraham saw Christ coming up. He saw his sins on Christ and was glad—he saw His day and was glad. All Abraham's seed lost their sins as much by Christ as we. For 4,000 years they were looking to the promise of his coming. They were not looking to the cross, but to the Messiah, and it was through Him all the nations of the earth were to be blessed. The difference is that we look back to Calvary, and they looked forward. Then again in Exodus xii. 13 we read—"And the blood shall be to you for a token upon the houses where you are; and when I see the blood I will pass over you, and the plague shall not be upon you to destroy you, when I smite the land of Egypt." I can imagine some of the lords and dukes and great men, as they rode through and saw the poor Egyptians sprinkling their dwellings, saying they never saw such foolishness, for they were spoiling their places. The blood was to be put upon the door-posts and lintels, and not upon the floor, for that is what many are doing now,

TREADING UPON THE BLOOD.

Wherever blood was upon the door-post death passed over, and that kept death out. It was not what they were. He did not say, "When I see your prayers, your good deeds, I will pass over you;" but "When I see the blood I will pass over you. A little child that night behind the blood in Goshen was as safe as Moses. People say, "If I was as good as that man who has been preaching for fifty years, or that mother in Israel who has long labored for Christ." But if you are behind the blood of the Lamb, you are as safe for heaven as any man living on earth. It ain't when I see how holy you are—how you go to church every Sabbath—how you say your prayers—how you pay your debts—but when I see the blood. Some one has said that the little fly in Noah's ark was as safe as the elephant. It was the ark that saved the fly and the elephant, and it is the blood that saves the weakest and the strongest. When death came that night with his sword, he entered the palace of the prince, and went into the houses of the great and mighty, and they all had to pay tribute to death, for the first-born in Egypt was smitten down that night. The only thing that kept death out was death itself. The only way that death can be met is by death. I have sinned, and must die, or get some one to die for me. Some people say it isn't the death of Christ, but His life. Suppose some one had said, I will have a live lamb; I will tie my little white lamb against the door. Death would have passed over that lamb, and into the house. The blood shall be a token, and the great question is, Have you got the token? If death should come after any one of us to-night, are we sheltered behind the blood? that is the point. It is the blood that atones. Not my good resolutions, or prayers, or position in society, or what I have done, but what has been done by another. God looks for the token.

HAVE YOU GOT THE TOKEN?

Some one has made use of this illustration. You go down to a railway station to start for Liverpool, and you get your ticket at the office, but the man doesn't care who you are. When I went down to Liverpool the other day, a man called, "Tickets." I have an idea the man could not tell whether I was a white man or a black man. All he looked for was the ticket, the token. If I hadn't got the token, he would have put me out; but, because I had the token, he passed me. God says, When I see the token—the blood—I will pass over you. If I am behind the blood, I am safe, and if I trample it under my feet, I must perish. These Egyptians made light of the Hebrews sprinkling their door-posts. The blood of Christ is worth more to us than all the world. It is that and that alone that can atone for sin.

STRONG AND SICKLY CHRISTIANS.

In the eleventh verse of the same chapter we read: "And thus shall ye eat it; with your loins girded, your shoes on your feet, and your staff in your hand; and ye shall eat it in haste; it is the Lord's Passover." Why you have not got more power is because you don't feed on the Lamb; and this is why there are so many weak Christians. The Lamb not only atones for our sins, but we are to feed upon the Lamb. We have got a wilderness journey before us, as the children of Israel had. After we are saved we are to feed upon Christ; He is the true bread from heaven. If I don't feed this soul with the true bread from heaven, I am sickly, and have not power to go and work for Christ. And that is the reason, I believe, why so few in the Church have power. Some people think if they have got one glimpse of Christ that is enough. You in England think much of your dinner, and why should not God's children think a good deal of their spiritual food? We should no more think of laying in spiritual food to last for ten years than we should bodily food. A good many people are living on stale manna. A man in Ireland said to his boy, "I want you to eat two breakfasts. Do you know why?" The boy said he understood one was for his body and the other for his soul. All Christians should take two breakfasts. Everything dated back to the passover night—to the time the blood was put upon the door-posts. All the time you are serving the world it goes for nought. If you have not come to Calvary you are losing time. Everything you do on the other side of the cross counts for nought; the first thing is to know we are saved, and then we commence our pilgrimage to heaven. We don't start, as some people suppose, from the cradle to heaven. We start from the cross. We have got a fallen nature that is taking us hellward. We must be born of the Spirit, and

SHELTERED BY THE BLOOD,

and then we become pilgrims for heaven. Turn to Exodus xxix. 16—"and thou shalt slay the lamb, and thou shalt take his blood and sprinkle it round about the altar." Even Aaron could not come to God until he sprinkled blood round about the altar. From the time Adam fell there has been no other way a man can approach God but by the blood. You cannot have an audience of God until you come by the way of the blood. So it has been for 6,000 years. It has never been otherwise, and never will be. Leviticus viii. 23—"And he slew it; and Moses took of the blood of it and put it upon the tip of Aaron's right ear, and upon the thumb of his right hand, and upon the great toe of his right foot." I had used to read a passage like this, and say it seemed absurd. I think I understand it now. The blood upon the ear, that a man can hear the voice of God. A man must be sheltered behind the blood before he can hear God's voice. The blood upon the hand, that a man may work for God. You cannot work for God until you are sheltered behind the blood, and until you are sheltered it all stands for naught. You may build churches, endow colleges, pay ministers and missionaries salaries, but it all goes for naught until you are sheltered behind the blood. Don't let any one deceive you on this point. Don't let Satan deceive you by telling you that you can get to heaven by some other way. They asked Christ, "What must we do that we may do the works of God?" Perhaps these men had got their pockets full of money, and were ready and willing to build churches. Christ told them that the work of God was that they believed. You cannot do anything to please God until you believe. As an illustration, suppose I should say to my little girl, "Emma, go and get me a glass of water;" and she was to say, "I don't want to do it, papa." She goes into another room and some one gives her a cluster of grapes, which she decides to give to her papa. Do you think these grapes would be acceptable if she did not want to get the water? I should say, "I do not want the grapes until you have brought the water. She goes out of the room again, and some one gives her an orange. If she brought the orange to me, do you think I should want it? Ten thousand times no, and that child cannot do anything to please me until I get the water. You cannot please God until you believe on His Son.

THE CHURCH NEEDS TO BE ON FIRE.

I wish the Church was on fire, and I wish all Christians were on fire. Don't let us set dead men to work. I don't believe in unconverted Sabbath-school teachers and unconverted men working in the Church, and I hope the line will ere long be drawn. God has given an unspeakable gift—the Son of His bosom—and if we reject that Son and won't follow Him, do you think anything we can do will please God? The blood upon the hand is that a man may work for God, and on the foot that a man may walk with God. When Adam fell he fell out of communion with God. Before he fell, he walked with God, but the moment he fell out of communion with Him, and from that time to this He has been trying to get men back into communion. God is full of truth and justice. His justice must be met, and after that has been met He is satisfied. God never walked with men until He put them behind the blood at Goshen. What could stand before them then? They passed through the Red Sea, and God said to Joshua, "Take this country, and no man shall be able to stand before you all the days of your life." Look at Joshua walking round Jericho, and as he does so a man stands before him with a sword in his hand. Joshua steps up to him and says, "Art thou for us, or against us?" He was to lead them on to victory, but God was testing Joshua's faith. When God gave them Saul as their king, they raised the cry, "God save the king!" and this cry has been raised ever since. They then, so to speak, voted God out when they had got a king. In the days of Joshua there were whole regiments of giants, but one stripling from the Lord's hosts defeated the giant of Gath. If God is with us, the giants will be like grasshoppers; but if God is not with us, it will be different. I would rather have ten men separated from the world, than ten thousand nominal Christians who go to the prayer-meeting to-night and the hall to-morrow. The Church and the world are mixed up into one. Now turn to Leviticus xvii. 11—"For the life of the flesh is in the blood; and I have given it to you upon the altar to make an atonement for your souls, for it is the blood that maketh an atonement for the soul." There may be some who are saying, Why does God demand blood? Some one said to me," I detest your God, He demands blood; I don't believe in such a God, for my God is merciful to all." I want to say, my God is full of mercy, but don't be so blind as to believe that God is not just, and that He has not got a government. Suppose Queen Victoria didn't like any man to be deprived of his liberty, and she threw all her prisons open, and was so merciful that she could not bear any one to suffer for guilt, how long would she hold the sceptre? how long would she rule this empire? Not twenty-four hours. Those very men who cry out about God being merciful would say, "We don't want such a Queen."

A REVOLT IN HEAVEN.

God is merciful, but He isn't going to take an unredeemed sinner into heaven. If He did, the redeemed would plant the banner of rebellion round the throne, and there would be a revolt. This verse tells why God demands blood. Atonement means at-one at-one-ment. God demands blood because He said to Adam, "On the day thou sinnest thou must die." Sin came into the world and brought death into the world. God's words must be kept. I must either die or get somebody to die for me, and in the fulness of time Christ comes forward to die for the sinner. He was without sin, but if He had committed one sin He would have had to die for His own sin. The life of the flesh is in the blood, and it is not blood He demands really, it is life, and life has been forfeited. We have sinned, and death must come, or justice must take its course. Glory to God in the highest to think He sent His Son, born of a woman, to take our nature and die in our stead, tasting death for every man. You take this blood out of this body of mine, and life is gone.

GOD DEMANDS BLOOD.

He demands life. Man has sinned, therefore life must be forfeited, and I must die or find somebody to die for me. My friends, I have only just touched this subject. If you read your Bibles carefully you will find the scarlet thread running through the Bible. It commenced in Eden and flows on to Revelation. I cannot find anything to tell me the way to heaven but by the blood. That book (holding up the Bible) wouldn't be worth carrying home if you take the scarlet thread out of it, and it don't teach anything else, for the blood commences in Genesis and goes on to Revelation. That is what this Book is written for. It tells its own story, and if a man should come and preach another gospel don't you believe him. If an angel should come and preach anything else don't belive it. And if you are in a church, either Dissenting or Established, and the minister doesn't preach the blood, you get out of it as Lot did out of Sodom. Don't trifle with this subject of the blood. In your dying hour you would give more to be sheltered behind this blood than for all the world. Christ died for us, and all I have to do is to accept Him. Christ said, "You take My life and I will take your sins." Don't you want to make this bargain? Death shall never have his hand on Christ again. Christ says, My life is yours. I will have it. Won't you? Isn't it the height of madness for any one of you to go out of this place and not accept it? Christ laid down His life that you and I might live, and now out of gratitude ought we not to serve Him? Some people think it is noble to lift up their voice against Christ, but it is a cowardly act.

THE MOTHER'S LOVE.

In the time of the Californian gold fever a man went to the diggings, and left his wife to follow him some time afterwards. While on her voyage with her little boy the vessel caught fire, and as there was a powder magazine on board the captain knew when the flames reached it the ship would be blown up. The fire could not be got under, so they took to the life-boats, but there was not room for all. As the last boat pushed off the mother and boy stood on the deck. One of the sailors said there was room for another. What did the mother do?—she gave up her boy. She kissed him, and told him if he lived to see his father to tell him she died to save her boy. Do you think when that boy grew up he could fail to love that mother who died to save him? My friends, this is a faint type of what Christ has done for you and me. He died for our sins. He left heaven for that purpose. Will you go away saying, I see no beauty in Him? May God break every heart here to-day! and may we become loyal to Him! You will need Him when you come to cross the swelling of Jordan. You will need Him when you go up to the bar of God. For death to come and find without Christ, and God, and hope,—may God forbid!

THE FIRST PREACHERS OF THE BLOOD.

I want to follow up the subject of yesterday, and those of you who were present then will remember I was speaking of "The Blood" especially in the Old Testament. This afternoon I will take up the subject from the New Testament. When I was in Dublin I gave a lecture on "The Blood," and a lady wrote me and said, if the blood was so important, why was it the early preachers, the apostles, and Christ Himself never referred to it?

I hadn't time to write to the lady, but I wish she was present to-day, for I will prove that the early Chrstians preached nothing else. Would you turn to Acts ii. 22–36. It was Christ and Him crucified the apostles preached, and nothing else. It was this preaching God blessed, and which brought so many in one day to the cross of Christ. [To further prove this, Mr. Moody quoted Acts iv. 10, Acts v. 29, Acts vii. 52, Acts viii. 32, Acts xvii. 2, 18, 31; Hebrews ix. 22; Matthew xxvi. 28, Revelations i. 5.] If a man makes light of the blood, how is he going to be washed in it? If he makes light of the blood, how is he going to get rid of his sins? and how is he going to stand before that pure God unless his sins are washed away? A great many people think God never loves them until they get rid of their sins, and because they are so vile they cannot come to God— "unto Him that loved us and washed us." He loves us in our sins, saves us from our sins, and washes us and clothes us with His own garment, and then we are able to have communion with Him. Satan says God will not love you because you are not pure. But let us keep to Scripture—"unto Him that loved us and washed us." Like the good Samaritan who went to the poor man who fell among thieves, so Christ comes to the sinner where he is. You cannot make yourselves clean, therefore stay where you are. Let Christ wash you, for our good intentions and prayers cannot atone for sin.

THE NEW AND LIVING WAY.

Heb. x. 19, 20—"Having, therefore, brethren, boldness to enter into the holiest by the blood of Jesus, by a new and living way which He hath consecrated for us, through the veil, that is to say, His flesh." When Christ expired on the cross the veil of the temple was rent, and as somebody said, not from the bottom to the top, but from the top down. Thanks be to God, we don't need any one to plead for us, for we can come right into His presence, for we are all priests now. I want to say to you who are running to this man and that man to plead for you, go right to the Master yourselves. Let us come by this new and living way Christ has made for us by rending the veil. Christ's flesh was nailed to that cross to open the living way. Before only the high priest could go into the holiest of holies, but the moment Christ expired He made us all priests. Some one may remark, "It says confess your sins one to another;" but if I have sinned against a man I must confess to him, and must also confess to God. If I have caused man to stumble I must go and remove that stumbling-block out of the way. The only man, it is said, who confessed his sins to man was Judas, and he hung himself. Peter confessed to God, and God forgave him. Numbers xxviii. 4—"The one lamb shall thou offer in the morning, and the other lamb shall thou offer in the even." That was done continually, and the priest could never take his seat in the holiest of holies because his work was never done. Now turn to Hebrews x. 12 —"But this man, after he had offered one sacrifice for sins, for ever sat down on the right hand of God." The blood of Christ is enough to cover every sin that was committed. All the lambs on the other side of the cross were typical. They were pointing to the true sacrifice, and they were all fulfilled in Him. We don't need to make any more sacrifice. He has made sacrifice Himself, and has made full atonement for every sin. All we have now to do is to trust the sacrifice. God says, I am satisfied with the finished work of Christ, and the moment the sinner is satisfied, God and the sinner are united. The

BLOOD HAS TWO CRIES.

It either cries for my condemnation, which means damnation,—excuse the strong expression,—or for my salvation. If I make light of the blood and trample it under my feet, then it cries out for God's condemnation; but if I am sheltered behind the blood, there is no condemnation for me. God dealt in judgment with Cain, and when Pilate wanted to know what to do with Christ, he washed his hands and said he was innocent. The Jews said, Let His blood be upon us and our children, not to save us, but to condemn us. Would they had said, Let His blood be upon us to save us and protect us. Nearly 1,900 years have rolled away, and the Jews are wanderers on the face of the earth without a king. Their having been scattered all these years, what a proof it is the Word of God is true! May our prayer be to-day, His blood be upon us and our children, not to condemn us, but to save us. Let that be our prayer, that we may know what it is to be sheltered behind the blood of God's dear Son. Colossians i. 20—"And having made peace through the blood of His cross, by Him to reconcile all things unto Himself; by Him, I say, whether they be things in earth or things in heaven." The blood of the cross speaks peace. If I am

SHELTERED BEHIND THE BLOOD

there is peace, and there is no peace until my sin is covered. If you had committed sin against a man you would get no peace until that was forgiven. Men are running after peace, and if it could be bought in the market, many would give hundreds of thousands of pounds to secure it. The blood of Christ speaks peace, and it will bring peace to every guilty conscience and aching heart to-day if you only seek it. John xix. 34—" But one of the soldiers with a spear pierced His side, and forthwith came thereout blood and water." There is a beautiful thought contained in this verse. The spear that went into the side of the Son of God was the crowning point of earth and hell. I don't see how they could have done a more cruel thing than that. You may say that was the crowning act of sin. And the blood came out and covered the spear, and a fountain was thus open in the house of David for sin. The blood touched the Roman spear, and it was not long before He had the Roman Government. The blood ran down from His side, and God will have the world by-and-by. He is the true Sovereign, and He will cast out the prince of darkness, and will sway His sceptre from end to end of the earth. He has redeemed the earth by His blood, and He will have all He has redeemed, bear that in mind.

HAS THE BLOOD TOUCHED YOU?

The blood of Christ makes us one, brings us into the family of God, and enables us to cry, " Abba, Father." During the days of slavery in America, when there was much political strife and strong prejudice against the black men, especially by Irishmen, I heard a preacher say when he came to the cross for salvation he found a poor negro on the right hand and an Irishman on the left hand, and the blood came trickling down upon them and made them one. There may be strife in the world, but every one Christ has redeemed He has made one. We are blood relatives. When I go before an audience there ain't hardly a person I have seen before; but as I begin to talk about the Kingdom their eyes light up, and I say they are kinsmen, they are blood relatives, and in a short time I become attached to them. A man may go into a town a perfect stranger, but how soon will he find out those who love God, and they will be one. I wish Christians had more of this oneness. I hope the time will soon come when these sectarian walls will be broken down, and people will not want to ask whether you belong to the Established, Wesleyan, or Baptist Churches. Here, mind, we are blood relatives. Thomas thrust his hand into the Master's side, and He was, afterwards seen by over 500 at once. The apostles saw Him go back to Heaven, but the blood which flowed on Calvary is there, and I believe

GOD IS GOING TO JUDGE THE WORLD BY THE BLOOD.

—" What did you do with the blood?" If we make light of that blood, and send back an insulting message, saying we don't want the blood of God's dear Son, we shall stand speechless before God's tribunal. What have you done with God's blood? If we make light of that blood what is going to become of our souls? Hebrews x. 28, 29—" He that despised Moses' law died without mercy under two or three witnesses. Of how much sorer punishment, suppose ye, shall he be thought worthy who hath trodden under foot the Son of God, and hath counted the blood of the covenant, wherewith he was sanctified, an unholy thing, and hath done despite unto the Spirit of grace?" To me these are very solemn verses. I don't see how any one can sit here and hear these verses read and not be saved. I don't care what you are or what your life has been, if you have rejected Christ up to the present time. Let me read these verses again. They died without mercy; but how much more sore will be the punishment of those who live in the age with an open Bible which tells how Christ died to redeem us, and make us heirs of heaven. Revelation xii. 11—" And they overcame him by the blood of the Lamb, and by the word of their testimony, and they loved not their lives unto the death." I don't believe there is a word in the Bible Satan is fearing more than the word blood. I shall receive a good many letters to-morrow attacking me for what I have said to-day. These letters will say it is heathenish to stand up and preach what would only do for an unenlightened age. May

God forgive anyone who would dare to teach such a thing. If you will read your Bible in the light of Calvary, you will find there is no other way of coming to heaven but by the blood. The devil don't fear 10,000 preachers who preach a bloodless religion. A man who preaches a bloodless religion is doing the devil's work, and I don't care who he is. It is said of old Dr. Alexander, of Princeton Seminary, that as the students left, he would take them by the hand, and say, "Young man, make much of the blood—

MAKE MUCH OF THE BLOOD."

As I have traveled up and down Christendom, I have found out that a minister who gives a clear sound upon this doctrine is successful. A man who covers up the cross, though he may be an intellectual man, and draw large crowds, there will be no life there, and his church will be like a gilded sepulchre. Those men who preach the doctrine of the cross, holding up Christ as the sinner's only hope of heaven, and as the sinner's only Substitute, and make much of the blood, God honors, and souls are always saved in the church where that is preached. I don't like to give advice to these gray-haired ministers, but make much of the blood. May God help us to make much of the blood of His Son. It cost God so much to give us this blood, and shall we try to keep it from the world which is perishing from the want of it? The world can get along without us, but not without Christ. Let us preach Christ in season and out of season. Let us go to the sick and dying, and hold up the Saviour who came to seek and save them, and died to redeem them. It is said of Julian, the apostate in Rome, that when he was trying to stamp out Christianity, he was pierced in the side by an arrow. He pulled the arrow out, and taking a handful of blood as it flowed from the wound, threw it into the air, shouting, "Thou Galileean, Thou hast conquered!" Yes, this Galileean is going to conquer. May God help us to give no uncertain sound on this doctrine. I would rather give up my life than give up this doctrine. Take that away, and what is my hope of heaven? Am I to depend upon my works? Away with it when it comes to the question of salvation. I must get salvation distinct and separate from work, for it is to him that worketh not and believeth. None will walk the celestial pavement of heaven but those washed in the blood. The first man that went up from this earth was Abel probably. You can see Abel putting his little lamb upon the altar, thus placing blood between him and his sin. Abel sang a song the angels could not join in. There must have been

ONE SOLO SONG IN HEAVEN,

because Abel had no one to join him. But there is a great chorus now, for the redeemed have been going up for 6,000 years, and they sing of Him who is worthy to receive honor because He died to save us from hell and damnation. Revelation vii. 14—"And I said unto him, Sir, thou knowest. And he said to me, These are they which came out of great tribulation, and have washed their robes, and made them white in the blood of the Lamb." Sinner, how are you going to get your robes clean if you don't get them washed in the blood of the Lamb? How are you going to wash them? Can you make them clean? I hope at last we shall get back to the paradise above. There they are singing the sweet song of redemption, and may it be the happy lot of each of us to join them. It may be a few months at the longest before we shall be there, and shout the song of redemption, and sing the sweet song of Moses and the Lamb. If you die without Christ, without hope, and without God, where will you be? Sinner, be wise! don't make light of the blood. An aged minister of the Gospel, when on his dying bed, said, "Bring me the Bible." Putting his finger upon the verse, "The blood of Jesus Christ, His Son, cleanses us from all sin," he said. "I die in the hope of this verse." It wasn't his fifty years' preaching, but the blood of Christ. When we stand before God's tribunal we shall be as pure as God, because we shall be washed in the blood of the Lamb. During the American war a doctor heard a man saying, "Blood, blood, blood!" The doctor thought this was because he had seen so much blood, and sought to divert his mind. The man smiled, and said, "I wasn't thinking of the blood upon the battle-field, but I was thinking how precious the blood of Christ is to me as I am dying." As he died, his lips quivered, "Blood, blood, blood!" and he was gone. It will be precious when we come to our dying bed—it will be worth more than all the world then.

LOSING THE BRAKE.

A stage-driver away on the Pacific coast —as I was told when I was there about three years ago—while lying on his dying bed, kept moving one of his feet up and down, saying, " I am on the down grade, and cannot reach the brake." As they told me of it I thought how many were on the down grade and could not reach the brake, and were dying without God and without hope. I plead with you as a fellow traveler; don't go out of this hall without saying, Heaven is my home and God is my Father. Don't let the scoffers laugh you into hell; they cannot laugh you out of it. The blood is upon the mercy-seat, and while it is upon the mercy-seat you go into the kingdom. God says, " There is the blood; it is all I have to give." The blood is there, and God says, ' As long as it is there, there is hope for you. I am satisfied with the finished work of My Son and will you be satisfied."

Don't go out until you can claim this as yours. Think of that atheist we have been praying for who is dying. I hope he will lay hold of Christ before he dies. How dark and sad it is to go to the bed-side of a dying infidel or atheist, or one who is dying without the light of the resurrection morn. I hope the light will burst in upon him before it is too late. If we trust to Christ, death has lost its sting and the grave its victory. You may have read of that good man in America, Alfred Cookman. While his friends were gathered round his dying couch his face lit up, and with a shout of triumph, he said, " I am sweeping through the gates, washed in the blood of the Lamb !" And this echoes and re-echoes through America to-day, " I am sweeping through the gates, washed in the blood of the Lamb !" May these be our last words, and there will be no trouble then about an entrance into the kingdom of God.

A NEW HAND-BOOK for the
BIBLE-READER AND STUDENT.
The Historic Origin of the Bible.

A HAND-BOOK of Principal Facts, from the best recent authorities, German and English. In three Parts, complete in One Volume. *Part I.*—The English Bible. *Part II.*—The New Testament. *Part III.*—The Old Testament. With APPENDICES: *I.*—Leading Opinions on Revision. *II.*—On the Apocrypha. By Rev. EDWARD CONE BISSELL, A.M. With an Introduction by Prof. ROSWELL D. HITCHCOCK, D.D., of the Union Theological Seminary, N. Y. One vol. small 8vo, 455 pp. $2.50.

Please Read this Description of the Book.

This is a complete manual of *Biblical Introduction*. It covers ground never before covered by any other volume. While giving, in a compact and practicable form, the gist of such important treatises as those of Bleek, Keil, Reuss, Credner, and De Wette, on the Old and New Testaments; it contains also a fresh and critical history of the English Bible, together with an exceedingly valuable Appendix of fifty closely printed pages, on the subject of revision. This Appendix furnishes, in their own language, the leading opinions of Christian scholars in England and America, both for and against revision. While showing the more important defects of our version, on which the plea for such a work is based, it brings down to date the history of the recent movement undertaken by the Convocation of Canterbury, and thus puts the general reader in possession of all needful facts for an enlightened judgment respecting its feasibility and expediency. The whole work has been written with extreme care. During its preparation, or while going through the press, it has passed under the eye of some of our most eminent Biblical scholars, who have expressed their gratification both with the general plan of the treatise, and its execution. While it is a book that might be expected to find a welcome place on the table of the minister and theological student, it is also particularly adapted to the wants of Sunday-School teachers, and is written in a style easily comprehensible by all intelligent readers of the Bible.

NOTICES OF THE PRESS.

"The results of a vast mass of Biblical learning are here presented in an intelligible and practical form. Although the work exhibits the fruits of wide research and mature scholarship, the author has evidently had in view the demands of common readers rather than the exigencies of professional students. He treats first of the history of the English Bible in its different versions and fragments of versions, giving a succinct account of the labors of Tyndale and other translators, before the so-called authorized version; then of the New Testament, the manuscripts, the ancient versions and printed text, the canon, and the different books in succession; closing with the Old Testament, under similar heads; and an appendix containing remarks on the Apocrypha and the revision of the current translation. *Probably no other work in the language contains so much accurate information on the subject, in so convenient a form.*"—*Tribune.*

"The chief value of the work, it seems to us, is in its clear arrangement of the facts collected by earlier writers as to the Historical origin of the Bible, in a concise yet sufficiently full and really popular form. It places at the command of every reader information which, heretofore, has been attainable only at the cost of such study and research as is possible only to the professed Biblical scholar. The author does not undertake to produce new evidence or new arguments; but, in the discussion of the materials before him, he develops a fine critical acumen, and in his independent reasoning is generally forcible. The work will be found very serviceable to clergymen of all denominations as a book of reference, and the general public will find in it the simplest and most intelligent account of the origin and history of the Bible, that, so far as we know, has ever been printed."—*Literary World.*

"Those who believe in the authenticity of the Sacred Scriptures, are by no means disheartened by the assaults of modern criticism. On the contrary, they have given themselves to increased study, and will meet book by book . . . . The present work bears evidence of scholarship and reverential study. The appendices are by no means the least valuable portion of the volume. An index of authorities, of passages of Scripture cited or illustrated, and a general index of subjects, completes a work which will be found of much value to students of the Bible, and for those who write on kindred themes and need a condensed book of reference."—*The Graphic.*

"It is an excellent companion to the Bible."—*Evening Mail.*

ANSON D. F. RANDOLPH & CO., 770 Broadway, cor. 9th Street, New York.

Sent by mail, prepaid, on receipt of the price, $2.50.

A NEW HAND-BOOK for the

BIBLE-READER AND STUDENT
The Historic Origin of the Bible.

A HAND-BOOK of Principal Facts, from the best recent authorities, German and English. In three Parts, complete in One Volume. *Part I.*—The English Bible. *Part II.*—The New Testament. *Part III.*—The Old Testament. With APPENDICES: *I.*—Leading Opinions on Revision. *II.*—On the Apocrypha. By Rev. EDWARD CONE BISSELL, A.M. With an Introduction by Prof. ROSWELL D. HITCHCOCK, D.D., of the Union Theological Seminary, N. Y. One vol., small 8vo, 455 pp. $2.50.

From the Rev. WM. A. STEARNS, D.D., LL.D.,
President of Amherst College.

"It is a highly interesting and valuable work. I do not know where else one can find so full a history of our English Bible, so well worked over, arranged, and condensed, as in this volume."

From the Rev. WILLIAM S. TYLER, D.D., LL.D.,
Williston Professor of the Greek Language and Literature, Amherst College.

"My examination confirms me in my opinion, formed from a cursory examination of the manuscript, of the great value and merits of the book. It covers a very wide field; much of it hitherto accessible only to scholars, but all of great interest and importance to every reader of the Scriptures. It meets a want widely felt by the intelligent Christian public, by answering, in a clear and satisfactory manner, a multitude of questions which they have hitherto had no means of answering. At the same time, there is so much accuracy, thoroughness, patience, and conscientiousness in the investigation and treatment of the subject, that the book is entitled to a high rank among the helps of educated men, ministers, and biblical scholars. The remarkable candor and fairness of the book is among its chief recommendations. The author seeks only to ascertain the truth, not to establish a theory, or support a tradition."

From Prof. GEORGE COOKE, Boston, Mass.

"This is just the book which many a student has been waiting for. Besides its professed value to theological and sabbath school study, it is, in my judgment, one of the most valuable of all contributions to the study of general history; giving as it does, in compact and thoroughly sustained forms of testimony, the vital pivotal facts of the world's progress in civilization. Such a work most legitimately belongs to every thorough plan or course of historical study; to the curriculum of every school or college affecting a thorough study of history; not only that superficial and vapid scepticism as to the claims of the Bible may be held in check, but that the springs and progress of human culture may be the most truthfully apprehended."

From the Rev. A. L. STONE, San Francisco, Cal.

"The book is a memorial of diligent study, patient research, accurate scholarship, and an ability to digest and reproduce, in orderly and effective method, the minutest material."

From the PRINCETON REVIEW.

"It is a very scholarly book, of decided value and interest, not only for ministers and theological students, but for intelligent Christians."

From the CHURCH AND STATE.

"It is with especial pleasure that we can recommend the work, which is at once thorough, exhaustive even, in its presentation of the results of Biblical investigation in its specific department, and, from beginning to end, represents wise discrimination and pains-taking research in the selection and presentation of its material."

From the CONGREGATIONALIST.

"For a book of reference, we have never seen one more conveniently prepared."

From the CHRISTIAN AT WORK.

"It is an intensely interesting book."

From the PACIFIC.

"We cordially recommend the book as a worthy, right honest and faithful piece of literary and Christian work. The printer and publisher have done themselves honor in its mechanical presentation."

ANSON D. F. RANDOLPH & CO., 770 Broadway, cor. 9th Street, New York.

Sent by mail, prepaid, on receipt of the price, $2.50.

[*Turn over.*]

"A perfect Harvest-nest of Good Things."

Recently Published in a handsome octavo vol., cloth extra, $2.50; gilt edges, $3.50; Half calf, $5.00; Mor. extra, $6.50.

Evenings with the Sacred Poets.

A Series of Quiet Talks about the Singers and their Songs.

BY THE AUTHOR OF

"FESTIVAL OF SONG," "SALAD FOR THE SOLITARY," ETC.

"Mr. SAUNDERS, whose "Salad for the Solitary" has delighted thousands of homes and readers, has just prepared another volume of far greater value, higher purpose and of captivating beauty. He calls it "Evenings with the Sacred Poets," and his exquisite taste, extensive reading and rare familiarity with bibliography, shine in these elegant pages. He roams through all the realms of Poesy, from the earliest times to our own; wanders among all nations, and through all climes, culling the sweetest flowers, and giving us all the most brilliant gems of song. It is a book to be kept near at hand, for refreshment and strength, for comfort and joy; and, when once read, is all the more attractive to be read again."—*New York Observer.*

"The requisites for the proper execution of such a work as this, are, good taste and large research, and these are abundantly manifested in this volume."—*Evening Post.*

"This volume is not only a library of religious poetry, but it is the best critical and historical essay upon the subject with which we are acquainted. The author has already won a reputation as one of the most scholarly and genial of critics, and as the most delightful purveyor of really fresh and entertaining literary gossip."—*N. Y. Round Table.*

"This volume does something more than string together a number of pieces of poetry more or less known; the author has evidently brought considerable research and study to the task of presenting a complete picture of the sacred poetry and hymnology of the Christian ages. Altogether we know of no selection of sacred poetry so suitable for Sunday reading as this. Not the least of its merits is, that it is not likely to become tiresome, for the comments of the author give it freshness and variety."—*N. Y. Times.*

"This book will be hailed with great satisfaction by all lovers of sacred song." *Christian Intelligencer.*

"The book in all its parts is well adapted to gratify taste and Christian feeling." *N. Y. Examiner and Chronicle.*

"Its style adapts it to popular use, and should ensure a wide circulation; especially when we remember the interest that every one takes in knowing something of the author of a favorite song. In this respect, we have all that could have been anticipated in a work of this character; the bits of biography and anecdote being charmingly given, and deftly incorporated in the glittering mosaic. The idea of the author is an excellent one; while the result fills a place in literature that has hitherto been unoccupied."—*Charlestown (Mass.) Adv.*

"The author has a quick eye for whatever is beautiful; his "quiet talks" are almost always interesting, and convey a pretty fair idea of the history of religious poetry, and the characteristics of different schools. We dismiss the work with cordial praise for its excellent spirit and general good workmanship."—*N. Y. Tribune.*

"This book is one for dreamy reading—a book to take up on a quiet Sunday afternoon, when the children are away at Sabbath school, and the house is still with that peculiar sacred stillness, which seems to imbue the very atmosphere of Sabbath—a book to read in snatches, here a little, and there a little, with much musing between."—*Independent.*

"Beginning with the poetry of the Bible, Mr. Saunders brings us down to the present day, culling from the sacred poets some of their choicest verses, and giving us just enough to provoke a most tantalizing appetite for more."—*Harper's Monthly.*

"Mr. Saunders has so ably and satisfactorily performed his task that we must not withhold our meed of praise. He has indeed done that for which the thanks of Christian readers are due. Looked upon in the light of a contribution to Christian literature, this book deserves to live, since it illustrates what has always been the most powerful and intimate phase of religion—the expression by song. Those who love pleasant things in books, should get and read this charming cluster of flowers of sacred poesy." —*N. Y. Evening Mail.*

Published by A. D. F. RANDOLPH & CO., 770 Broadway, N. Y.

Sent by mail, prepaid, on receipt of price.

"A domestic Story, which may be taken into the family with the feeling that it is entirely free from the objectionable features which too often characterizes the modern novel."

JANET'S LOVE AND SERVICE. By MARGARET M. ROBERTSON, author of "Christie; or, the Way Home." 12mo. 586 pages. Bound in cloth, $1 75.

It is rarely that we read a story which gives us so much satisfaction and so little cause of complaint as *Janet's Love and Service*, by MARGARET M. ROBERTSON (A. D. F. Randolph). We should characterize it as a book of emphatically *quiet* power. There are no strivings after sensational effects; there are no sudden scene-shiftings; no marvellous transformations; no astute villains; no innocent angels caught in their toils; no intricate mysteries through which reader and heroine flounder alike, to be alike extricated only at the full of the curtain ; no elaborate misunderstandings and tortured hearts ; in a word, there is no plot and but little incident. The characters are drawn by a vigorous hand; but they are just such characters as you meet with in daily life. You may see Mr. Snow in any New England parish. Graeme is just such a young girl as might be wrought out of ordinary material, by the experiences of premature responsibility which constituted her schooling. Janet is indeed a rare servant; but she does not remain servant long, and whoever is familiar with Scotch character could find her prototype among his own circle of acquaintances. In a word, the portraits are all photographs. If we were told that the authoress were Graeme, and that she had given us only her actual journal, we should not be surprised. The story is as natural as the actors. It is the story of just such a life as has occurred again and again in the past, and will again and again in the future. In form and style and method of treatment the volume is as simple as it is select in the rather scant materials of which it is composed. It is not without humor; but there are no clowns introduced for the purpose of cracking poor jokes ; no cheap burlesques of Yankee farmers and New England deacons. It is rather genial than witty or even humorous. It is characterized by unusual pathetic power. We pity the reader who peruses its pages with undimmed eyes. But there is no weeping heroine, or sickly sentimentalism. It is a religious novel—its story that of a minister's life. But there are no dogmas to be advocated; no moralizing to be skipped ; no mawkish and tawdry piety to belie the cause of true, healthful religion. So far as there is a moral, it is indicated in the title. So far as there is a heroine, it is Janet, the Scotch servant, who leaves mother and son to care for the poor motherless bairns of her master, and whose service is always that of love—never that of mere self-seeking—never based on wages. The author will, indeed, not be apt to become a popular novel-writer. But we trust she may find a sufficiently large circle of appreciative readers to encourage her to future work. In literature, as on the stage, the more unnatural the drama the more popular. It would be a satisfaction to know that there is a public who prefer the romance of real life to that of the cheap fire and the sheet-iron thunder of a third-rate melodrama.—*Harpers' Magazine.*

"A simple tale of the struggles and life incidents of a Scotch minister and his family, who emigrate to a new settlement in America ; and the writer vividly depicts the habits and manners of the village, and the struggles of the family, with a directness and picturesque effect that are deeply interesting."—*Express.*

"It will be read with interest by those who do not like the sensational trash of many popular fictions."—*Rochester Union.*

"Janet's life presents an example worthy of note and imitation. It is marked with many little incidents which try the faith and patience of one in her sphere of active service, and contains salutary lessons of instruction for a numerous class of readers engaged in the ordinary avocations of life."—*Christian Observer.*

"*This is a story of family life, tending to show that, as a rule, we men need not go beyond the home circle to find an appropriate sphere of labor and influence, and that the daily routine of domestic care and duty, accepted as God-given work, and pursued in a right spirit, permits the full and symmetrical development of woman's nature, and tends 'to establish her goings' in paths, which must ever be to the greater number, the best and safest.*"

"The faithful Janet devotes her life to the welfare and protection of the seven orphan children. She is the life and genius of the whole work, and the moral of the tale is to show that the most humble in life may be useful and truly invaluable to their fellow-creatures."—*Episcopalian.*

Published by A. D. F. RANDOLPH & CO., 770 Broadway, N. Y.

Sent by mail, prepaid, on receipt of price.

www.ingramcontent.com/pod-product-compliance
Lightning Source LLC
Chambersburg PA
CBHW021359230426
43666CB00006B/580